Clean Energy Production Technologies

Series Editors
Neha Srivastava, Department of Chemical Engineering and Technology,
IIT (BHU) Varanasi, Varanasi, India
P. K. Mishra, Department of Chemical Engineering and Technology,
IIT (BHU) Varanasi, Varanasi, India

The consumption of fossil fuels has been continuously increasing around the globe and simultaneously becoming the primary cause of global warming as well as environmental pollution. Due to limited life span of fossil fuels and limited alternate energy options, energy crises is important concern faced by the world. Amidst these complex environmental and economic scenarios, renewable energy alternates such as biodiesel, hydrogen, wind, solar and bioenergy sources, which can produce energy with zero carbon residue are emerging as excellent clean energy source. For maximizing the efficiency and productivity of clean fuels via green & renewable methods, it's crucial to understand the configuration, sustainability and techno-economic feasibility of these promising energy alternates. The book series presents a comprehensive coverage combining the domains of exploring clean sources of energy and ensuring its production in an economical as well as ecologically feasible fashion. Series involves renowned experts and academicians as volume-editors and authors, from all the regions of the world. Series brings forth latest research, approaches and perspectives on clean energy production from both developed and developing parts of world under one umbrella. It is curated and developed by authoritative institutions and experts to serves global readership on this theme.

Santosh Kumar • Vivek Kumar Singh
Editors

Renewable Energy Development: Technology, Material and Sustainability

Springer

Editors
Santosh Kumar
Department of Chemistry, School of Basic and Applied Sciences
Harcourt Butler Technical University
Kanpur, Uttar Pradesh, India

Vivek Kumar Singh
Energy for Sustainability Initiative
University of Coimbra
Coimbra, Portugal

ISSN 2662-6861 ISSN 2662-687X (electronic)
Clean Energy Production Technologies
ISBN 978-981-97-9625-0 ISBN 978-981-97-9626-7 (eBook)
https://doi.org/10.1007/978-981-97-9626-7

This Springer imprint is published by the registered company Springer Nature Singapore Pte Ltd.
The registered company address is: 152 Beach Road, #21-01/04 Gateway East, Singapore 189721, Singapore

Preface

Energy consumption is growing around the globe, and there might be various reason behind on that as; economic development, growing population and advancement of technological developments. Environmental issues directly related to energy production and the combustion of fossil fuels are also the leading cause of greenhouse gas emissions. Renewable energy is growing, as innovation brings down cost and starts to supply on the promise of a clean energy. Renewable energy is often brought up as clean energy that comes from natural sources or processes. Renewable energy resources are available from various sources, including solar energy and wind energy.

This book aims to highlight the current renewable energy technologies: solar, wind, hydrogen, hydropower, wave power, biomass, and biofuel. It will offer a unique perspective, opening a new window on traditional renewable energy sources and new technological developments, advanced material innovations, and sustainability aspects. Key topics will include energy storage, transmission, institutional and economic factors, renewable energy applications in smart cities and buildings, solar thermal applications, environmental impacts, electrical energy generation, and climate change mitigation potential to achieve net-zero targets.

Additionally, the book will explore the integration of renewable energy into existing grids, the role of policy and regulatory frameworks, and the importance of international cooperation in accelerating the transition to a sustainable energy future. By examining both established and emerging technologies, the book aims to provide a comprehensive understanding of the current landscape and future potential of renewable energy, offering valuable insights for researchers, policymakers, and industry professionals dedicated to advancing clean energy solutions.

Chapter 1 discusses on renewable energy development sources and technology overview, Chap. 2 discusses on renewable energy insights: sources and technological advancements, Chap. 3 discusses on solar energy and environmental application, Chap. 4 discusses the role of solar energy in sustainable environment, Chap. 5 discusses on solar energy and smart cities: a holistic approach for sustainable development, Chap. 6 discusses on solar energy and its utilization in smart cities, Chap. 7 discusses on integration of renewable energy in buildings, Chap. 8 discusses on sustainable electricity generation through solar energy technologies, Chap. 9 discusses on biomass energy conversion technologies to produce electrical power, green hydrogen, bio-oil, and biochar through biomass gasification and pyrolysis, Chap. 10 on discusses hydrogen energy: innovation in production, storage, and diverse applications, Chap. 11 discusses on hydrogen horizons: advancements, applications, and global initiatives, Chap. 12 discusses on porous organic polymers for renewable energy development, Chap. 13 discusses on advanced biobased materials for renewable energy development, Chap. 14 discusses on achieving net-zero emission: a sustainable future with renewable energy, Chap. 15 discusses on renewable energy in focus: development trends, challenges, and policy responses, Chap. 16 discusses on environmental aspects and electrical energy generation, and Chap. 17 discusses environmental aspects in electrical energy generation: a comprehensive review.

Kanpur, Uttar Pradesh, India — Santosh Kumar
Coimbra, Portugal — Vivek Kumar Singh

Contents

Editors and Contributors

About the Editors

Santosh Kumar is currently working as an Associate Professor in the Department of Chemistry, Harcourt Butler Technical University, Nawabganj Kanpur, Uttar Pradesh, India. He also served as Research Professor in Department of Organic and Nano System Engineering, Division of Chemical Engineering, Konkuk University, Seoul, South Korea. Dr. Kumar received his B.Sc, M.Sc., and D.Phil. in Chemistry from University of Allahabad. He worked as a Research Associate (CSIR) at Motilal Nehru National Institute of Technology (MNNIT), Allahabad. He was awarded CAS-TWAS Postdoctoral Fellowship in 2008 and worked in Changchun Institute of Applied Chemistry, Chinese Academy of Sciences, Changchun, China. Later, he joined as a KU Brain Pool Professor and researcher at the KU, Seoul, South Korea (2010–2013). He was awarded FCT Postdoctoral Fellowship 2012 and later worked at University of Coimbra, Coimbra, Portugal (2013–2017). Dr. Kumar has published more than 111 publications in reputed journals and books along with 19 book chapters and 3 edited books. His current research interests are chitosan biopolymer, nanotechnology, biomedical, food packaging, energy and environmental chemistry.

Vivek Kumar Singh is the Founding Director of Emerald Gears Initiative Foundation (EGIF) Global, located in Greater Noida, Uttar Pradesh, India. His tenure expanses as a researcher in the Energy for Sustainability Initiative at the University of Coimbra in Coimbra, Portugal, and a postdoctoral researcher at the Institute for Systems Engineering and Computers in Coimbra (INESC Coimbra). Dr. Singh has worked with prominent organizations such as the Energy and Resources Institute (TERI), the Center for Study of Science, Technology and Policy (CSTEP), and the United Nations Development Program (UNDP) in New Delhi. Dr. Singh earned his doctoral degree with distinction in Sustainable Energy Systems from the University of Coimbra as part of the Massachusetts Institute of Technology, USA-Portugal sandwich program. He holds an M.Tech degree in Energy & Environmental Engineering from Vellore Institue of Technology (VIT) University, Vellore, Tamil Nadu, India. With a robust academic background, he has published over 53 articles in reputed journals and contributed 4 book chapters. His research interests are in policy research in renewable energy, energy and climate economics modeling, sustainability, and many more. Dr. Singh has accumulated 18 years of experience in industry consultancy, academia, and research across Asia and Europe.

Contributors

Shoaib Akhtar Department of Chemical Engineering, Birsa Institute of Technology, Sindri, Jharkhand, India

Mohammad Izhar Alam Department of Environmental Studies, Delhi University, New Delhi, New Delhi, India

Neeru Anand University School of Chemical Technology, Guru Gobind Singh Indraprastha University, New Delhi, New Delhi, India

Shraddha Awasthi Department of Environmental Studies, Motilal Nehru College (E), Delhi University, New Delhi, New Delhi, India

Anurag Bajpai GreenTree Global, Noida, Uttar Pradesh, India

Ankur Bhardwaj GreenTree Global, Noida, Uttar Pradesh, India

Arun Kumar Choudhary Ministry of New and Renewable Energy, New Delhi, New Delhi, India

G. L. Devnani Department of Chemical Engineering, Harcourt Butler Technical University, Kanpur, Uttar Pradesh, India

Kruthika Eswaran Lead Sustainability and Climate Consultant, Net Zero Think Pvt. Ltd., Bengaluru, Karnataka, India

Rajiv Ganguly Department of Civil Engineering, Harcourt Butler Technical University, Kanpur, Uttar Pradesh, India

Arijit Ghosh S.A.P. Automations (India) Private Limited, New Delhi, New Delhi, India

Mukul Kumar Gupta Department of Electronics and Instrumentation Engineering, MJP Rohilkhand University, Bareilly, Uttar Pradesh, India

Nazrul Hsan Department of Materials Science and Engineering, Konkuk University, Seoul, Republic of Korea

Vandana Kaushik Department of Physics, Sanskaram University, Jhajjar, Haryana, India

Joonseok Koh Department of Materials Science and Engineering, Konkuk University, Seoul, Republic of Korea

Avanish Kumar Department of Chemical Engineering, Harcourt Butler Technical University, Kanpur, Uttar Pradesh, India

Manoj Kumar Department of Physics, Harcourt Butler Technical University (HBTU), Kanpur, Uttar Pradesh, India

Santosh Kumar Department of Chemistry, School of Basic and Applied Sciences, Harcourt Butler Technical University, Kanpur, Uttar Pradesh, India

Subhajit Majumder Department of Chemical Engineering, National Institute of Technology Hamirpur, Hamirpur, Himachal Pradesh, India

Ambneesh Mishra Faculty of Sciences and Technology, University of Algarve, Faro, Portugal

Nirupama Department of Chemical Engineering, Birsa Institute of Technology, Sindri, Jharkhand, India

Dan Bahadur Pal Department of Chemical Engineering, Harcourt Butler Technical University, Kanpur, Uttar Pradesh, India

P. Raman Energy Efficiency and Environment P. Ltd, New Delhi, New Delhi, India

Shivani Rani GreenTree Global, Noida, Uttar Pradesh, India

Amit Kumar Rathoure Department of Chemical Engineering, Harcourt Butler Technical University, Kanpur, Uttar Pradesh, India

S. K. Sharma Department of Physics, Harcourt Butler Technical University (HBTU), Kanpur, Uttar Pradesh, India

Swadha Shree Department of Chemical Engineering, Birsa Institute of Technology, Sindri, Jharkhand, India

Ira Singh Department of Chemistry, School of Basic and Applied Sciences, Harcourt Butler Technical University, Kanpur, Uttar Pradesh, India

Roushan Kumar Department of Mechanical Cluster, School of Advanced Engineering, UPES, Dehradun, Uttarakhand, India

Satyendra Singh Department of Chemical Engineering, Harcourt Butler Technical University, Kanpur, Uttar Pradesh, India

Shalinee Singh Department of Chemistry, School of Basic and Applied Sciences, Harcourt Butler Technical University, Kanpur, Uttar Pradesh, India

Varun Pratap Singh Department of Mechanical Cluster, School of Advanced Engineering, UPES, Dehradun, Uttarakhand, India

Vivek Kumar Singh Energy for Sustainability Initiative, University of Coimbra, Coimbra, Portugal
Emeraldgears Initiative Foundation (EGIF) Global, Greater Noida, Uttar Pradesh, India

Gyanendra Singh Sisodia Ajman University of Science and Technology, College of Business Administration, Ajman University, Ajman, UAE

Sweta Swaraj Department of Chemical Engineering, Birsa Institute of Technology, Sindri, Jharkhand, India

Aruna Thakur Department of Mechanical & Manufacturing Engineering, National Institute of Advanced Manufacturing Technology, Ranchi, Jharkhand, India

Pooja Thakur Department of Chemical Engineering, National Institute of Technology Hamirpur, Hamirpur, Himachal Pradesh, India

Uplabdhi Tyagi University School of Chemical Technology, Guru Gobind Singh Indraprastha University, New Delhi, New Delhi, India

Chapter 1
Renewable Energy Development Sources and Technology: Overview

Avanish Kumar and Dan Bahadur Pal

Abstract The chapter systematically examines various types of renewable energy resources and explains the current scenario of both nonrenewable and renewable energy sources, emphasizing the significance of renewable energy amidst global concerns regarding climate change and sustainability including wind energy, solar energy, biomass energy, geothermal energy, ocean thermal energy conversion (OTEC), hydroelectric power, and tidal and wave energy. Each renewable energy type is explored in detail, with its technological advancements, potential for power generation, and associated challenges. For instance, wind energy discussions encompass onshore and offshore wind farms, while solar energy discussions delve into photovoltaic power plants and solar ponds. Biomass energy exploration covers combustion, cogeneration, and biofuels such as ethanol and biodiesel. Similarly, geothermal energy production techniques and OTEC technologies are comprehensively analyzed. The paper also scrutinizes the status of renewable energy deployment, highlighting capacity growth, cost reductions, competitiveness, and the pivotal role of renewable energy in addressing global warming concerns. However, it also acknowledges challenges related to renewable energy development, including intermittency, grid integration issues, technological barriers, and the imperative need for energy storage and grid modernization. This chapter offers a comprehensive understanding of renewable energy sources, technologies, challenges, and policy frameworks, serving as a valuable resource for researchers, policymakers, and stakeholders invested in advancing renewable energy transition in India and beyond. Furthermore, it examines government incentives, subsidies, and renewable portfolio standards (RPS) aimed at renewable energy adoption. It concludes by providing insights into current electricity policies in India, underscoring the importance of integrated approaches to sustainable energy development. In summation, this chapter offers a holistic understanding of renewable energy sources, technologies,

A. Kumar · D. B. Pal (✉)
Department of Chemical Engineering, Harcourt Butler Technical University, Kanpur, Uttar Pradesh, India

S. Kumar, V. K. Singh (eds.), *Renewable Energy Development: Technology, Material and Sustainability*, Clean Energy Production Technologies,
https://doi.org/10.1007/978-981-97-9626-7_1

challenges, and policy frameworks, catering to researchers, policymakers, industry practitioners, and other stakeholders invested in advancing renewable energy transitions globally.

Keywords Renewable energy · Solar energy · Biomass energy · Geothermal energy

1.1 Introduction

India's energy consumption is continuously increasing with population growth and rising living standards. Presently, India's energy planning is mostly dependent on thermal power plants, which contribute nearly 70% of the total installed capacity of Indian conventional resources of energy (Tripathi et al. 2016). This heavy reliance imposes significant strain on fossil fuel resources, raising concerns about their preservation for future generations while concurrently exploring alternative energy sources to foster robust and sustainable economic growth. Usually, thermal power plants exert partially adverse impacts on the surrounding environment, due to the large number of emissions of carbon dioxide (CO_2) ranging from 0.9 to 0.95 kg/kWh, as well as emissions of SO_x and NO_x, contributing to global warming and climate change (Miah et al. 2010). Research shows that carbon dioxide CO_2 concentration increased by 28% after the Industrial Revolution came into existence, resulting in a global temperature increase of 0.3–0.6 °C and a sea level increase of up to 10–15 cm over the last few decades. Environmentalist warns that temperature may rise to 13.5 °C and sea level may rise to 15–95 cm if greenhouse gas emissions are not checked (Barnard and Morgan 2000). Notably, the industrial sector requires electricity consumption, which increases approximately 37% of global greenhouse gas emissions. India, ranking 145th globally in per capita equity, emits 1.25t CO_2 per annum (Rao and Rao 2019). These all-cause effects are adverse on the environment which prompted India to adopt the favor of renewable energy sources.

1.2 Current Scenario of Nonrenewable and Renewable Energy in India

The leading consumer of energy globally, after China, Russia, and the United States (Ahmad and Zhang 2020), India's per capita total electricity consumption remains around 917.2 kWh as per data of March 2013 (Dubey 2015), 4500 BKWH by 2031–2032 (Kumar and Pal 2014). The result indicates that electricity consumption in India will increase by approximately 2280 BKWH by 2021–2022, and the power sector in India is renowned for its remarkable diversity on a global scale. Over time, it has made consistent strides in augmenting its generation capacity, and the total availability of conventional resources of power plants in the country is 253.39 GW. Thermal power constitutes the largest share at 69.5% as per data indicated from the source of electricity in India as of November 31, 2014, while renewable

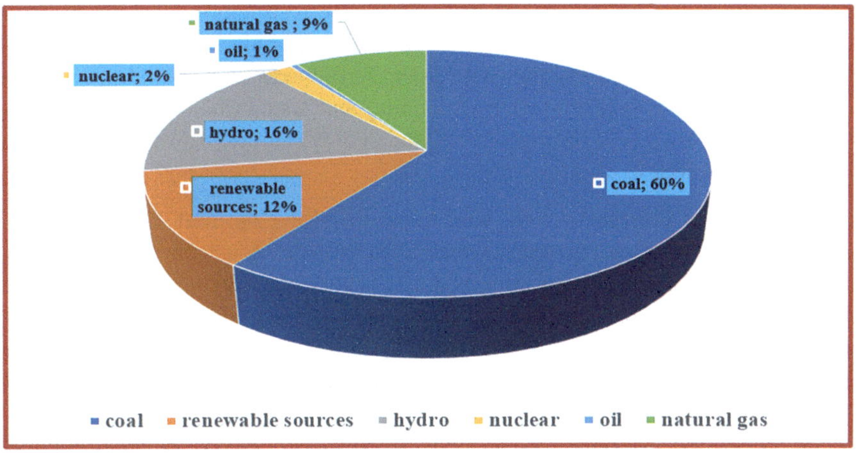

Fig. 1.1 Installed capacity as per source of electricity: November 11, 2014

energy sources contribute 12.5%. Nuclear power holds a 2% share, and hydroelectric power accounts for 16% of the total installed capacity (Tripathi et al. 2016). According to the annual report of the Ministry of Renewable Energy (MNRE) for 2017–2018, India's renewable energy potential is considerable. The report indicates that, at a 100-meter mast height, the potential of wind power was assessed at 302.251 GW. Additionally, small hydropower was estimated at 19.749 GW, biomass power at 17.536 GW, bagasse cogeneration at 5 GW, waste to energy (WTE) at 2.554 GW, and solar at 748.990 GW. The cumulative estimated renewable potential amounted to 1096.080 GW. Figure 1.1 shows the installed capacity as per source of electricity.

Despite India's diverse geographical and climatic conditions as a tropical country, it boasts a wide array of nonconventional energy resources, including solar energy, wind energy, hydropower, geothermal energy, and biomass (Mohammed et al. 2017). Renewable energy resources are sources of energy that are naturally replenished. These resources are considered sustainable because they are derived from processes or mechanisms that are continuously occurring and can be harnessed without depleting the Earth's finite resources. The different sources of renewable energy are wind energy, hydropower, biomass, solar energy geothermal energy, etc.

1.3 Significance of Renewable Energy

The critical emissions of greenhouse gases like CO_2, CH_4, N_2O, ozone, and peroxyl acetyl nitrate trap heat from the Earth's surface, increasing surface temperatures. This necessitates the enhancement of the world's power infrastructure and reduced human reliance on them (Fitzsimmons 1999). With rising global temperatures and energy demands in India, there's a need for sustainable renewable energy sources and improved energy efficiency measures. India possesses significant renewable energy potential to fulfil its energy needs. However, restructuring its energy infrastructure to meet socioeconomic goals amidst increasing electricity demands presents challenges. Therefore, renewable energy resources are pivotal for human survival, addressing environmental, economic, and social challenges. The indispensable role of such type of nonconventional energy sources in sustainable human development includes:

Independence in Energy Security and Economic Development Dependence on a variety of renewable energy sources decreases dependency on conventional energy resources, enhancing energy security by diversifying the energy mix. The renewable energy sector is a prominent field to create jobs in the manufacturing, installation, maintenance, and research and development field. This may change economic growth and support local communities.

- *Reduced GHC emissions and the effect of climatic change:* Nonconventional energy sources produce power with lower or zero value of greenhouse gas emissions compared to conventional resources. This helps mitigate climate change and reduce air pollution.
- *Global collaboration, off-grid solutions:* Investing in renewable energy encourages driving advancements in clean energy technologies, and fostering a culture of sustainability renewable energy technologies, solar and wind, can provide decentralized and off-grid solutions, which contributes to social development by improving living conditions and supporting education and healthcare services. The adoption of renewable energy is often part of global efforts to address environmental challenges. International cooperation and agreements promote the exchange of knowledge, technology, and resources to promote to move for sustainable energy systems.

1.4 Types of Renewable Energy Resources

These are the energy sources that are naturally replenished, reducing the dependence on finite resources on fossil fuel. This helps in preserving biodiversity and ecosystems. Mostly, all renewable energy technologies require low land and water requirement, and important renewable energy resources are classified as Fig. 1.2 shows the schematic diagram: (a) wind power plant, (b) photo voltaic, and (PV) solar power plant.

1.4.1 Wind Energy

This type of resource uses the kinetic potential of wind into electrical energy, which may be applicable for various purposes. Wind turbines consist of rotor blades attached to a central hub. As the wind blows, the rotor blades rotate around the hub.

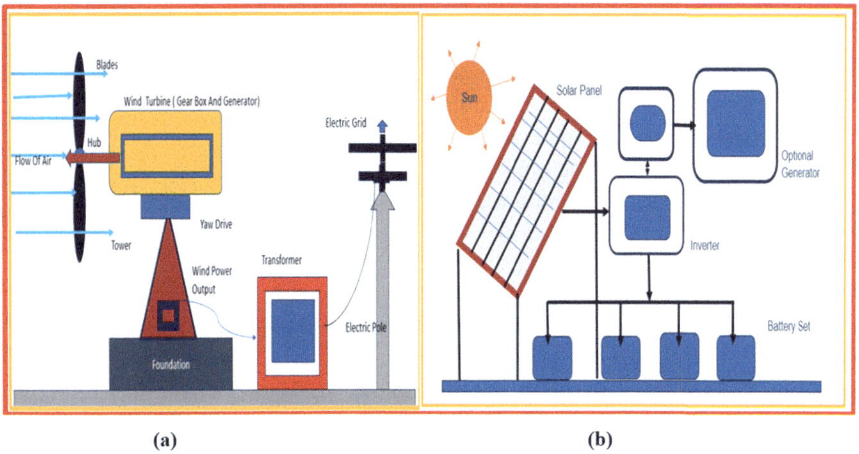

Fig. 1.2 Schematic diagram: (**a**) Wind power plant. (**b**) Photo voltaic (PV) Solar power plant

The rotating blades transfer the kinetic power of the wind to the rotor shaft. The rotor shaft is connected to a gearbox, which increases the rotational speed of the shaft to optimize the generator's efficiency. The attached rotor shaft to an electric generator is located inside the nacelle, which is the housing atop the tower. As the rotor shaft rotates, it drives the generator, which changes mechanical energy into electrical energy. The generator usually produces alternating current (AC) electricity. The electricity generated by the wind turbine is transmitted through cables down the tower and eventually into the electrical grid. It may go through transformers to adjust voltage levels for efficient transmission over long distances. Wind turbines are attached to sensors and control systems to monitor wind speed, direction, and other parameters. These systems optimize the turbine's performance by adjusting the orientation of the blades and other operational parameters to maximize energy output and ensure safe operation.

1.4.1.1 Wind Energy Potential and Their Challenges

The total installed Indian wind power capacity is 21,136.40 MW (MNRE), which makes it an important contributor to the global energy market. Wind power contributes to approximately 75% of the grid-connected renewable energy power installed capacity in the states such as Tamil Nadu, Maharashtra, Karnataka, Rajasthan, and Gujarat, which have been prominent wind energy contributors in India, owing to favorable wind conditions and conducive policies. Despite all favorable conditions in India, the actual utilization of wind power in India is relatively low, which indicates that policymaker should focus to prepare policy incentives for the commissioning of wind mill rather than the operation of plants. But, in reality, only 1.6% of total actual power production in India is generated from wind energy, despite the available potential being around 8.6% (Kumar and Pal 2014; Tripathi et al. 2016).

Wind power generation is a challenge because it depends entirely on climatic conditions such as direction and wind speed. Lack of wind and excessively high wind speeds can affect the reliability and consistency of power generation from wind turbines (Hansen et al. 2012). Most of the wind turbines generate mechanical sound from the gearbox and generator. However, modern turbines are required to modify the design to minimize noise emissions. Even the noise pollution caused by wind turbines is not considered important, because it falls within acceptable limits recommended by environmental authorities (Guski et al. 2017).

1.4.1.2 Different Wind Energy Production Technology with Their Benefit

There are various wind energy transformation technologies, which may be classified based on different ranges of turbine designs and configurations, including horizontal-axis turbines, vertical-axis turbines, and hybrid systems (Aşçıoğlu 2019). These technologies vary in their efficiency, scalability, and applicability to different wind conditions and geographic locations; then primary methods of harnessing wind energy include onshore wind turbines, offshore wind farms, and different wind energy conversion technologies:

Onshore Wind Turbines

Onshore wind turbines consist of tall towers with rotor blades mounted on top. As wind moves over the blades, it start to rotate. The rotational motion drives a generator, converting mechanical energy into electrical energy. Onshore wind turbines play a prime role in combating climate change by producing electricity with low greenhouse gas emissions (Wang et al. 2019). They optimize land use efficiently, leaving ample space for agricultural activities, grazing, or other land purposes. Wind farms, when strategically located, can harmoniously coexist with wildlife, safeguarding habitats and biodiversity. Moreover, onshore wind energy stands out as a cost-effective renewable energy source, outperforming traditional options. Additionally, wind energy initiatives spur job growth across various sectors like manufacturing, construction, maintenance, and operation, invigorating local economies and fostering employment prospects (Von Oertzen 2018).

Offshore Wind Farms

Offshore wind farms deploy large wind turbines in bodies of water, typically in coastal areas or offshore wind parks. The turbines operate similarly to onshore turbines, utilizing the kinetic energy or mechanical energy of wind to change electricity (Von Oertzen 2018). It provides energy surety to the nation and decreases dependency on imported fossil fuels, thus enhancing resilience. Coastal areas experience economic growth through investments, job creation, and the revitalization of port infrastructure driven by offshore wind projects. Integration of offshore wind farms into existing offshore infrastructure lowers transmission expenses and enhances grid stability. Moreover, placing offshore wind farms away from populated zones mitigates visual and aesthetic concerns commonly associated with onshore wind projects.

1.4.2 Solar Energy

Solar technology requires sunlight to generate electricity or heat, offering a sustainable and abundant source of renewable energy. The topography of some states in India is versatile, which makes it an ideal location for solar power generation. The Indian solar potential is estimated to be around 6000 million GWh annually. Indian states have receive solar radiation from about 7 kWh per square meter per day. This high level of solar irradiance contributes to the country's vast solar energy potential (Guski et al. 2017). The target of the National Solar Mission to achieve several key targets by 2022 is about 20 MW grid-connected solar power capacity, an off-grid solar power capacity of 2000 MW, and a solar thermal collector area of 20 million square meters (Ogunmodimu and Okoroigwe 2018). The successful implementation of the National Solar Mission would significantly contribute to India's energy security, environmental sustainability, and economic development. It will reduce greenhouse gas emissions, create employment opportunities, and affordable energy sources for common people,

1.4.2.1 Photovoltaic (PV) Power Plant

Photovoltaic systems convert sunlight into electricity using photovoltaic cells, which are made of semiconductor materials. When sunlight strikes on PV cells, it excites electrons and produces direct current (DC), which may be changed into alternating current (AC) by inverters for use for various purposes in industries, in household, in businesses, and in the grid (Dahiya 2018). Solar energy, being renewable and abundant, decreases dependence on finite resources while curbing greenhouse gases and pollutants, thereby mitigating pollution and combating climate change. Photovoltaic (PV) systems yield long-term cost savings on electricity bills since sunlight is freely available and abundant. Additionally, they stimulate economic growth and employment opportunities through manufacturing, installation, maintenance, and research endeavors associated with solar energy technology.

1.4.2.2 Solar Pond

A solar pond is a type of solar energy collector system that utilizes the sun's radiation to heat water or other fluids for various applications, such as electricity generation, heating, solar refrigeration, etc. The working principle of a solar pond involves the exploitation of the natural phenomenon of salinity stratification to trap and store solar energy efficiently. A solar pond consists of a large shallow basin or pond filled with water and salt. The salt concentration gradually increases from the surface to the bottom, creating distinct layers of varying salinity. The layers of water in the solar pond stratify due to differences in salinity. The top layer, known as the surface zone, has the lowest salt concentration and is exposed to direct sunlight. The middle layer, called the gradient zone, experiences a gradual increase in salinity with depth. The bottom layer, known as the storage zone, has the highest salt concentration and acts as a thermal storage medium. Sunlight penetrates the surface zone of the solar pond and heats the water. The salt content of the surface zone prevents the heated water from convective mixing with the cooler layers below, thus trapping the heat within the top layer. As the surface water heats up, it transfers thermal energy to the underlying gradient and storage zones through conduction. The high salt concentration in the storage zone enhances heat retention, allowing the solar pond to store thermal energy for extended periods, even after sunset. Heat exchangers or heat pumps are used to extract thermal energy from the solar pond for various applications such as space heating, water heating, electricity generation through steam turbines, or desalination processes. Solar ponds harness abundant solar energy as their primary heat source, rendering them a renewable and sustainable energy solution. They offer cost-effectiveness in construction and operation compared to other solar energy technologies, demanding minimal maintenance and boasting low operational expenses once established. Moreover, solar ponds operate without emitting greenhouse gases or air pollutants, fostering cleaner air and aiding in climate change mitigation. Furthermore, solar ponds can be seamlessly integrated with desalination processes, enabling the production of fresh water from seawater or brackish water, thereby addressing water scarcity concerns in coastal regions. Through the utilization of solar energy, solar ponds diminish reliance on fossil fuels, fostering energy independence for both communities and industries.

1.4.3 Biomass Energy

This is a significant energy source, primarily derived from natural resources such as plants, crop residues, agricultural waste, wood, and animal dung. As of mid-2013, reported by the (MNRE) ministry of new renewable energy, the nation has sufficient resources for biomass electricity generation, and it is estimated at 23 GW for biomass including bagasse, an agro waste power generation system (Negi et al. 2023). Biomass energy plays a crucial role in meeting the energy needs of rural and remote areas in India. It has facilitated the provision of electricity to sections of the population previously unreached by conventional grid systems. Globally, biomass fuels account for approximately 13% of the total potential of energy. Biomass offers a sustainable supply of energy through various forms such as biogas, producer gas, vegetable oil, biodiesel, and direct combustion of biomass. Biomass combustion produces CO_2, as the prime greenhouse gas. However, the carbon released during combustion is a portion of the natural carbon cycle, as plants absorb CO_2 during growth. Therefore, biomass energy is considered carbon-neutral over its life cycle. Biomass contains low levels of sulfur and nitrogen, resulting in lower emissions of sulfur oxide and nitrogen oxide compared to thermal power plants (Erlach 2014). However, the power plants based on biomass need water for the boiler and cooling system, which can lead to warmer water discharge into lakes or rivers, potentially harming aquatic ecosystems. Biomass power generation in India attracts significant investments and contributes to rural employment, generating over 5000 million units of electric power annually and producing employment opportunities in rural areas (Deshmukh 2009).

1.4.3.1 Different Technology for Biomass to Energy Production

Biomass energy utilizes organic materials derived from plants and animal waste to produce heat, electricity, or biofuels. Figure 1.3 shows the schematic diagram of different biomass conversion processes. Here's an overview of different biomass energy technologies:

Combustion and Cogeneration

Biomass combustion involves burning wood, agricultural residues, and urban waste to produce heat or electricity. Cogeneration, also known as combined heat and power (CHP) generation, is considered an effective source of biomass energy where both heat and electricity are generated simultaneously in a single process (Karampinis et al. 2015; Igbum et al. 2014).

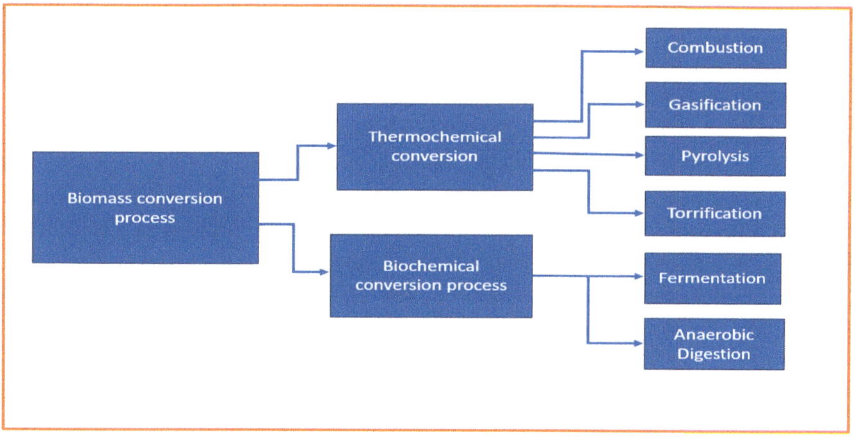

Fig. 1.3 Schematic diagram of different biomass conversion process

Biomass stands as a significant renewable energy source that can be managed and harvested sustainably, lessening reliance on fossil fuels. Through biomass combustion, agricultural residues, forestry by-products, and organic waste materials can be utilized, diverting them from landfills and curbing methane emissions. Cogeneration systems further enhance energy efficiency by harnessing waste heat from electricity generation for heating or industrial applications, thereby optimizing energy utilization and lowering emissions.

Biofuels (Ethanol, Biodiesel)

Biofuels are liquid fuels produced from biomass feedstocks, such as sugarcane, corn, soybeans, and different types of algae. Ethanol is manufactured by the fermentation of the juice of sugarcane, while biodiesel is derived from agro seeds or animal fats, which is an important green fuel because it generates low SO_X and NO_X (Igbum et al. 2014). Additionally, they play a pivotal role in bolstering energy security by diversifying the transportation fuel supply, lessening reliance on imported oil, and fostering domestic energy production and security. Moreover, biofuel production fosters rural development by supporting agricultural economies, generating employment opportunities in farming, processing, and distribution, and revitalizing rural communities through heightened economic activity.

1.4.4 Geothermal Energy

India plays an important role in the global geothermal energy landscape due to its significant potential in this field. The Ecological Survey of India projected India's geothermal energy potential to be around 10,000 MW, which could make it an economical and economical leader in power generation. Chandra Sakharam estimates the present installed capacity of geothermal energy in India to be 203 MW (Shelare et al. 2023). This figure indicates the existing infrastructure and utilization of geothermal resources for power generation. Unlike wind and solar energy sectors, geothermal energy in India has seen fewer benefits or incentives to attract investment. However, efforts are being made to promote geothermal energy, such as plans to set up the first geothermal power plant in Puga, Jammu, and Kashmir, with a capacity of 2–5 MW (Nisar and Monroy 2012). Geothermal energy generation has minimal emissions because no fuels are combusted during the process. However, there may be potential environmental impacts related to drilling wells and withdrawing hot water or steam, or groundwater contamination. Proper management techniques, such as reinjecting used water back into the ground, can help mitigate these impacts and prevent the introduction of pollutants into surface waters.

1.4.4.1 Different Technology Used for the Production of Geothermal Energy

Geothermal energy converts the available internal heat of the Earth's surface for various applications into heat or electricity, direct heating, and cooling (Barber 2002). It is generated by the decay of radioactive materials, like uranium, thorium, and potassium, and the movement of tectonic plates. Here are some key points regarding geothermal energy and its development in India geothermal energy contributes around 6.5% of electricity generation worldwide.

Geothermal Power Plants

Geothermal power plants extract heat energy from reservoirs of hot water or steam from the ground of the earth. The heat is applied to produce steam, which gives force to run turbines connected to generators for producing electricity. There are many types of geothermal power plants, e.g., dry steam, flash steam, and binary cycle plants, depending on the temperature and fluid characteristics of the steam or hot water. This type of energy is recognized as a renewable resource that offers consistent and dependable baseload power, diminishing reliance on fossil fuels and lowering the emission of greenhouse gas. Geothermal power plants exhibit minimal emissions of air pollutants and greenhouse gases in contrast to traditional fossil fuel power plants, thereby promoting cleaner air and aiding in the mitigation of climate change (Akorede et al. 2012). Moreover, geothermal power plants boast low operating costs and extended operational lifespans, ensuring stable and predictable electricity prices over the long term. This stability benefits both consumers and utilities alike; however, despite such advancements, ocean energy has vast untapped potential for future.

Direct Applications of Geothermal Energy

Geothermal energy involves utilizing heat directly from underground reservoirs for heating, cooling, and industrial processes with or without electricity generation (Kabeyi 2019). This can include district heating systems, greenhouse heating, aquaculture, and spa and recreational facilities. Direct-use applications of geothermal energy exhibit exceptional energy efficiency by harnessing heat directly from the source, bypassing intermediate conversion processes and thus minimizing energy losses. From an environmental perspective, these applications significantly decrease reliance on fossil fuels for heating and industrial processes, leading to reductions in greenhouse gas emissions and environmental pollution (Kabeyi 2019). Moreover, direct-use projects yield notable economic benefits by invigorating local economies through job creation, lowering energy expenses for businesses and households, and enticing investments in the development and infrastructure of geothermal resources.

1.4.5 Ocean Thermal Energy Conversion (OTEC)

This type of power generation system utilizes the temperature gradient from surface to lower side temperature difference within the ocean to operate a heat engine and generate useful work to generate electric energy it needs to explain that the National Institute of Ocean Technology is taking a pioneering initiative to establish an ocean thermal energy conversion unit in the coast of Tamil Nadu, approximately 60 km from Tuticorin. With a gross generation capacity of 1 MW, while net power is obtained at 500 kW, this endeavor marks the world's inaugural floating plant in the realm of OTEC technology (Shelare et al. 2023; Mitra et al. 2023).

1.4.5.1 Different Types of OTEC Technologies

There are mainly three different kinds of OTEC technologies: open-cycle, closed-cycle, and hybrid systems restarts. Open-cycle systems are simpler than closed-cycle systems but are less efficient and may cause environmental concerns due to the release of warm seawater back into the ocean in closed-cycle OTEC systems; hot surface seawater is used to vaporize a working fluid with a highly volatile gas such as ammonia. The vaporized fluid drives a turbine, which generates electricity (Anderson and Mayer 1979). The vapor is then condensed using cold, deep seawater, and the process repeats. Closed-cycle systems are efficient and can generate electricity continuously; hybrid systems combine elements of both closed-cycle and open-cycle OTEC. They aim to improve efficiency and address environmental concerns by minimizing the release of warm seawater into the ocean thermal energy conversion (OTEC) offers a comprehensive array of advantages. Figure 1.4 shows the schematic diagram: (a) geothermal plant, (b) ocean thermal energy conversion (OTEC) plant.

It reduces reliance on fossil fuels and bolsters energy security by harnessing a renewable and plentiful ocean energy resource while concurrently creating employment opportunities across various sectors, including engineering, construction, and maintenance. OTEC facilities also boast minimal environmental impact compared to traditional power generation methods like fossil fuel combustion. Moreover, OTEC platforms double as artificial reefs, attracting marine life and enhancing biodiversity in surrounding oceanic areas (Quirapas and Taeihagh 2021). Countries with access to OTEC technology can capitalize on its potential for electricity exportation, stimulating economic growth and fostering international trade. Additionally, OTEC generates electricity. It avoids fossil fuel combustion, thus significantly decreasing greenhouse gas emissions and actively reducing the effect of climate change.

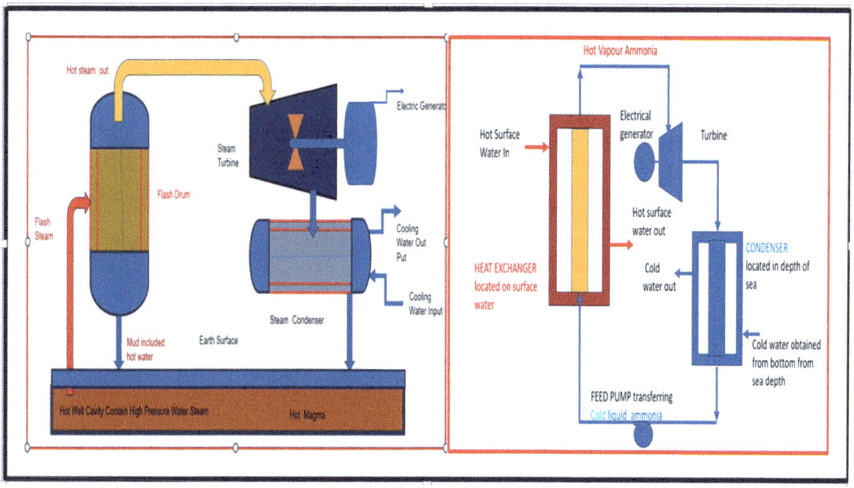

(a) (b)

Fig. 1.4 Schematic diagram: (**a**) Geothermal plant, (**b**) ocean thermal energy conversion (OTEC) plant

1.4.6 Hydroelectric Power Plant

A hydroelectric plant produces electricity using the gravitational force of falling water from an elevated place such as the peak of a mountain. A hydroelectric plant requires a reliable source of water, typically a river, dammed reservoir, or diverted stream. The water source must have sufficient elevation and flow to generate significant energy. A dam is built on the river to create a reservoir to store water. The dam's primary function is to raise the water level, creating potential energy or gravitational potential energy. So, a separate structure containing intake gates or pipes that control the flow of water into the power plant is developed. The intake pipes lead to a large, high-pressure pipeline called a penstock. The penstock directs the water downhill toward the turbines. As the pressurized water flows through the penstock, it strikes the blades of a turbine, causing them to spin. The turbine is connected to a generator, usually through a shaft, so the rotational energy of the turbine is converted into electrical energy by the generator. The spinning turbine turns the generator's rotor, which is surrounded by coils of wire within a magnetic field. This movement induces an electric current in the wire coils, generating the electricity generated by the generator typically in the form of alternating current (AC). It is sent to a transformer, which increases the voltage for efficient long-distance transmission. The electricity is transmitted through high-voltage transmission lines to substations, where the voltage is reduced for distribution to homes, businesses, and industries. The total available electric power generated by small Indian hydropower plants is about 3803.68 MW. This sector is also growing rapidly over the last decade.

Hydroelectric power is generated from the energy of flowing water. It is one of the important global electric energy sources for many decades (Bilgili et al. 2018).

1.4.6.1 Different Technology Used for Hydro Electric Power Plant

These are different technologies used for hydroelectric power production, which may be divided based on their available resources and design.

Conventional Hydroelectric Plants

Conventional hydroelectric plants use dams to store water in reservoirs. When water is released from the reservoir with the help of penstock, it flows over the turbines, which are connected to generators. The force of the flowing water spins the turbines, converting mechanical energy into electrical energy (Güney and Kaygusuz 2010) hydroelectric power stands as a renewable energy source, harnessing the natural water cycle for its energy generation. Conventional hydroelectric plants offer consistent and reliable baseload power, effectively meeting a substantial portion of electricity demand. Furthermore, dams constructed for hydroelectric power generation serve the dual purpose of flood control by regulating water flow and mitigating flood risks downstream. Additionally, hydroelectric reservoirs frequently create recreational opportunities like boating, fishing, and wildlife habitats, thus enhancing local tourism and economies.

Pumped-Storage Hydroelectricity

This type of hydroelectricity facility consists of two reservoirs situated at different heights. During periods of low electric demand from the user side. Surplus electricity is used to pump water from the lower reservoir to the upper reservoir. When electricity demand is high, this surplus water is used from the upper reservoir through turbines to generate electricity. Pumped-storage hydroelectricity plays a crucial role in energy storage by offering valuable capabilities to balance supply and demand, integrate intermittent renewable energy sources, and bolster grid stability (Quirapas and Taeihagh 2021). Moreover, pumped-storage hydroelectricity facilities contribute ancillary services such as frequency regulation, voltage control, and grid stabilization, thereby enhancing overall grid reliability and resilience.

Run-of-the-River Hydroelectric Systems

This type of system does not require large reservoirs or significant changes to natural water flow. Instead, they divert a portion of the river's flow through turbines to generate electricity (Aşçıoğlu 2019). These systems typically have smaller environmental footprints compared to conventional hydroelectric plants. Run-of-the-river hydroelectric systems boast minimal impact on river ecosystems, thereby safeguarding natural habitats and aquatic biodiversity. These projects can be crafted to minimize ecological disruption and honor local communities' social and cultural values, fostering sustainable development. Moreover, run-of-the-river hydroelectric systems offer distributed generation capabilities, making them suitable for deployment in remote or rural areas. By providing reliable electricity access, they stimulate local economic development and contribute to community prosperity.

1.4.7 Tidal and Wave Energy

The tidal energy is generated by the flow of low and high tides. Tidal energy is generated from the kinetic energy from water bodies, particularly the movement of tides caused by gravitational forces exerted by the moon and the sun. Tidal energy technologies are designed to capture and convert the energy present in tidal movements into electricity. These technologies typically involve tidal barrages, tidal stream generators, and tidal turbines. In India, potential sites for tidal power plants include the Gulf of Kutch and the Gulf of Cambay in Gujarat, as well as the delta of the Ganges in the Sundarbans region of West Bengal. In India, the tidal power potential is estimated to be around 8000 MW, in the area of Gulf of Cambay and Gulf of Kutch (Mitra et al. 2023).

1.4.7.1 Different Technologies of Tidal and Wave Energy

Tidal and wave energy are forms of renewable energy derived from the natural motion of the oceans. They offer significant potential for generating electricity with minimal greenhouse gas emissions. Here's an overview of the main technologies used to harness tidal and wave energy.

Tidal Barrages

Tidal barrages are structures built across estuaries or tidal rivers to capture the energy of rising and falling tides. As the tide flows in and out, water passes through turbines installed within the barrage, generating electricity. These systems produce electricity without air pollutants and emission of greenhouse gases during operation, thereby aiding in climate change mitigation and enhancing air quality. Additionally, tidal barrages play a crucial role in flood control by regulating tidal flows and mitigating flood risks in coastal areas. This helps protect communities and infrastructure from inundation, ensuring their safety and resilience against natural disasters.

Tidal Stream Generators

This type of power plant operates similarly to underwater wind turbines but harnesses the flow power of tidal currents instead of wind. They are anchored to the seabed and use rotor blades to capture the energy of flowing water and drive generators to produce electricity. Tidal currents serve as a highly predictable energy source, enabling precise forecasting of electricity generation and seamless grid integration (Nicholls-Lee and Turnock 2008). This predictability enhances the reliability and stability of energy systems. Unlike tidal barrages, tidal stream generators exert minimal environmental impact, safeguarding marine habitats and biodiversity in coastal and estuarine ecosystems. Furthermore, tidal stream projects foster job opportunity in different sectors as production, maintenance, and operation. This support for local economies and communities in coastal regions contributes to sustainable development and prosperity.

Wave Energy Converters (WECs)

Wave energy converters harness the up-and-down flow of ocean waves to generate electricity power. WECs come in various designs, including point absorbers, oscillating water columns, and attenuators, each utilizing different mechanisms to capture wave energy. Wave energy represents an abundant resource that is widely distributed across coastal regions worldwide, presenting a substantial potential source of renewable electricity. Additionally, wave energy has the capacity to enhance grid stability by complementing other energy sources. It can provide a reliable and complementary energy supply that assists in balancing fluctuations in electricity demand and generation, thereby contributing to a more stable and resilient electrical grid infrastructure. Wave energy represents an abundant resource that is widely distributed across coastal regions worldwide, presenting a significant potential source of renewable electricity. Furthermore, wave energy has the capability to complement other energy sources such as wind and solar power. By providing a reliable and complementary energy supply, wave energy helps to balance fluctuations in electricity demand and generation, thus enhancing grid stability and contributing to a more sustainable energy ecosystem.

1.5 Renewable Energy Status

The condition of renewable energy in the present and previous scenarios in comparison to fossil fuels' presence and their different issues may reported as.

1.5.1 Renewable Energy Capacity Growth

Over the last decade, the capacity of renewable energy has tripled, signaling a substantial surge in its adoption and investment worldwide. Among all types of renewable energy sources, wind energy has consistently maintained a leading position, underscoring its prominence in the nonconventional energy sector. Additionally, solar photovoltaic (PV) capacity has witnessed rapid expansion following the initiatives spearheaded by the Jawaharlal Nehru National Solar Mission in 2009, further diversifying the renewable energy landscape.

1.5.2 Reductions in Cost and Competitiveness

The rate of wind turbine installations has remained relatively stable since the 1990s and has shown consistency over the past decade, contributing to the continuous progress of wind energy capacity. Moreover, the average capacity factor of wind turbines has progressively increased, signifying enhanced efficiency in electricity generation. Rapid declines in solar PV costs, with a notable 50% reduction between 2010 and 2014, have made solar power increasingly competitive with conventional energy resources. Furthermore, the widespread prospect of grid connectivity for solar power globally is emerging, indicating that renewable energy sources, particularly solar PV, are becoming economically viable options in many countries. Additionally, biomass power generation technologies, particularly those utilizing low-value agricultural, forest residues, or crop waste, offer economically feasible options for power generation, further diversifying the renewable energy landscape.

1.5.3 Concern Regarding Global Warming

Many countries are growing increasingly concerned about global warming, prompting heightened attention to decreasing greenhouse gas emissions. All renewable energy sources present a cleaner alternative to fossil fuels, offering countries a means to mitigate their contributions to global warming. In regions like Punjab and Haryana, where environmental issues arise from the burning of agricultural residues, there is a pressing need to adopt sustainable alternatives, such as biomass energy for electricity generation. This thought to renewable energy sources not only addresses environmental concerns but also contributes to increased efforts to combat climate change in the world.

1.5.4 Hydropower Growth and Constraints

While hydropower is considered an economical and sustainable electricity generation technology, so its growth is lower compared to solar PV and wind due to geographical constraints and various construction requirements.

1.5.5 Growth in Renewable Energy Capacity

Over the last decade, the capacity of renewable energy has tripled, reflecting significant growth and investment in renewable energy infrastructure globally. Wind energy stands out as the predominant renewable energy source, maintaining a substantial percentage share due to the stable prices of wind turbines since the 1990s. Efficiency improvements in wind energy, evidenced by the gradual growth in average capacity factors, have resulted in increased electricity generation and reduced costs. Solar photovoltaic (PV) has experienced rapid expansion following the introduction of the Jawaharlal Nehru National Solar Mission in 2009, driven by decreased costs of solar PV panels. However, hydropower faces challenges due to geographical constraints and construction requirements, despite offering the cheapest electricity among other generation technologies. Biomass power generation, particularly from low-value agricultural or forest waste, proves to be economical, with the combustion of agricultural residue observed in regions like Haryana and Punjab, indicating potential for biomass electricity generation. These trends underscore the dynamic landscape of renewable energy adoption, with each technology facing unique opportunities and challenges.

1.6 Challenges with the Renewable Energy Resource Development

The sources of renewable energy hold essential potential for addressing global energy needs while mitigating climate change. However, several challenges need to be addressed to fully realize their benefits. Let's explore these challenges and future prospects:

1.6.1 Intermittency and Grid Integration Challenges

The intermittent nature of renewable resources like wind and solar energy poses challenges for grid operators due to fluctuations in energy generation caused by weather conditions. To address this issue, grid infrastructure needs to be flexible and adaptable, capable of accommodating these fluctuations (Sundaravadivelu et al. 2017). This entails implementing advanced grid management techniques, integrating energy storage solutions, deploying demand response programs, and offering grid-scale balancing services to ensure grid stability and meet electricity demand effectively. Moreover, regional and seasonal variability in renewable energy resources necessitates the development of diverse portfolios of renewable energy technologies and the interconnection of regional grids to balance supply and demand across different geographical areas. These strategies are needed for optimizing the reliability and efficiency of renewable energy integration into the grid infrastructure.

1.6.2 Technological Barriers and Research Needs

Improving the efficiency and performance of nonconventional energy technologies, for example, solar photovoltaic, wind turbines, and biomass conversion processes, is paramount for maximizing energy yields and reducing costs. Additionally, the development of cost-effective and scalable energy storage technologies, including batteries, pumped water storage, and hydrogen storage systems, is crucial for addressing intermittency challenges, facilitating renewable energy integration, and enhancing grid stability and resilience (Sundaravadivelu et al. 2017). Furthermore, research into advanced materials, manufacturing techniques, and innovative designs can result in lighter, more durable, and cost-effective renewable energy components, thereby enhancing the overall performance and durability of renewable energy systems.

1.6.3 Role of Energy Storage and Grid Modernization

Since energy storage solutions are difficult to manage and smooth out intermittent renewable energy generation, optimizing energy dispatch, and providing grid ancillary services, continued advancements in energy storage technologies and grid-scale deployment are essential for maximizing renewable energy penetration and grid stability. Additionally, modernizing grid infrastructure with advanced monitoring, control, and automation capabilities enhances grid flexibility, reliability, and resilience. Smart grid technologies, demand-side management, and grid-scale energy storage systems enable efficient integration of renewable energy resources and support the transition to a decentralized and dynamic energy system. Clear and consistent policy support, market incentives, and regulatory frameworks are also vital for encouraging investment in renewable energy deployment, energy storage projects, and grid modernization initiatives. Collaborative efforts among governments, utilities, and industry stakeholders are necessary to create conducive environments for renewable energy adoption and innovation to thrive.

1.6.4 Government Incentives and Subsidies

The government of India offers several financial incentives to encourage renewable energy deployment. These incentives include capital subsidies, generation-based incentives (GBI), and viability gap funding (VGF) for project developer, which help reduce the initial investment costs and improve project feasibility. Additionally, renewable energy projects are eligible for accelerated depreciation benefits, allowing developers to claim higher depreciation rates on project assets, thereby reducing taxable income and enhancing project economics. Furthermore, tax incentives such as income tax exemptions, customs duty exemptions on imported equipment, and excise duty exemptions on specific renewable energy components are offered to incentivize investment in the sector. These financial incentives play a vital role in accelerating the growth of renewable energy projects and fostering investment in sustainable energy infrastructure across India.

1.6.5 Renewable Portfolio Standards (RPS)

The National Solar Mission (NSM) seeks to advance solar power generation in India and requires the development of 100 GW of solar capacity by 2022 through initiatives like utility-scale solar projects, rooftop solar installations, and solar parks nationwide (Adhana 2015). Similarly, India's National Wind Energy Mission targets require to develop wind energy up to 60 GW of wind capacity by 2022 (Kumar et al. 2022). The government extends policy support and incentives to bolster wind energy development, employing measures like competitive bidding mechanisms and feed-in tariffs to stimulate growth in the sector. These missions underscore India's commitment to develop renewable energy sources, with solar and wind power considered broader sustainable energy resources.

1.6.6 Current Policies of Electricity in India

Several pivotal policies and acts have steered India's electricity sector and renewable energy endeavors. The National Electricity Policy 2005 aims to ensure high-quality power supply at affordable rates, augment per capita electricity availability, and render the electricity sector commercially viable. The Electricity Act 2003 consolidates prior regulations, prioritizing rural transmission and distribution of electricity while advocating for cogeneration and electricity generation from renewable sources (Elavarasan et al. 2020). Addressing connectivity with the grid, electricity sales, tariff principles, and the restructuring of state electricity boards, it lays a comprehensive foundation. The Tariff Policy 2006 establishes minimum energy purchase percentages and encourages distribution companies to procure energy from renewables at preferential tariffs. The National Rural Electrification Policies 2006 strive for reliable and economical electricity access for all households by 2009, particularly through grid solutions in rural areas. The National Plan to Mitigate Climate Change 2008 encompasses various missions, notably the Jawaharlal Nehru National Solar Mission (JNNSM) and the National Mission for Enhanced Energy Efficiency (NMEEE), which aim to make solar energy sustainable and economical while focusing on energy management and demand-side savings. These policies collectively reflect India's commitment to enhancing energy accessibility, fostering renewable energy adoption, and tackling climate change challenges through strategic planning and regulatory frameworks.

1.7 Conclusion

The promotion and development of renewable energy sources in India are critical for mitigating greenhouse gas from the power sector and addressing the challenges posed by fossil fuel scarcity. As renewable energy becomes increasingly integrated into the energy mix, it will play a pivotal role in bridging power supply deficits and expanding energy access to remote areas. It's projected that greenhouse gas emissions could be reduced by 4–45% in the coming years, with up to 14% achievable through the substitution of coal with renewable energy sources. However, realizing this potential requires a combination of appropriate technologies and changes in human behavior, considering the limitations inherent in each technology. To give pace to the adoption of renewable energy, increased investment in research and development, innovation, and benchmarking are essential for improving efficiency and overcoming technological limitations. Moreover, robust government support mechanisms, such as subsidies, tax exemptions, feed-in tariffs, green certificates, and various financial incentives, are crucial for encouraging local entrepreneurs and driving renewable energy technologies forward. By leveraging these strategies and fostering a supportive policy environment, India can harness its abundant renewable energy potential to build prosperous energy future while addressing environmental challenges and promoting economic development.

References

Adhana D (2015) Solar Energy Mission: Paving the Way for India's Transformational Future. *International Journal of Advanced Research in Management and Social Sciences*, *4*(12)

Ahmad T, Zhang D (2020) A critical review of comparative global historical energy consumption and future demand: the story told so far. Energy Rep 6:1973–1991

Akorede MF, Hizam H, Ab Kadir MZA, Aris I, Buba SD (2012) Mitigating the anthropogenic global warming in the electric power industry. Renew Sust Energ Rev 16(5):2747–2761

Anderson JH, Mayer DF (1979) Sea thermal power: competitive electricity and chemicals from the sea. In: Advances in energy systems and technology. Academic Press, New York, pp 49–100

Aşçıoğlu B (2019) Improving the energy conversion stability of hybrid solar photovoltaic/wind turbine renewable energy systems. Master's thesis, Middle East Technical University
Barber E (2002) Geothermal energy technology and current status: an overview. Renew Sust Energ Rev 6(1–2):3–65
Barnard RC, Morgan DL (2000) The Basic Science Issue in Global Warming is Whether Human Activity is Forcing Changes in Climate. *Inside EPA's Risk Policy Report*, *7*(5), 33–35
Bilgili M, Bilirgen H, Ozbek A, Ekinci F, Demirdelen T (2018) The role of hydropower installations for sustainable energy development in Turkey and the world. Renew Energy 126:755–764
Dahiya AK (2018) Z Source Inverter Based Solar Power Generation Fed Induction Motor Drive
Deshmukh A (2009) The role of decentralized renewable energy for rural electrification. Maharashtra case study, India. Assessed from http://lup.lub.lu.se/luur/download
Dubey DK (2015) Issues and challenges in electricity sector in India. Bus Manag Rev 5(4):132
Elavarasan RM, Shafiullah GM, Padmanaban S, Kumar NM, Annam A, Vetrichelvan AM, Mihet-Popa L, Holm-Nielsen JB (2020) A comprehensive review on renewable energy development, challenges, and policies of leading Indian states with an international perspective. IEEE Access 8:74432–74457
Erlach B (2014) Biomass upgrading technologies for carbon-neutral and carbon-negative electricity generation. https://doi.org/10.14279/depositonce-4224
Fitzsimmons C (1999) The use of non-thermal plasma systems for VOC destruction and air quality improvement. The University of Manchester, Manchester
Güney MS, Kaygusuz K (2010) Hydrokinetic energy conversion systems: a technology status review. Renew Sust Energ Rev 14(9):2996–3004
Guski R, Schreckenberg D, Schuemer R (2017) WHO environmental noise guidelines for the European region: a systematic review on environmental noise and annoyance? Int J Environ Res Public Health 14(12):1539
Hansen KS, Barthelmie RJ, Jensen LE, Sommer A (2012) The impact of turbulence intensity and atmospheric stability on power deficits due to wind turbine wakes at horns rev wind farm. Wind Energy 15(1):183–196
Igbum OG, Eloka-Eboka AC, Ubwa ST, Inambao FL (2014) Evaluation of environmental impact and gaseous emissions of biodiesel fuels and blends of selected feed-stocks. Int J Glob Warm 6(1):99–112
Kabeyi MJB (2019) Geothermal electricity generation, challenges, opportunities and recommendations. Int J Adv Sci Res Eng 5(8):53–95
Karampinis E, Kourkoumpas DS, Grammelis P, Kakaras E (2015) New power production options for biomass and cogeneration. Wiley Interdiscip Rev Energy Environ 4(6):471–485
Kumar S, Pal A (2014) A review of bio-diesel as an alternative fuel for vehicles. Int J Eng Technol Manag Appl Sci 2(7):163–177
Kumar A, Pal D, Kar SK, Mishra SK, Bansal R (2022) An overview of wind energy development and policy initiatives in India. Clean Techn Environ Policy 24(5):1–22
Miah MD, Masum MFH, Koike M (2010) Global observation of EKC hypothesis for CO2, SOx and NOx emission: a policy understanding for climate change mitigation in Bangladesh. Energy Policy 38(8):4643–4651
Mitra M, Singha NR, Chattopadhyay PK (2023) Review on renewable energy potential and capacities of south Asian countries influencing sustainable environment: a comparative assessment. Sustain Energy Technol Assess 57:103295
Mohammed YS, Mustafa MW, Bashir N, Ibrahem IS (2017) Existing and recommended renewable and sustainable energy development in Nigeria based on autonomous energy and microgrid technologies. Renew Sust Energ Rev 75:820–838
Negi H, Suyal DC, Soni R, Giri K, Goel R (2023) Indian scenario of biomass availability and its bioenergy-conversion potential. Energies 16(15):5805
Nicholls-Lee RF, Turnock SR (2008) Tidal energy extraction: renewable, sustainable and predictable. Sci Prog 91(1):81–111
Nisar A, Monroy CR (2012) Potential of the renewable energy development in Jammu and Kashmir, India. Renew Sustain Energy Rev 16(7):5260–5267

Ogunmodimu O, Okoroigwe EC (2018) Concentrating solar power technologies for solar thermal grid electricity in Nigeria: a review. Renew Sust Energ Rev 90:104–119
Quirapas MAJR, Taeihagh A (2021) Ocean renewable energy development in Southeast Asia: opportunities, risks and unintended consequences. Renew Sust Energ Rev 137:110403
Rao KR, Rao KR (2019) Global wind energy and power generation options: socioeconomic factors. In: Wind energy for power generation: meeting the challenge of practical implementation. Springer, Cham, pp 703–828
Shelare S, Kumar R, Gajbhiye T, Kanchan S (2023) Role of geothermal energy in sustainable water desalination—a review on current status, parameters, and challenges. Energies 16(6):2901
Sundaravadivelu S, Norman MSR, Johnsi Stella I (2017) Solar photovoltaic power systems: principles, design and applications. Notion Press, Chennai
Tripathi L, Mishra AK, Dubey AK, Tripathi CB, Baredar P (2016) Renewable energy: an overview on its contribution in current energy scenario of India. Renew Sust Energ Rev 60:226–233
Von Oertzen D (2018) Economic impacts of the deployment of renewable energy technologies in Namibia. Konrad Adenaeur Stiftung, Bonn
Wang S, Wang S, Liu J (2019) Life-cycle green-house gas emissions of onshore and offshore wind turbines. J Clean Prod 210:804–810

Chapter 2
Renewable Energy Insights: Sources and Technological Advancements

Ankur Bhardwaj and Vivek Kumar Singh

Abstract Renewable energy derives from continuously replenished natural resources including sunlight, wind, waves, biomass, tides, and geothermal heat. These sources are inherently sustainable, capable of indefinite utilization without depletion, due to their natural replenishment cycles. This abstract explores the diverse sources and technologies of renewable energy, their progress, global implications, and sustainable applications in powering our world. Cutting-edge technologies like floating solar panels, perovskite solar cells, offshore wind energy, and production of green hydrogen are shaping the future of renewable energy. Smart grid integration plays a critical role in this advancement. In developing countries, renewable energy presents pathways to prosperity, offering sustainable energy solutions, enhanced energy security, climate action, rural electrification, empowerment, and economic growth. However, the global shift toward renewable energy faces unique challenges, particularly in developing countries, including intermittency, land acquisition, and financing. Overcoming these challenges necessitates innovative solutions and strategic planning, which are essential for achieving net-zero emissions and promoting sustainable development. This book chapter outlines the opportunities and considerations for advancing renewable energy integration, highlighting its potential as an engine for economic growth and climate responsibility.

Keywords Green hydrogen · Renewable energy · Sustainable energy · Rural electrification · Net-zero emissions

A. Bhardwaj
GreenTree Global, Noida, Uttar Pradesh, India

V. K. Singh (✉)
Energy for Sustainability Initiative, University of Coimbra, Coimbra, Portugal

Emeraldgears Initiative Foundation (EGIF) Global, Greater Noida, Uttar Pradesh, India

S. Kumar, V. K. Singh (eds.), *Renewable Energy Development: Technology, Material and Sustainability*, Clean Energy Production Technologies,
https://doi.org/10.1007/978-981-97-9626-7_2

2.1 Introduction

Renewable energy encompasses energy derived from continuously replenished including natural resources sunlight, wind, waves, tides, and geothermal heat and biomass (including plant and animal matter). These sources are inherently sustainable, meaning they can be utilized indefinitely without depletion. The sustainability of renewable energy sources indeed lies in their natural replenishment cycles. Unlike finite fossil fuels, these resources can be harnessed indefinitely, ensuring long-term availability (Suhrab et al. 2023).

Renewable energy sources not only help mitigate climate change and enhance energy security but also provide substantial environmental and health benefits. Unlike fossil fuels, which emit particulate matter, sulfur dioxide (SO_2), nitrogen oxides (NOx), and other pollutants, renewable energy technologies generally have minimal adverse impacts on air quality (Gyimah et al. 2024). This reduction in pollution helps to improve public health outcomes by mitigating respiratory conditions and other health concerns caused by poor air quality. Furthermore, renewable energy production typically involves less habitat destruction and ecosystem disruption compared to the extraction and combustion of fossil fuels. For instance, hydropower, while altering river ecosystems to some extent, generally has a lower impact on land use and biodiversity compared to coal mining or oil drilling operations. Similarly, solar and wind energy installations can coexist with agriculture or wildlife habitats, minimizing direct environmental impacts.

However, the sustainability of renewable energy sources is not without challenges (White 2024). While renewable resources are naturally replenished, their sustainability can vary depending on several factors. Effective resource management practices are essential to ensure that renewable sources should be used in a way that preserves their availability and quality over time. For instance, while biomass energy from agricultural waste or forest residues is theoretically renewable, it can become unsustainable if harvesting practices exceed the rate at which organic materials regenerate naturally. Excessive harvesting of biomass can result in soil degradation, diminished biodiversity, and decreased ecosystem services. Additionally, technological advancements are vital for improving the sustainability and efficiency of renewable energy systems. Continuous research and development are crucial for improving the performance and cost-efficiency of solar photovoltaic cells, wind turbines, energy storage systems, and other renewable technologies. Advances in materials science, energy storage, and grid integration are key to managing intermittent energy production and boosting the reliability of renewable energy sources.

Furthermore, advancing toward a renewable energy future necessitates the implementation of supportive policies and significant investments in infrastructure, and public acceptance. Governments, businesses, and communities must collaborate to accelerate the Implementing of renewable energy technologies and supporting the shift from fossil fuels are crucial. Policy incentives, including subsidies for renewable energy, tax credits, feed-in tariffs, and carbon pricing mechanisms, can drive investment in clean energy and ensure a fair competitive environment for renewable technologies in the energy market. The renewable energy represents a cornerstone of sustainable development and a critical pathway to achieving global climate goals. Utilizing solar, wind, hydro, geothermal, and biomass energy allows societies to cut greenhouse gas (GHG) emissions, bolster energy security, and advance environmental conservation. However, to fully capitalize on renewable energy's potential, it is essential to tackle the technological, economic, and policy challenges involved (Gielen et al. 2019). Through innovation, collaboration, and commitment to sustainability, we can build a cleaner, more resilient energy system that serves todays and tomorrow's needs while protecting the planet for future generations.

2.2 Renewable Energy: Exploring Diverse Sources and Technologies

In today's world, the dual amid the growing challenges of increasing energy demands and climate change, the transition to renewable energy technologies offers a promising solution. These cutting-edge technologies harness naturally replenishing sources—like sunlight, wind, water, geothermal heat, and biomass—providing a sustainable and eco-friendly alternative to conventional fossil fuels (Ellabban et al. 2014).

Unlike finite resources like coal, oil, and natural gas are finite resources, renewable energy sources are replenished continuously by natural processes. This inherent renewal ensures long-term energy security and significantly reduces environmental impact, as these sources emit minimal to no greenhouse gases during their operation (Osman et al. 2023). This pivotal characteristic not only mitigates the exacerbation of global warming but also fosters a more resilient energy infrastructure capable of sustaining future generations. Renewable energy technologies encompass a diverse range of options, each providing its own set of benefits and applications. Some of the most prominent include:

- *Solar Energy*: Solar technologies capitalize on the sun's abundant energy, with options ranging from photovoltaic panels that directly convert sunlight into electricity to concentrated solar power systems that use mirrors or lenses to focus and harness sunlight for generating heat used in electricity production.
- *Wind Energy*: Wind turbines, emblematic of renewable energy, capture the kinetic energy of the wind. They convert this energy into electricity through the rotation of blades linked to a generator, offering scalable solutions from small turbines for individual homes to vast offshore wind farms.
- *Hydropower*: Harnessing the gravitational energy of flowing water, hydropower systems generate electricity using turbines installed in dams or rivers. This technology, ranging from large-scale hydroelectric dams to small run-of-river installations, provides reliable and predictable power generation with minimal environmental emissions once the infrastructure is in place.

- *Geothermal Energy*: Exploiting geothermal energy from within the Earth, geothermal power plants convert thermal energy from deep within the Earth into electricity. This renewable resource is particularly valuable in regions with high geothermal activity, where heat can be accessed through wells drilled into hot rock formations or shallow ground sources.
- *Bioenergy*: Produced from organic materials like agricultural residues, forestry by-products, and specialized energy crops, bioenergy encompasses a versatile range of technologies. These include biofuels for transportation and heating, as well as biomass power plants that burn organic matter to generate electricity or produce biogas through anaerobic digestion processes.

2.3 Renewable Energy: Progress, Trends, and Global Implications

As of 2023, the global capacity of renewable energy has experienced remarkable growth, marking a pivotal shift toward sustainable energy sources. The International Energy Agency (IEA) reported (IEA 2024) an unprecedented addition of 500 GW in renewable energy capacity, representing a significant 50% increase from the previous year in Fig. 2.1. This surge underscores the rapidly increasing momentum in the global adoption of cleaner energy technologies. Projections by IEA indicate that by 2030, global renewable energy capacity is poised to reach 7300 GW, signaling a trajectory that could triple current levels. Such advancements are not only driven by technological innovations and policy frameworks but also highlight a collective

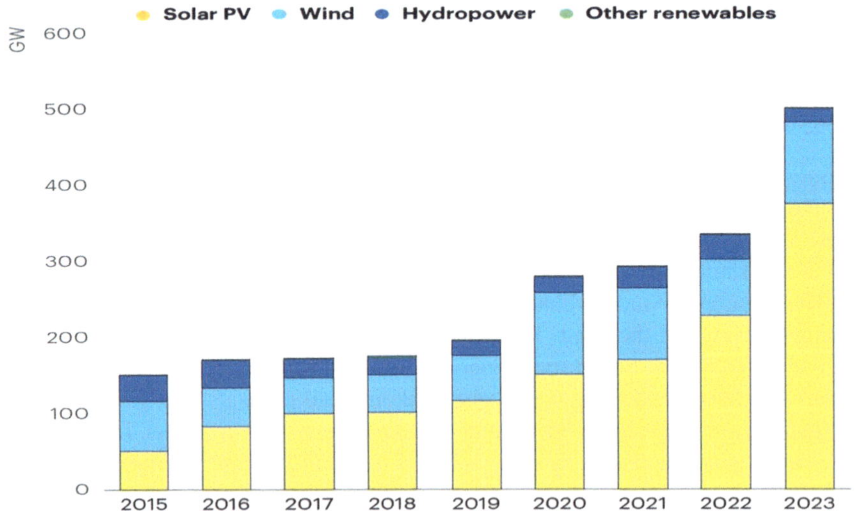

Fig. 2.1 Global renewable energy additional capacity 2015–2025. (IEA 2024)

commitment to achieving ambitious climate goals set forth in international agreements like COP28. The ongoing expansion of renewable energy has shown a promising path toward a stronger and more robust global energy landscape. The exploration and advancement of renewable energy technologies are fueled by ongoing research, technological innovation, and policy support across the globe. These efforts are crucial in combating climate change, reduce dependency on fossil fuels, and promote energy security. Renewable energies such as solar, wind, hydropower and biomass are at the forefront of this change and are providing large-scale solutions with minimal environmental impact compared to traditional energy sources.

In 2023, the substantial increase of 500 GW in global renewable energy capacity is a testament to the speedy scale-up of renewable energy systems wide-reaching. This expansion is driven by various issues, incorporating of declining price of renewable technologies, favorable country specific policies, and growing public and private sector investments in clean energy projects. The increase in capacity not only addresses the growing global energy demand but also plays a crucial role in reducing greenhouse gas emissions and enhancing air quality. IEA projections for 2030, estimating a global renewable energy capacity of 7300 GW, highlight the robust growth anticipated in the sector. This forecast demonstrates a significant acceleration compared to historical trends, underscoring the growing viability and competitiveness of renewable energy sources. The aim to triple current capacity by 2030 aligns with international agreements such as the Paris Agreement and COP28, showcasing the global dedication to sustainable development and climate resilience. This ambitious trajectory highlights the increasing adoption of renewable energy as a key strategy in addressing climate change and fostering a sustainable future for all.

The expansion of renewable energy capacity is not limited to a few regions but is a global phenomenon. Countries across continents are ramping up their renewable energy investments and installations, driven by both environmental imperatives and economic opportunities. Regions such as China, Europe, the United States, and India have made significant progress in the use of renewable energy sources. This progress is driven by the main goal and policies of decarbonizing energy. Technology development plays a significant role in this progress, with the development of solar photovoltaic (PV) systems, wind turbine construction, energy storage, and collaborative projects to increase the reliability and efficiency of renewable energy solutions. These advances are critical to solving traditional interoperability issues with renewable energy sources and making them more reliable for the energy mix.

Moreover, the declining costs of renewable energy technologies have been instrumental in driving their widespread adoption. The levelized cost of electricity (LCOE) for renewable resources such as solar and wind has significantly decreased in recent years, positioning them as viable alternatives to fossil fuel-based generation in many parts of the world. As economies of scale and technological improvements continue to drive down costs, renewable energy is becoming the preferred choice for new power generation capacity globally. Policy frameworks and government incentives have been instrumental in accelerating the deployment of renewable energy. Measures such as feed-in tariffs, renewable portfolio standards, tax incentives, and carbon pricing mechanisms have driven investments in clean energy projects and promoted the shift away from carbon-intensive fuels. Additionally, international collaborations and partnerships have enhanced knowledge sharing and capacity building, supporting countries in achieving their renewable energy goals. The expansion of renewable energy capacity not only reduces climate risks but also drives economic growth and job creation. The renewable energy sector has become a major global employment source, providing opportunities across the value chain—from manufacturing and installation to maintenance and research (World Economic Forum 2023). As countries invest in renewable energy infrastructure, they diversify their energy sources while simultaneously building resilient and sustainable economies for the future.

Looking ahead, the continued momentum in renewable energy deployment is essential for achieving long-term climate goals and ensuring energy security. Governments, businesses, and civil society must continue to collaborate and innovate to overcome remaining barriers and accelerate the transformation to a sustainable, low-carbon energy infrastructure. Investments in research and development, supportive policies, and international cooperation are essential to fully realize the transformative potential of renewable energy to create a robust and sustainable energy landscape, safeguarding the well-being and prosperity of future generations (Dey et al. 2022).

2.4 Applications of Renewable Energy: Powering Our World Sustainably

Renewable energy stands at the forefront of global initiatives to combat climate change, strengthen energy security, and promote sustainable development (Osman et al. 2023; Akella et al. 2009). With a diverse array of technologies harnessing renewable resources such as wind, sunlight, water, and geothermal heat, renewable energy proposals multifaceted solutions that extend beyond electricity generation, permeating various sectors of the economy. Let's explore in-depth how renewable energy is driving transformative change across key areas:

2.4.1 Electricity Generation: Powering Our Future

Renewable energy has transformed electricity generation, ushering in an era of cleaner and more sustainable power sources:

- Large-Scale Integration: Solar and wind farms, alongside geothermal and hydropower plants, are pivotal in national and regional energy grids. These facilities are pivotal in diminishing reliance on fossil fuels, resulting in significant reductions in greenhouse gas emissions and substantial improvements in air quality within urban and industrial centers.

- Empowering Remote Communities: Decentralized renewable energy systems are vital in supplying electricity to remote and underserved communities, effectively bridging the energy access gap and promoting equitable and sustainable development (Ha and Kumar 2021). Technologies such as solar home lighting kits and micro-hydropower installations ensure reliable energy access, enhancing livelihoods, and promoting economic development where grid connectivity is limited.

2.4.2 *Transportation: A Green Shift in Motion*

Renewable energy is reshaping the transportation sector, offering sustainable alternatives to longstanding fossil fuel-powered vehicles:

- Electric Vehicles (EVs): The proliferation of EVs powered by renewable electricity marks a significant shift toward cleaner mobility solutions. From personal cars and buses to maritime vessels and airplanes, the widespread adoption of electric vehicles (EVs) significantly diminishes dependence on fossil fuels, drastically cuts tailpipe emissions, and substantially improves air quality in urban environments.
- Sustainable Biofuels: Renewable biofuels derived from biomass offer sustainable alternatives to traditional gasoline and diesel fuels. Bioethanol and biodiesel contribute to reducing lifecycle emissions and provide viable solutions in sectors where full electrification is challenging, such as aviation and heavy freight transport.

2.4.3 *Heating and Cooling: Innovating Comfort*

Renewable energy technologies are redefining how we heat and cool buildings, enhancing energy efficiency and reducing carbon footprints:

- Solar Thermal Systems: Solar water/air heaters effectively harness solar energy to provide hot water and space heating. By reducing dependence on conventional heating methods, these systems lower energy costs and promote sustainable building practices using abundant solar resources.
- Geothermal Heat Pumps: Geothermal energy leverages the Earth's stable subsurface temperatures to efficiently heat and cool buildings throughout the year. This innovative technology provides a cost-effective and sustainable alternative to traditional HVAC systems, significantly enhancing energy savings and reducing greenhouse gas emissions in residential, commercial, and industrial applications.

2.4.4 Industry: Toward Sustainable Practices

Renewable energy is driving sustainability across industrial sectors, optimizing energy use and reducing environmental impacts:

- Industrial Applications: Renewable energy resources such as solar and wind power are progressively more integrated into industrial operations, from manufacturing and mining to food processing. This integration helps industries reduce carbon footprints, enhance energy security, and comply with stringent environmental regulations.
- Combined Heat and Power (CHP): CHP systems cogenerate electricity and useful heat from a single fuel source, often renewable. By capturing and utilizing waste heat, CHP technologies significantly improve overall energy efficiency in industrial processes, thereby minimizing energy costs and environmental impacts.

2.4.5 Agriculture and Rural Development: Empowering Communities

Renewable energy solutions show a crucial role in promoting sustainable agriculture and rural development:

- Solar-Powered Irrigation: Solar irrigation pumps provide reliable water access for agricultural activities in remote areas with limited grid connectivity. These systems enhance crop productivity, support food security, and promote sustainable water management practices by harnessing solar energy effectively.
- Biogas: Transforming Waste into Fuel: Biogas systems convert organic waste, like agricultural residues and animal manure, into biogas for cooking, lighting, and electricity generation in rural communities. This decentralized energy solution not only reduces organic waste and methane emissions but also fosters energy independence and economic resilience in rural areas.

Looking ahead, the future of renewable energy is poised for remarkable growth and innovation, driven by advancements in technology, supportive policies, and expanding market dynamics. Emerging technologies are at the forefront of this evolution, with breakthroughs in energy storage, smart grid solutions, and advancements in renewable resource management are laying the groundwork for more reliable and scalable renewable energy systems. Coupled with breakthroughs in materials science and digital technologies, these innovations are driving down costs and expanding access to renewable energy solutions across diverse geographic regions. Policy support plays a critical role in accelerating the adoption of renewable energy technologies. Supportive policies, incentives, and regulatory frameworks—including carbon pricing, renewable energy mandates, and green finance initiatives—are pivotal in stimulating investments in clean energy infrastructure. These mechanisms not only spur technological innovation but also catalyze sustainable economic growth by creating new markets and job opportunities within the renewable energy sector. Through these combined efforts, we are witnessing a transformative shift towards a cleaner, more resilient energy future.

2.5 Cutting-Edge Technologies Paving the Way for the Renewable Energy Future

Devolving country, with its vast geographical expanse and diverse resources, holds immense potential for renewable energy generation. As the country strives to meet its ambitious clean energy goals, several innovative technologies are emerging to harness this potential efficiently and sustainably. One of the most promising areas is solar energy, where advanced photovoltaic (PV) technologies are leading the way. Innovations such as bifacial panels, which can capture sunlight on both sides, and perovskite solar cells, celebrated for their exceptional efficiency and low manufacturing costs, are transforming the solar energy sector. These advancements are particularly beneficial in regions with high solar irradiance, allowing devolving country to maximize its solar potential and generate significant amounts of clean energy.

Wind energy is another critical segment of the renewable energy mix in devolving country. Recent developments in wind turbine design, including larger blades and taller towers, have greatly boosted the efficiency and capacity of wind farms. Furthermore, offshore wind technology is increasingly gaining popularity, facilitated by floating turbines that allow deployment in deeper waters. This advancement harnesses wind resources that were previously inaccessible, thus expanding the horizons of wind energy production. The combination of onshore and offshore wind energy technologies positions devolving country to harness its substantial wind energy potential effectively. Energy storage solutions are essential for addressing the intermittent nature of renewable energy sources. Progress in battery technology, particularly with lithium-ion and the development of solid-state batteries, is improving energy-storing capabilities and enabling the storage of excess energy generated during peak production phases and its utilization during high-demand periods. These storage solutions are critical for safeguarding a constant and unfailing energy supply, enabling devolving country to integrate more renewable energy into its grid. Moreover, advancements in grid-scale storage systems and new smart grid-connected technologies are significantly boosting the overall efficiency and resilience of energy infrastructure. Hydropower remains a significant player in Devolving Country's renewable energy portfolio. Innovations in small-scale and micro-hydro systems are expanding the potential for hydropower generation, particularly in remote and rural areas. These systems can provide reliable and consistent power, contributing to local energy security and supporting sustainable development. Additionally, advancements in turbine technology and the addition of hydropower with other renewable energy resources, such as solar and wind, are optimizing energy production and distribution. Finally, bioenergy is gaining momentum as a versatile and sustainable energy source, increasingly recognized for its potential across various applications. Innovations in biofuel production, including advanced bio-refining processes and the use of non-food biomass, are enhancing the efficiency and sustainability of bioenergy. These technologies provide an effective means of lowering greenhouse gas emissions and promoting energy self-sufficiency.

The deployment of cutting-edge technologies in devolving country is laying the new path for renewable energy sustainability. By leveraging advancements in solar, wind, energy storage, hydropower, and bioenergy, the country can harness its diverse resources effectively and sustainably. These innovations are not only helping Devolving Country meet its clean energy goals but also positioning it as a leader in the global transition to renewable energy. Here are a few noteworthy examples:

2.5.1 Floating Solar Panels

Floating solar panels, also referred to as floating photovoltaic (FPV) systems, are installed on water bodies, such as reservoirs, lakes, ponds, and offshore sites. These panels are typically mounted on buoyant structures that keep them afloat. Floating solar panels offer several significant benefits that make them an appealing option for expanding renewable energy sources. One major advantage is land optimization (Kowsar et al. 2023); by utilizing underutilized water bodies like reservoirs and lakes, these systems avoid competing for valuable land resources, which are often scarce in densely populated areas. Furthermore, the cooling effect of water improves the proficiency of the panels, as the water helps maintain lower temperatures compared to land installations, improving electricity generation. Floating solar panels also provide environmental benefits by reducing water evaporation from reservoirs, a crucial factor in arid and semiarid regions where water conservation is essential. Moreover, they offer enhanced performance and longevity by staying cleaner with less dust and dirt accumulation, which reduces maintenance needs. Their compatibility with hydropower infrastructure allows them to complement existing facilities, sharing resources like transmission lines and water bodies. Lastly, the flexibility and scalability of floating solar installations enable them to be easily expanded or relocated, adapting to various geographical and environmental conditions. These benefits collectively make floating solar panels a versatile and effective solution for addressing energy and water conservation challenges.

Despite these benefits, floating solar also presents challenges such as higher installation costs compared to ground-mounted systems, potential impacts on aquatic ecosystems, and logistical challenges in maintenance and installation. Floating solar panels represent a promising avenue for expanding renewable energy capacity while addressing land constraints and water conservation challenges.

As technologies advance and more projects are implemented globally, they are likely to play a crucial role in the energy mix of various countries working to achieve their clean energy targets.

2.5.2 Perovskite Solar Cells

Perovskite solar cells are an emerging generation of photovoltaic technology that utilize perovskite materials, offering several advantages over traditional silicon-based solar cells (Bati et al. 2023). These advantages include higher efficiency, lower production costs, and greater flexibility in applications. Perovskite materials can be produced using simpler manufacturing processes, which can significantly reduce the cost of solar panel production. Additionally, perovskite solar cells can be made into lightweight, flexible panels, expanding their potential uses beyond traditional rigid panels. Their ability to achieve high efficiency at a lower cost makes them a promising advancement in renewable energy technology. Perovskite solar cells offer several key benefits that variety them an elegant substitute to traditional silicon-based solar cells. First, they exhibit enhanced efficiency, with the potential to achieve over 30% efficiency in the future, surpassing the efficiency of conventional silicon cells. Second, the lower cost of perovskite materials and their simpler manufacturing processes result in more affordable solar panels. Finally, their flexibility and lightweight nature allow perovskite cells to be used in diverse applications, including rooftops, building facades, and wearable electronics. These advantages position perovskite solar cells as a promising technology in the renewable energy landscape.

2.5.3 Offshore Wind Energy

Offshore wind farms capitalize on the more powerful and steady winds present over ocean, compared to onshore locations. This results in higher energy output and more reliable power generation. Additionally, offshore wind farms can be built on a larger scale and face fewer space constraints, making them a significant contributor to renewable energy sources. Their location away from populated areas also minimizes visual and noise impacts on communities.

Offshore wind farms offer several notable benefits. They have a higher capacity factor due to the stronger and steadier winds over oceans, allowing for more efficient and consistent electricity generation. Additionally, the reduced visual impact of offshore wind farms, being located further from populated areas, minimizes their effect on coastlines and residential areas. This form of development also provides access to new resources for clean energy generation, particularly valuable for countries with limited land availability, expanding the potential for renewable energy production.

2.5.4 Green Hydrogen Production

Green hydrogen is produced through electrolysis, which uses electricity generated from renewable sources such as solar, biomass, or wind energy to split water into hydrogen and oxygen (Ikuerowo et al. 2024). This method underscores its eco-friendly credentials by removing the carbon emissions during the traditional hydrogen production methods. The process is emission-free and uses renewable energy to split water into hydrogen and oxygen. Green hydrogen has the potential to be a clean fuel for many applications, contributing to global decarbonization efforts by providing an alternative to fossil fuels.

Green hydrogen could provide significant benefits to many sectors. First, it reduces greenhouse gas emissions by acting as a clean fuel to replace fossil fuels in transportation, steelmaking, chemical manufacturing and other industries (Islam et al. 2024). Second, green hydrogen plays an important role in energy storage by providing a way to store additional energy from sources such as solar and wind. This stored hydrogen can be released at times when demand is high or renewable energy is low, increasing grid security and flexibility. Green hydrogen can also manage the shift in renewable energy production by storing renewable energy to support balancing projects and releasing it when needed. The ability to store it for long periods without significant energy loss makes it particularly suitable for seasonal energy storage, providing strong energy all year round. The versatility of green hydrogen also extends beyond transportation and industry. It can also generate electricity from fuel cells and provide heat for buildings, increasing its many uses in industry. By promoting the use of green hydrogen, society can reduce its dependence on foreign fossil fuels, improve energy security and promote greater independence. Furthermore, the development of green hydrogen technology and infrastructure supports job growth and future empowerment by creating new jobs in the production, construction, and maintenance of hydrogen production facilities.

2.5.5 Smart Grid Integration

Smart grids combine technologies to improve the efficiency, reliability, and performance of the grid. Smart grids can optimize renewable energy sources such as solar and wind energy by integrating digital communication and control (Khalid 2024). This technology helps instantly monitor, identify, and respond to changes in energy demand and supply, optimize operational plans, and ensure fire management is stable. Smart grids also increase grid security, enable communication between utilities and consumers, and support a more robust and interactive energy ecosystem.

Smart grids offer significant benefits to modernize and optimize energy infrastructure (Kabeyi and Olanrewaju 2023). This technology helps instantly monitor, identify and respond to changes in energy demand and supply, optimize operational plans, and stabilize fire management. Smart grids also increase grid security, enable communication between both utilities and consumers, and support a more robust and interactive energy ecosystem. Control during this period alone increases grid flexibility and reliability; This is important for stable maintenance of power when there are changes in new electricity supply. Second, smart grids facilitate demand management, allowing customers to use electricity based on current price or grid conditions. This not only optimizes energy consumption but also reduces peak demand and reduces the height of the grid during high demand. Finally, smart grid technology increases overall project security through integrated monitoring and advanced management. When it comes to access to renewable energy, this capacity is vital to ensuring continued sustainability even during the transition period. The use of smart energy systems allows countries like India to move towards an energy efficient future, promote energy independence, reduce carbon emissions, and help mitigate climate change.

2.6 Pathways to Prosperity: Harnessing Renewable Energy for Development in Developing Countries

Developing countries, with their rapidly growing energy demands, face a dual challenge: meeting these needs while committing to sustainable development. Renewable energy emerges as a strategic solution, offering numerous benefits across multiple dimensions:

- *Sustainable Energy Solution*: Renewable energy resources, such as biomass, solar, wind, and hydropower, provide low-carbon, inexhaustible substitutes to fossil fuels. These sources are pivotal in addressing the increasing energy requirements of developing countries. Unlike fossil fuels, renewables do not deplete over time and have minimal environmental impact, making them ideal for sustainable long-term energy strategies.
- *Enhanced Energy Security*: By expanding domestic renewable energy generation, developing nations can greatly diminish their reliance on imported fossil fuels. This shift mitigates vulnerability to external market fluctuations and supply chain disruptions. Greater energy independence enhances national security and economic stability, as countries are less affected by geopolitical tensions and volatile fuel prices.
- *Climate Action and Responsibility*: Transitioning to renewable energy is central to achieving ambitious emission reduction targets. For instance, India has set ambitious goals to minimize carbon emissions and maximize the share of renewables in national energy mix. Adopting renewable energy demonstrates a commitment to global climate action and positions nations as responsible leaders in combating climate change. This transition is crucial for meeting nationally determined contributions (NDCs) under international agreements like the Paris Agreement and for mitigating the adverse effects of global warming.

- *Rural Electrification and Empowerment*: The decentralized nature of renewable energy systems, especially with solar panels and wind forms, facilitates electrification in remote and underserved areas. This decentralization improves the quality of life by providing reliable and sustainable electricity access, crucial for health, education, and economic activities. Electrification empowers rural communities, bridging the energy access gap and promoting inclusive development.
- *Economic Growth Engine*: Investments in renewable energy infrastructure, manufacturing, and research spur economic growth. Developing renewable energy sectors creates substantial numbers of sustainable jobs, from construction and installation to operation and maintenance of renewable energy systems. These investments foster innovation and stimulate local economies, contributing to overall economic development. Moreover, renewable energy projects often economic benefits.

Renewable energy is not just an alternative but a strategic imperative for developing countries. It addresses their burgeoning energy demands while fostering sustainable development, enhancing energy security, mitigating climate change, empowering rural areas, and driving economic growth. Embracing renewable energy lays the foundation for a resilient and thriving future.

2.7 The Way Forward: Opportunities and Considerations

The way forward for developing countries involves harnessing the opportunities presented by renewable energy while addressing the associated challenges. Renewable energy is integral to achieving net-zero emissions and contributing to global climate action. Advancements in technology, economic growth, and overcoming obstacles through coordinated efforts will drive this transition. Countries like India have the potential to lead the clean energy revolution, bringing about significant positive impacts on a global scale.

- *Integral to Net-Zero* Renewable energy is indispensable for developing countries (Islam et al. 2023) to achieve their long-term climate goals and contribute responsibly to global climate action. As nations like India commit to net-zero emissions (Renné 2022), integrating renewable energy into their energy mix becomes a crucial strategy. This shift not only addresses climate change but also supports global efforts to limit temperature rise and reduce carbon footprints.
- *Next-Generation Technologies* Advancements in scaling up new generation technologies, including green hydrogen (Marouani et al. 2023), energy storage, and smart grids, are essential for optimizing renewable energy integration. Green hydrogen produced from renewable energy sources provides a clean fuel for applications such as industry and transportation. The best solutions, such as next-generation batteries, can ensure the sustainability of renewable energy by reducing interconnection problems. In addition, smart grids, when combined with technology, can increase grid security and efficiency and support renewable energy management.

- *Economic Driver* Renewable energy value chain presents vast potential for domestic innovation, manufacturing growth, and sustainable job creation. Developing countries can leverage this opportunity to establish robust renewable energy industries. Investments in new research and development can foster innovation, foremost to creating a new technologies and business models (World Economic Forum 2021). Manufacturing renewable energy components domestically reduces reliance on imports and stimulates local economies. Additionally, the renewable energy sector generates numerous jobs, from installation and maintenance to research and administration, contributing to overall economic prosperity.
- *Overcoming Obstacles* While the outlook appears promising, addressing several issues to realize the full potential of renewable energy. Grid integration is a significant challenge, requiring upgrades to existing infrastructure to accommodate renewable energy sources. Land availability in large-scale projects can be a constraint, necessitating strategic planning and efficient land use policies. Project financing remains a hurdle, as renewable energy projects often require substantial upfront investment. Overcoming these obstacles requires coordinated policy efforts, public-private partnerships, and innovative financing mechanisms.

India, a developing country with significant renewable energy ambitions, is poised to lead the global clean energy revolution. The nation's sustained commitment to renewable energy development, bolstered by strategic investments and progressive policies like the International Solar Alliance, positions it at the forefront of this transformation. India's leadership in renewable energy promises far-reaching economic, environmental, and social benefits, including strengthening energy security, reducing GHG emissions, and improving people's quality of life.

2.8 Challenges and Considerations

As the global transition to renewable energy accelerates, developing countries face some unique and challenging challenges. Renewable energy sources such as wind and solar hold great promise for sustainable development, but their integration with existing power systems also presents unique challenges to solve. Key issues include intersections, land acquisition, and financing, all of which require innovative solutions and strategic planning.

Intermittency One of the most significant challenges with renewable energy is interconnection. Unlike fossil fuels, which provide stable and controlled energy, renewable energy sources such as solar and wind are unique in their ability to generate electricity. This interaction plays a significant role in ensuring the reliability and stability of the supply chain. While solar energy relies on sunlight that varies from day to day and seasonally, wind energy relies on unpredictable wind patterns. Such changes can impact grid management by causing periods of excess power followed by periods of supply shortages. Strong strategic planning and advanced solutions are essential to manage this interaction. Grid infrastructure needs to be improved to cope with different types of renewable energy. This includes simple grid development and efficient power distribution. Energy storage systems, such as batteries, play a key role in reducing interaction by storing a large portion of the energy produced during peak production periods and releasing it during low production periods. This ability increases the reliability and sustainability of renewable energy and facilitates its integration into the overall electricity grid. To solve these problems, technologies such as water pumps, compressed air energy storage, and emerging batteries have been researched and used. In addition, smart energy technology can improve energy distribution and energy efficiency and provide stable and reliable energy.

Land Acquisition New large-scale energy projects in renewable, particularly ground-mounted solar techparks and wind farms, need the hugest land areas. Securing suitable land for these projects can be a complex and contentious process. In densely populated regions or areas with competing land uses, finding available and appropriate land can pose a significant hurdle. This challenge is exacerbated by issues related to land rights, environmental concerns, and community opposition. To overcome land acquisition challenges, strategic planning and efficient land use policies are necessary. Identifying and prioritizing areas with high renewable energy potential and minimal competing land uses can facilitate project development. In some cases, repurposing degraded or underutilized land can provide viable sites for renewable energy projects. Engaging with local communities and stakeholders early in the planning process is vital for addressing concerns and ensuring the equitable distribution of benefits from renewable energy projects. Transparent and fair land acquisition processes, coupled with appropriate compensation and benefit-sharing mechanisms, can help mitigate opposition and foster community support.

Financing Continuous investment is vital for supporting renewable energy projects and developing new technologies. The considerable initial capital required for renewable energy infrastructure, coupled with long payback periods, can deter investors. Moreover, developing countries may face additional challenges in accessing financing due to perceived risks, limited financial markets, and competing budget priorities. Addressing the financing challenge requires innovative financing mechanisms and supportive policy frameworks. Public-private partnerships can mobilize private sector investment and share risks. Green bonds, climate funds, and concessional financing can provide affordable capital for renewable energy projects. Governments can perform a crucial task by offering marginal incentives, subsidies, and tax breaks to attract investment and reduce financial barriers. International cooperation and support from multilateral development banks and donor agencies can also help bridge financing gaps and provide technical assistance.

Renewable energy presents a transformative opportunity for sustainable development, particularly in developing countries. However, unlocking this potential requires overcoming key challenges associated with intermittency, land acquisition, and securing financing. These factors are crucial in ensuring the sustainable deployment and scalability of renewable energy solutions. By investing in robust grid infrastructure and energy storage solutions, adopting strategic land use policies, and leveraging innovative financing mechanisms, developing countries can overcome these obstacles, and to hasten their shift to a renewable energy future, collaboration between governments, the private sector, and local communities is essential. Such cooperation ensures that the benefits of renewable energy are widely realized, fostering economic growth, enhancing energy security, and promoting environmental sustainability.

References

Akella AK, Saini RP, Sharma MP (2009) Social, economical and environmental impacts of renewable energy systems. Renew Energy 34:390–396. https://doi.org/10.1016/j.renene.2008.05.002

Bati ASR, Zhong YL, Burn PL et al (2023) Next-generation applications for integrated perovskite solar cells. Commun Mater 4:1–24. https://doi.org/10.1038/s43246-022-00325-4

Dey S, Sreenivasulu A, Veerendra GTN et al (2022) Renewable energy present status and future potentials in India: an overview. Innov Green Dev 1:100006. https://doi.org/10.1016/J.IGD.2022.100006

Ellabban O, Abu-Rub H, Blaabjerg F (2014) Renewable energy resources: current status, future prospects and their enabling technology. Renew Sust Energ Rev 39:748–764. https://doi.org/10.1016/J.RSER.2014.07.113

Gielen D, Boshell F, Saygin D et al (2019) The role of renewable energy in the global energy transformation. Energ Strat Rev 24:38–50. https://doi.org/10.1016/J.ESR.2019.01.006

Gyimah J, Batasuma S, Yao X, Wauk G (2024) The adoption of renewable energy towards environmental sustainability: evidence from partial least square structural equation modelling (PLS-SEM). PLoS One 19:e0299727. https://doi.org/10.1371/JOURNAL.PONE.0299727

Ha YH, Kumar SS (2021) Investigating decentralized renewable energy systems under different governance approaches in Nepal and Indonesia: how does governance fail? Energy Res Soc Sci 80:102214. https://doi.org/10.1016/J.ERSS.2021.102214

IEA (2024) Massive expansion of renewable power opens door to achieving global tripling goal set at COP28. News. https://www.iea.org/news/massive-expansion-of-renewable-power-opens-door-to-achieving-global-tripling-goal-set-at-cop28. Accessed 1 July 2024

Ikuerowo T, Bade SO, Akinmoladun A, Oni BA (2024) The integration of wind and solar power to water electrolyzer for green hydrogen production. Int J Hydrog Energy 76:75–96. https://doi.org/10.1016/J.IJHYDENE.2024.02.139

Islam E, Shah A, Karim TA (2023) Role of renewable energy policy in ensuring net-zero carbon emissions and energy sustainability: a Bangladesh perspective. In: Moving toward net-zero carbon society, Springer climate, pp 59–77. https://doi.org/10.1007/978-3-031-24545-9_4/TABLES/1

Islam A, Islam T, Mahmud H et al (2024) Accelerating the green hydrogen revolution: a comprehensive analysis of technological advancements and policy interventions. Int J Hydrog Energy 67:458–486. https://doi.org/10.1016/J.IJHYDENE.2024.04.142

Kabeyi MJB, Olanrewaju OA (2023) Smart grid technologies and application in the sustainable energy transition: a review. Int J Sustain Energy 42:685–758. https://doi.org/10.1080/14786451.2023.2222298

Khalid M (2024) Smart grids and renewable energy systems: perspectives and grid integration challenges. Energ Strat Rev 51:101299. https://doi.org/10.1016/J.ESR.2024.101299

Kowsar A, Hassan M, Rana MT et al (2023) Optimization and techno-economic assessment of 50 MW floating solar power plant on Hakaluki marsh land in Bangladesh. Renew Energy 216:119077. https://doi.org/10.1016/J.RENENE.2023.119077

Marouani I, Guesmi T, Alshammari BM et al (2023) Integration of renewable-energy-based green hydrogen into the energy future. Processes 11:2685. https://doi.org/10.3390/PR11092685

Osman AI, Chen L, Yang M et al (2023) Cost, environmental impact, and resilience of renewable energy under a changing climate: a review. Environ Chem Lett 21:741–764. https://doi.org/10.1007/S10311-022-01532-8/METRICS

Renné DS (2022) Progress, opportunities and challenges of achieving net-zero emissions and 100% renewables. Solar Compass 1:100007. https://doi.org/10.1016/J.SOLCOM.2022.100007

Suhrab M, Soomro JA, Ullah S, Chavara J (2023) The effect of gross domestic product, urbanization, trade openness, financial development, and renewable energy on CO2 emission. Environ Sci Pollut Res 30:22985–22991. https://doi.org/10.1007/s11356-022-23761-8

White J (2024) Renewable vs non-renewable energy: understanding the differences. EcoFlow US Blog renewable energy vs non renewable energy. https://blog.ecoflow.com/us/renewable-energy-vs-non-renewable-energy/. Accessed 2 July 2024

World Economic Forum (2021) 5 ways to boost renewable energy investment in developing nations. https://www.weforum.org/agenda/2021/06/boost-renewable-energy-investment-in-developing-economies/. Accessed 2 July 2024

World Economic Forum (2023) How renewable energy transition is creating a green jobs boom. https://www.weforum.org/agenda/2023/01/renewable-energy-transition-green-jobs/. Accessed 1 July 2024

Chapter 3
Solar Energy and Environmental Application

Subhajit Majumder, Pooja Thakur, and Aruna Thakur

Abstract Solar energy is a significant source of renewable energy that plays a crucial role in succeeding sustainable development energy requirement. It is a clean and eco-friendly technology that offers an attractive energy supply for generating electricity. Solar photovoltaic technologies and concentrated solar power are continuously evolving to meet our energy needs. The significant installed capacity of solar energy applications globally supports the energy sector and provides business opportunities for people. This chapter aims to emphasize the critical role of solar energy in sustainable development, along with its environmental issues, concerns, and applications of solar energy. It provides an in-depth analysis of solar energy sustainability, comprising its environmental and commercial impact. The paper confidently identifies India's contributions to the field of solar energy applications toward sustainable development. It highlights how solar energy is meeting people's energy needs, creating employment, and augmenting environmental protection. Overall, solar energy is a reliable and efficient source of energy that is crucial for achieving a sustainable future. Therefore, it is essential to prioritize its development and implementation in global energy solutions.

Keywords Photovoltaic (PV) systems · Concentrating solar power (CSP) systems · Solar water heating (SWH) · Environmental application and Impacts

S. Majumder · P. Thakur (✉)
Department of Chemical Engineering, National Institute of Technology Hamirpur, Hamirpur, Himachal Pradesh, India

A. Thakur
Department of Mechanical & Manufacturing Engineering, National Institute of Advanced Manufacturing Technology, Ranchi, Jharkhand, India

S. Kumar, V. K. Singh (eds.), *Renewable Energy Development: Technology, Material and Sustainability*, Clean Energy Production Technologies,
https://doi.org/10.1007/978-981-97-9626-7_3

3.1 Introduction

Energy represents the fundamental concept of the capacity to perform work and is an essential aspect in the fields of physics and engineering. Energy can be enumerated in various classes, such as mechanical, chemical, thermal (radiant), electrical, and nuclear energy. Energy can be generally categorized into two forms as per the method of production and obtainability, namely, renewable and nonrenewable (Østergaard et al. 2023). The renewable energy sources, such as solar, wind, hydro, geothermal, and biomass energy, can be harnessed indefinitely, whereas nonrenewable energy sources, such as fossil fuels (coal, oil, and natural gas) and nuclear energy, are finite and become depleted over time. It is essential to comprehend the diverse forms of energy and their classification to enable businesses, policymakers, and researchers to make informed decisions regarding sustainable energy use, conservation, and development. The move toward renewable energy sources is a critical global challenge that requires interdisciplinary efforts from various sectors to ensure a sustainable and secure energy future (Sukhatme and Sukhatme 1996; Tiwari and Mishra 2012).

Renewable energy sources are natural energy sources that are replenished continuously, making them freely produced and almost inexhaustible. Examples of these sources include solar, wind, biomass, wood, geothermal, tidal, and ocean energy. In contrast, nonrenewable sources of energy, such as coal, petroleum, and natural gas, have been met naturally over a long period and are not replenished within human life spam. The depletion of nonrenewable sources is imminent due to overuse; thus, there is a pressing need for alternative sources of energy that are abundant and environmentally friendly. Renewable energy sources provide a promising solution to this problem by providing almost limitless energy. The benefits of renewable energy sources are numerous. Apart from being almost limitless, they are also eco-friendly, helping to reduce the impact on the environment. Renewable energy sources are also cost-effective and provide a stable and reliable source of energy. By harnessing the potential of renewable energy sources, we can meet the current energy demands while also ensuring the sustainability of our planet for future generations. The current demand for energy requires alternative sources that are both capable of meeting this demand and are environmentally sustainable. Renewable energy sources have emerged as a viable solution to address the energy crisis, as these sources are eco-friendly and nearly inexhaustible (Malik et al. 1982; Zahedi 2007). This shift in energy consumption patterns emphasizes the importance of transitioning from no-renewable sources to renewable energy sources to meet current and future energy demands while also preserving our planet's environment. The proliferation of renewable energy is anticipated to escalate significantly, primarily due to its cost competitiveness in numerous regions. It is currently the most economically viable option for new power generation, and by 2030, it is expected to contribute between 45% and 50% of global generation. This contribution is projected to increase to between 65% and 85% by 2050, regardless of the scenario. Solar energy is expected to be the leading renewable energy source, followed by wind energy (Ruelas et al. 2013; Stevovic et al. 2021).

Solar energy is transmitted through electromagnetic waves, which allows it to travel freely in a vacuum without any medium restrictions. The conversion of heat energy from the sun into beneficial heat and electrical energy is achieved through devices like solar photovoltaic and solar thermal collectors. Based on the technology employed for collection and distribution, solar energy can be further classified into two primary methods: active and passive. The active method involves the direct utilization of this radiation energy through the use of photovoltaic cells (PV cells), solar water heaters (SWS), and other devices. Meanwhile, passive methods function to support the active method by orienting the solar devices for optimal exposure to sun (Tiwari et al. 2009). To showcase the feasibility of renewable energy sources, the Government of India (GOI) has set ambitious targets to be achieved by the year 2022. This chapter confidently highlights the unparalleled potential of solar energy as the most prolific, versatile, and sustainable renewable energy source. It provides a comprehensive review of various cutting-edge solar technologies and their applications in converting raw solar energy into usable forms, such as electrical and thermal energy. The article cogently elucidates the compelling rationale behind adopting solar energy as the predominant renewable energy source and the various state-of-the-art technologies employed to harness this form of energy. The discourse also confidently delves into the advantages, limitations, feasible applications, and potential challenges of different solar technologies, as well as their futuristic scope and potential (Maxmut O'g'li 2023). The primary objective of this astute exposition is to confidently provide a profound considerate of advanced solar techniques and its potential sectors to use, over and above to confidently explore futuristic features of solar technology.

3.2 Advantages of Solar Power

The government, business sector, and consumers have taken the initiative to develop a renewable energy sourcing method that harnesses natural resources such as solar, biomass, wind, geothermal, hydroelectric, among others. Unlike conventional sources, these renewable sources are abundant in quantity and more affordable (Kurtz et al. 2013; Laine et al. 2019). Presently, solar power generation is thriving due to rapid technological advancements. This work focuses on the advantages of solar power generation, which are as follows and depicted in Fig. 3.1:

(a) Solar energy production surpasses fossil fuels as it relies on radiation energy emitted by the sun, an abundant and limitless source, in contrast to other conventional sources that are finite.
(b) Solar power generation is eco-friendly, having a nearly insignificant impact on the environment compared to other forms of energy invention.
(c) Solar technology has advanced to the point where it can meet peak time energy demands of consumers, through the collection and supply of energy during periods of high demand.

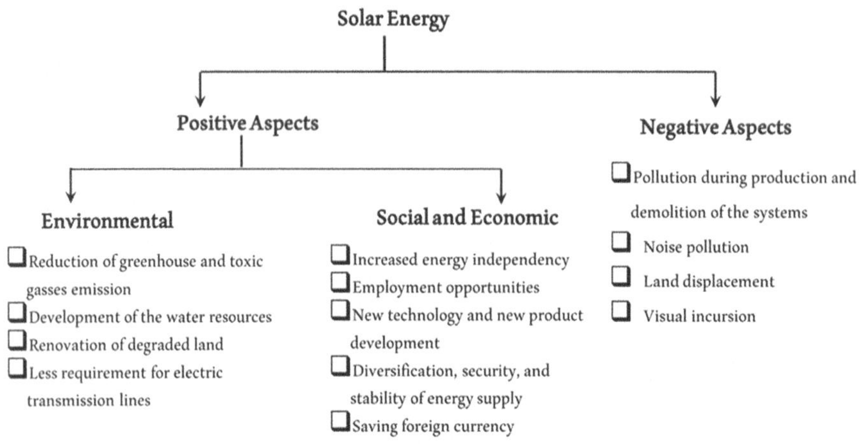

Fig. 3.1 Various aspects of solar energy

(d) The size and scalability of solar technologies have improved significantly, reducing the infrastructure required for power generation when compared to other forms of power generation, such as power plants.
(e) Solar power producing units are versatile and can be installed at the location of the user's choice, reducing investment in production and transmission.
(f) Large number of governments worldwide has increased their solar energy production and are advocating for numerous programs to encourage consumers to install solar photovoltaic and other solar equipment.

3.3 Solar Energy Harvesting Technologies

Solar power can be harnessed through various technologies, including standalone and subsidiary systems, which enable the generation of useful energy, such as electrical, from solar system (Dawn et al. 2016; Jacobson 2009). There are three key technologies employed to harvest energy from solar energy. These technologies include photovoltaic (PV) systems, concentrating solar power (CSP) systems, and solar water heating (SWH) systems. Each of these technologies is designed to capture solar radiation and convert it into a usable form of energy. PV systems use photovoltaic panels to convert sunlight directly into electricity. CSP systems, on the other hand, use mirrors or lenses to concentrate sunlight onto a small area, which heats up a fluid and generates steam to turn a turbine and produce electricity. SWH systems use solar collectors to absorb solar radiation and heat water for domestic or industrial use. Overall, solar energy is an encouraging source of renewable energy that can help reduce our dependence on nonrenewable sources of energy. However, the choice of technology used to harness solar energy will depend on factors such as location, climate, and energy needs.

3.3.1 Photovoltaic (PV) Technologies

Photovoltaic (PV) technologies are extensively employed to directly convert light energy from sun into electrical current. The solar photovoltaic (PV) arrangement is promptly advancing and is deemed to be bigger to further systems. One of the noticeable structures of the PV system is that it operates noiselessly, without generating any pollution, and can function in harsh climatic situations with minimal human participation in operations and maintenance (O&M) (Ma et al. 2014). The PV solar systems have the potential for comprehensive electricity generation. The process of solar energy generation is presented in Fig. 3.2. The PV system operates by utilizing the photovoltaic effect, a novelty witnessed by Alexandre E. Becquerel in 1839 (Kurtz et al. 2013; Nowak 2005). The arrangement comprises a photovoltaic unit, which consists of numerous solar cells prepared of semiconductor substance. While solar emission (photons) impinges on the semiconductor substance, an electron disengages at the atomic level of that material, which, when directed to flow, produces a power-driven current. The competence of the solar cell relies on various factors, including the dimension of the PV unit, the landscape where it is deployed, construction, and congress techniques, among others. The efficiency of various materials used in PV cells ranges from approximately 10–40%. Some common materials used are monocrystalline silicon, polycrystalline silicon, amorphous silicon, cadmium telluride, and copper indium gallium selenide. Current research efforts are focused on improving the efficiency of photovoltaic (PV) cells through the use of novel materials and experimental designs. Materials such as

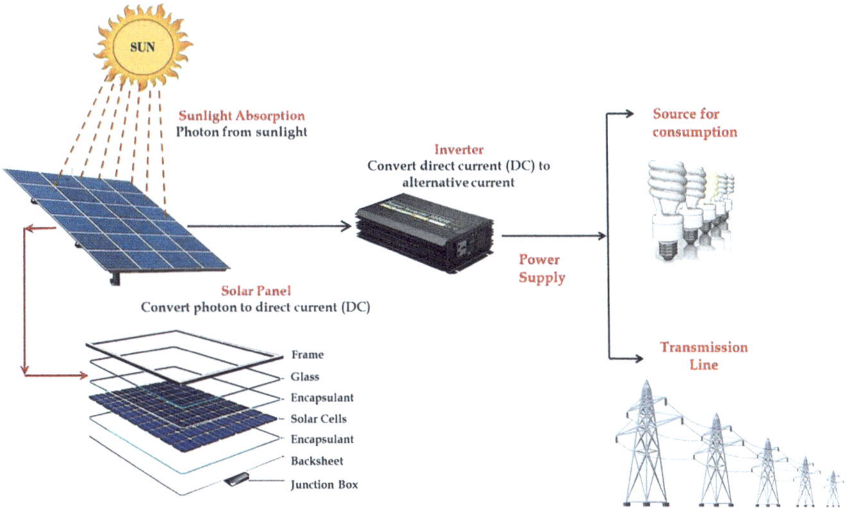

Fig. 3.2 Energy generation mechanism from the PV solar system

Table 3.1 Merits and demerits of photovoltaic (PV) technology based solar system

S. No.	Merits	Demerits
1.	PV systems convert sunlight into electricity and are becoming popular due to their decreasing cost, making them an attractive investment for renewable energy	One of the most significant problems is that they require a lot of land space to generate a substantial extent of electricity. This can be extremely challenging, particularly in densely populated areas where land is expensive and hard to find
2.	Due to their widespread popularity, there are many suppliers available globally	PV panels need batteries for energy storage because they rely on sunlight to produce electricity, which can be expensive and bulky
3.	PV panels can be installed on roofs, walls, or large fields to create solar farms, and they don't require water to function	PV systems experience a decline in efficiency as they age, resulting in a reduction in power output due to degradation caused by various factors. Therefore, stakeholders and decision-makers should consider its impact when evaluating the long-term economics of PV systems
4.	Improved technology has made solar panels cheaper and more efficient, allowing them to be used in arid regions without affecting local water supplies. This is a crucial step toward clean and sustainable energy sources	Manufacturing photovoltaic (PV) panels requires caution to avoid environmental harm. It is vital to consider the entire lifecycle, including production processes and waste disposal. To minimize the impact, adopt eco-friendly practices and use renewable energy such as solar power

radiation-sensitive nanoparticles and gallium arsenide have demonstrated superior competence when compared to conventional silicon-based solar cells (Green et al., 2018, 2022). The challenge with PV systems lies in their static nature, as sunlight is subject to variability and changes hourly throughout the day, as shown in Table 3.1. To maximize system performance, it is necessary to increase the amount of time that PV cells are exposed to direct sunlight (Vick and Clark 2005). Solar trackers are employed to track the Sun's trajectory, while stationary mounted systems can be analyzed using sun pathway data. The challenges and advantages of PV panels are mentioned in Table 3.1. Panels are often inclined at an angle equal to the latitude of the installation site. The use of inexpensive perovskite is being explored as a replacement for costly crystalline silicon in the production of solar cells (Saliba et al. 2018).

3.3.2 Concentrated Solar Power (CSP)

CSP technology is founded upon a concentrated heat system, which leverages a multitude of mirrors to concentrate sunlight onto a particular point. The resultant concentration of sunlight produces a substantial amount of heat, which is then utilized to produce steam. This steam, in turn, runs turbines to generate electricity (Dixit 2020; Liu et al. 2016). The CSP production is bifurcated into two integral sections. One section is dedicated to accumulating and transforming solar energy

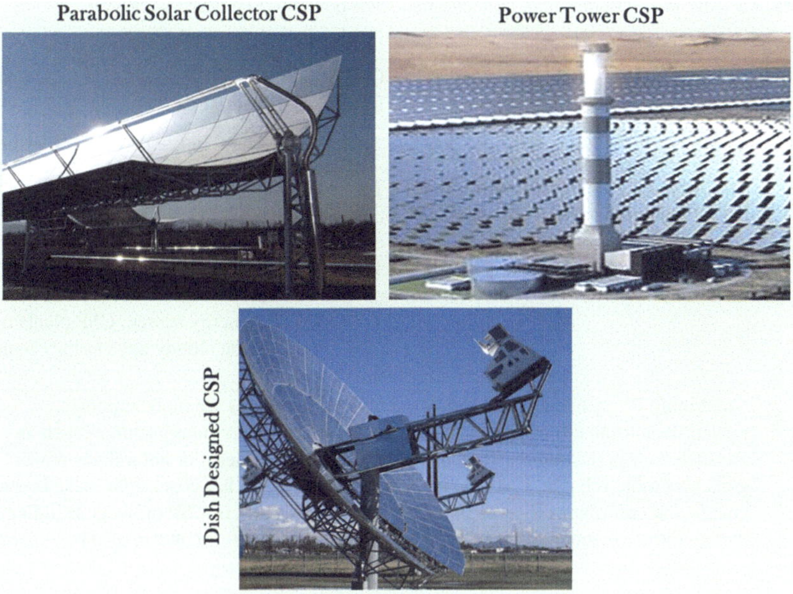

Fig. 3.3 Different designs of CSP solar system

into heat, while the next section is responsible for converting said thermal energy into an utilizable form of energy via the secondary method. For instance, the heat is transferred to a boiler to produce steam, and power is produced by successively connected turbines attached with generators. CSP technology confers a plethora of benefits when it comes to comprehensive electricity manufacture. However, it necessitates a significant amount of space to set up the plant. Technologically speaking, CSP can be categorized into three methodologies for their uses (Bean and Diver 1993). CSP can be implemented using three distinct approaches: power towers, parabolic trough systems, and dish/engine systems, as shown in Fig. 3.3. Each of these methods has its unique advantages and restrictions that must be cautiously evaluated before selecting the most suitable option for a particular application, discussed in Table 3.2. Therefore, it is critical to have an in-depth understanding of the technological aspects of CSP to make informed decisions regarding its implementation.

3.3.2.1 Power Towers

The technology based on concentrated solar power (CSP) has made significant advancements with the advent of the latest version, which utilizes a centrally placed high receiver tower (Alami et al. 2023; Lovegrove and Csiro 2012). This advanced version allows for the focused sunlight from a large number of mirrors to be

Table. 3.2 Merits and demerits of concentrated solar power (CSP) based solar system

S. No.	Merits	Demerits
1.	CSP systems can store thermal energy, allowing for electricity production even during times without sunlight, such as at night or on cloudy days	CSP is facing a challenging market due to the low prices of natural gas
2.	This innovative energy storage solution has the potential to revolutionize the solar industry	The technology needs the right weather conditions and a lot of space to operate efficiently, which can be difficult to achieve. These factors present significant obstacles that need to be addressed to make the most of CSP as a renewable energy source. CSP plants may face difficulties on cloudy days as they require direct sunlight
3..	Combining PV systems with solar power plants is an effective way to enhance their performance. This firm generates more reliable, sustainable energy and contributes to a resilient energy infrastructure	CSP technologies are more expensive compared to other energy sources such as photovoltaic systems or natural gas power stations. This higher cost might make investors hesitant to invest in CSP projects. Building and maintaining a CSP plant can also be expensive

concentrated on the tower. The fluid, which is flowing in the receiver, is heated up to 565 °C with the help of this concentrated sunlight. The use of molten salt as the fluid enhances the capabilities of the plant by storing the heat efficiently and allowing for electricity production even after sunset, on cloudy days, and during peak hours when load demand is high. The stored heat can be used for a long time, thereby increasing the overall efficiency of the plant (Islam et al. 2018).

3.3.2.2 Parabolic Trough System

This particular concentrated solar power (CSP) approach utilizes U-shaped (parabolic) reflectors to gather and concentrate the sun's thermal energy onto a focal point (mirror). At this point, the oil contained tube is fixed along the axial dimension. The use of U-shaped mirrors allows for the creation of high-heated oil, reaching temperatures approximately 400 °C. This oil is then conveyed to a heat exchanger to create steam from water. The steam, in turn, powers turbines that are attached to generators, producing electrical energy (Fernández-García et al. 2010; Price et al. 2002).

3.3.2.3 Dish/Engine Systems

The dish/engine system is an effective method of generating electricity using mirrored dishes. The system is made up of three main components: the concentrator (mirrored dishes), receiver, and generator. The mirrored dishes concentrate sunlight on a specific position where the receiver is located, resulting in a concentration of around 92% of the sunlight. Each dish has its own receiver and can independently generate between 5 kw and 50 kW of electricity. By connecting them in series, their capacity can be significantly increased (Ghosh and Prelas 2011). To optimize the system's output, it can be transformed into an active solar device that tracks the sun's position to maintain maximum focus systems. Furthermore, the assembly includes a heat engine (Stirling engine) that is linked with the receiver, generating electricity directly. This system is highly efficient and dependable, making it an excellent choice for electricity generation (Ghavami et al. 2022).

3.3.3 Solar Heating and Cooling

The solar thermal energy technique is employed to harness heat energy emanating from the sun rays, and subsequently stored heat is utilized for the purpose of heat exchanging as a coolant or heater both domestic and commercial systems. The various collectors utilized include flat plate, evacuated tube, thermosiphon, and concentrating collectors. The principle behind this technique is that solar thermal collector panels accumulate heat which is then conveyed directly to generate residential hot water (Balaras et al. 2007; Pietruschka et al. 2006). Essentially, the solar heating and cooling process involves a transfer of heat from the collectors to regulate cold water and convert it to hot water.

3.4 Environmental Applications of Solar Energy

Solar energy, harnessed from the radiant light and heat emitted by the sun, stands as a pivotal solution in addressing environmental concerns and promoting sustainable development (Zhang et al. 2018). As the global community grapples with the adverse effects of climate change and seeks cleaner alternatives to traditional energy sources, solar power emerges as a frontrunner in mitigating environmental impacts (McVeigh 2013). Solar energy has numerous environmental applications that contribute to sustainable practices and the reduction of environmental impact. Solar energy can be harnessed and applied in various ways within agricultural greenhouses to enhance efficiency, reduce costs, and promote sustainable farming practices (Hassanien et al. 2016). Several applications of solar energy are delineated below:

3.4.1 Solar Water Pumping

Solar-powered water pumps can be used to irrigate crops within the greenhouse. These pumping systems have emerged as a sustainable and efficient alternative to traditional electric or diesel-powered water pumps (Hassanien et al. 2016). As the world grapples with environmental challenges and seeks to transition toward cleaner energy sources, solar water pumping has gained significant attention (Benghanem et al. 2014). Solar water pumping systems operate on a straightforward principle: sunlight is converted into electricity, which powers a water pump. The photovoltaic panels generate direct current (DC) electricity, which is subsequently directed toward the controller and inverter for further processing. The controller regulates the power flow, and the inverter converts it to AC power suitable for the pump (Aliyu et al. 2018). The pump, in turn, circulates water to the desired location. Solar water pumping systems represent a sustainable and innovative approach to water management, providing a reliable and eco-friendly solution for various applications (Hilarydoss 2023). As technology advances and costs decrease, the adoption of solar water pumping is expected to grow, contributing to a more sustainable and resilient water supply system globally (Al-Mamun et al. 2023).

3.4.2 Solar Ventilation Systems

In the pursuit of sustainable and energy-efficient building designs, solar ventilation systems have gained prominence as an environmentally friendly solution for maintaining indoor air quality and temperature control (Lei et al. 2023; Nasri 2023; Sharon 2023). Solar ventilation systems are employed in residential buildings to enhance indoor air quality and comfort. They can be integrated into new constructions or retrofitted into existing homes, providing an energy-efficient solution for homeowners. In larger structures, such as offices, warehouses, or industrial facilities, solar ventilation systems contribute to sustainable building practices. They help reduce reliance on grid electricity for ventilation, leading to energy cost savings and environmental benefits. In agriculture, solar ventilation systems are utilized in greenhouses to regulate temperature and humidity. These systems provide a controlled environment for plant growth, enhancing crop yields and reducing the need for conventional heating and cooling methods. Solar ventilation systems significantly reduce reliance on grid electricity, utilizing renewable solar energy for their operation. This leads to increased energy efficiency and lower carbon footprints associated with building ventilation. Despite the higher upfront costs, solar ventilation systems are an economically viable alternative for businesses that seek long-term cost savings. By significantly reducing energy bills and operational expenses, businesses can recoup their investments in a relatively short time frame. This makes solar ventilation systems an attractive proposition for businesses that are looking to reduce their carbon footprint and save on energy expenses. Many governments and organizations also provide incentives and rebates to promote the adoption of renewable energy systems (Amanowicz 2023). By harnessing solar energy for ventilation, these systems contribute to sustainable building practices, aligning with global efforts to reduce environmental impact and promote eco-friendly technologies. The installation expenses associated with integrating a solar ventilation system may pose a potential challenge for some individuals. The initial cost allocation and required investment may become a hindrance factor that impedes the user's decision to opt for this type of ventilation system. It is crucial to consider the long-term benefits of such a system, which includes cost savings through reduced energy consumption, as well as environmental sustainability. Therefore, it is recommended that users weigh the initial expenses against the long-term benefits before making a decision. The return on investment may take several years, and this can be a challenge for those seeking immediate financial benefits. Solar ventilation systems are weather-dependent, with their efficiency influenced by sunlight availability. Cloudy days or periods of low solar radiation can impact their performance, necessitating supplementary methods for continuous ventilation.

3.4.3 Solar Thermal Heating

Solar thermal heating is a renewable energy technology that harnesses sunlight to produce heat for various applications, ranging from space heating and hot water production to industrial processes (Andharia et al. 2023; Sehrawat et al. 2023; Yu et al. 2023). Solar thermal heating systems operate on the principle of collecting and utilizing solar radiation to generate heat. There are two primary types of solar thermal systems (Alshuhail et al. 2023): concentrating collectors and non-concentrating collectors. Concentrating collectors employ mirrors or lenses to focus sunlight onto a small area, thereby augmenting the intensity of solar radiation. The concentrated energy is subsequently channeled to heat a fluid or substance, such as water or oil, which facilitates the transfer of heat for a variety of applications. This technology has gained significant traction in recent years due to its potential to increase energy efficiency and reduce carbon emissions, making it an attractive prospect for businesses and researchers alike. Non-concentrating collectors, on the other hand, absorb sunlight across a larger surface area without focusing it. These collectors are typically flat-plate collectors, and the absorbed solar energy is transferred directly to a fluid or material circulating within the collector.

Solar thermal heating is commonly used for space heating in residential and commercial buildings. It reduces the reliance on conventional heating methods, leading to energy savings and reduced carbon emissions (Alshuhail et al. 2023). Solar water heating systems are widely used to provide hot water for domestic purposes, such as bathing, cooking, and cleaning (Patel 2023; Rodriguez-Pastor et al. 2023; Singh et al. 2021). They are cost-effective and environmentally friendly alternatives to traditional water heaters. Solar thermal heating finds applications in various industrial processes, including food drying, textile manufacturing, and chemical production (Kumar et al. 2021). The high-temperature heat generated by concentrating collectors is particularly valuable in these applications.

Solar thermal heating relies on sunlight, a renewable and clean energy source, reducing dependence on fossil fuels and mitigating greenhouse gas emissions. By harnessing solar energy for heating purposes, users can significantly reduce their energy bills and operational costs, especially in applications like space heating and hot water production. Solar thermal systems have a low environmental impact compared to conventional heating methods, contributing to sustainable and eco-friendly practices (Gebreyohannes et al. 2020). However, the effectiveness of solar thermal heating is influenced by weather conditions and the intermittent nature of sunlight (Lingayat et al. 2022). Cloudy days or periods of low solar radiation can affect system performance. The initial cost of installing solar thermal heating systems, especially in the case of concentrating collectors, can be relatively high. However, this is often offset by long-term savings in energy bills. Solar thermal heating systems play a vital role in the transition to sustainable and renewable energy solutions. As technology continues to advance and costs decrease, the widespread adoption of solar thermal heating is expected to grow, contributing to a more sustainable and resilient energy future (Ismail et al. 2022).

3.4.4 Photovoltaic System for Electricity

At the core of solar power generation lies photovoltaic (PV) technology (Sayigh 2012; Soumya et al. 2021). Photovoltaic cells, commonly referred to as solar cells, are highly efficient devices that have the ability to convert sunlight directly into electricity with the use of the photovoltaic effect. The photovoltaic effect, whose discovery can be attributed to Alexandre-Edmond Becquerel in 1839, remains at the core of photovoltaic technology and is responsible for generating electrical current when photons from sunlight interact with certain materials, such as silicon. In the realm of materials science, certain materials, specifically semiconductors such as silicon, have been demonstrated to be proficient in converting sunlight into electrical energy. This process is achieved by exposing these materials to sunlight, which in turn dislodges electrons from their atomic state, leading to the creation of an electric current. This direct conversion of sunlight into electricity forms the basis of photovoltaic cells. Silicon, due to its unique properties, is the most widely used material in PV cells (Fernández et al. 2018). Monocrystalline silicon cells are made from a single crystal structure, providing high efficiency but at a higher cost (Zhang et al. 2018). Polycrystalline cells, being composed of multiple crystal structures, present a cost-effective alternative to their monocrystalline counterparts. In contrast, thin-film cells utilize layers of semiconductor material to offer increased flexibility and adaptability for various applications. These distinctions in the composition and structure of photovoltaic cells are crucial in determining their suitability for specific requirements.

PV systems begin with solar panels, which consist of interconnected PV cells (Reinders et al. 2017). Solar panels are essential components in the process of generating electricity from sunlight. They function by converting the sun's energy into direct current (DC) electricity. However, the electricity generated from solar panels is not immediately suitable for use in homes or businesses as it is in the form of DC electricity. To make it usable, the DC electricity must be converted into alternating current (AC) electricity. This is where inverters play a crucial role. They convert the

DC electricity generated by the solar panels into AC electricity, which is safe and suitable for use in homes and businesses. Therefore, the combination of solar panels and inverters is necessary for a successful solar energy system (Stanojevic 2021). Inverters also enable grid-tied systems to synchronize with the local electricity grid (Rathore et al. 2021). Mounting structures support the solar panels and ensure optimal orientation and tilt for maximum sunlight exposure. Tracking systems, either single-axis or dual-axis, dynamically adjust the position of the panels to follow the sun, increasing energy yield (Reinders et al. 2017).

Grid-tied PV systems are connected to the local electricity grid, allowing for the seamless integration of solar-generated electricity with the existing infrastructure (Bagher et al. 2015). During periods of high solar production, excess electricity can be fed back into the grid. Grid-tied systems often incorporate net metering, a mechanism that enables users to receive credits for excess electricity generated (Gorjian and Shukla 2020). This encourages the adoption of solar power and supports a more efficient use of energy (Goetzberger and Hoffmann 2005). Off-grid PV systems operate independently of the electricity grid and are essential for remote locations or areas with unreliable grid access (Sampaio and González 2017). The present-day energy systems incorporate energy storage solutions, such as batteries, to accumulate surplus energy for utilization during periods of limited solar availability. This approach has been instrumental in mitigating the energy shortfall issues and ensuring a consistent supply of power to cater to the energy demands of the populace. Perovskite solar cells are an area of active research, offering high efficiency and lower manufacturing costs (Green et al. 2014). These cells, made from organic-inorganic hybrid materials, hold the potential to revolutionize the solar industry. Bifacial solar panels capture sunlight from both the front and rear sides, increasing overall energy yield. This technology is gaining popularity, especially in ground-mounted solar installations (Guerrero-Lemus et al. 2016).

3.4.5 Solar Drying Systems

Solar energy can be utilized for drying crops and herbs. Solar dryers can be installed within or around the greenhouse to efficiently dry harvested produce, preserving their quality for storage and sale. Solar drying systems harness the power of the sun to remove moisture from various materials, transforming the traditional drying process into an energy-efficient and sustainable practice (Lingayat et al. 2022; Udomkun et al. 2020). Solar drying systems rely on the absorption of solar radiation by the collectors. The absorbed energy is then transferred to the air or a heat transfer fluid circulating through the system. The heated air is directed into the drying chamber, where it comes into contact with the material to be dried. As the air absorbs moisture from the material, it becomes laden with water vapor. The moisture-laden air is then expelled from the drying chamber, and fresh air is introduced. This process continues until the desired moisture content in the material is achieved (Kong et al. 2020). Solar drying systems are widely used in the preservation of fruits, vegetables, herbs, and other food products. By removing moisture, these systems extend the shelf life of produce, preventing spoilage and reducing the need for chemical preservatives (Kong et al. 2020; Lingayat et al. 2021). In agricultural settings, solar drying systems are employed to dry crops such as grains, nuts, and seeds. This enhances their storability and market value while reducing post-harvest losses. Solar drying is also applied in various industrial processes, including the drying of textiles, ceramics, and wood. The controlled and energy-efficient nature of solar drying makes it an attractive option for industries seeking sustainable practices (Lingayat et al. 2022). Solar drying systems utilize renewable solar energy, reducing dependence on conventional energy sources. This leads to lower operational costs and a smaller environmental footprint compared to traditional drying methods. Whilst the initial investment required to implement solar drying systems may be considerably higher, the long-term operational expenses are significantly reduced due to the elimination of fuel costs and reduced electricity consumption. This can lead to a lower total cost of ownership over time, making solar drying systems an attractive option for businesses seeking to achieve cost savings in their operations. By harnessing the power of the sun, solar drying systems offer a sustainable and cost-effective alternative to conventional drying methods, with the added benefit of reducing the carbon footprint of a business or academic institution. Overall, the long-term benefits of solar drying systems make them a wise investment for any enterprise seeking to optimize their operations while minimizing their environmental impact. Solar drying systems promote environmental sustainability by reducing greenhouse gas emissions associated with traditional drying methods. They align with global efforts to adopt eco-friendly technologies and decrease reliance on nonrenewable resources. The efficiency of solar drying systems is weather-dependent. Cloudy days or limited sunlight can affect the drying process, potentially necessitating backup methods for continuous operation. The upfront cost of installing a solar drying system can be a barrier for some users, particularly in regions where financial resources are limited. Solar drying systems offer a sustainable and energy-efficient solution for moisture removal across various industries (Ismaeel and Yumrutaş 2020; Menon et al. 2020). As technology advances and awareness of environmental sustainability grows, the adoption of solar drying systems is likely to increase, contributing to more efficient and eco-friendly drying practices globally.

3.4.6 Solar Desalination for Water Treatment

Desalination is a vital technology in the face of water scarcity. This process involves the removal of salt and other impurities present in seawater. This technology has become a crucial solution in addressing the rising demand for freshwater. Desalination is a complex process that involves several stages, including pretreatment, membrane filtration, and posttreatment. The process requires state-of-the-art technology and highly skilled personnel to ensure the production of high-quality water that meets regulatory standards. The increasing adoption of desalination technology has made it possible to provide freshwater to populations that previously had limited access to it. As such, desalination has become an essential tool in promoting sustainable development and ensuring a reliable water supply for future generations. Solar desalination is a sustainable and environmentally responsible approach to freshwater production, which involves integrating desalination processes with solar energy. This approach relies on solar collectors, including flat-plate collectors or concentrating collectors, which are designed to capture sunlight and generate heat to power the desalination process. By utilizing solar energy, this method offers a practical and efficient way to produce freshwater that is both cost-effective and eco-friendly. Furthermore, the integration of renewable energy sources in the desalination process can significantly reduce the carbon footprint of the operation, making it an attractive option for businesses and industries seeking to minimize their environmental impact. In solar desalination systems, these collectors provide the energy needed to drive the desalination process. The desalination unit is the core of the system and can use various technologies, including distillation (multistage flash distillation, multiple-effect distillation), reverse osmosis, or electrodialysis. Each technology has its unique mechanism for separating salt and impurities from water. A heat exchanger transfers the solar-generated heat to the seawater, increasing its temperature before entering the desalination unit. This preheating step enhances the overall efficiency of the desalination process. Solar desalination systems using solar collectors employ solar thermal energy to heat seawater. This heated seawater is then introduced to the desalination unit, where the elevated temperature facilitates the separation of freshwater from salt and impurities through the chosen desalination process. Another approach involves using solar photovoltaic (PV) panels to generate electricity. The utilization of solar energy for the generation of electricity has opened up new avenues for desalination processes such as reverse osmosis. In this process, seawater is separated from salt and impurities through a semipermeable membrane, producing clean drinking water. The application of solar desalination has been identified as a viable solution for regions that face the challenge of water scarcity due to a lack of freshwater sources. By harnessing the power of the sun, this technology has the potential to provide a sustainable and long-term solution to this global issue. It can serve as a decentralized and sustainable option for communities facing water scarcity issues. In agriculture, especially in regions with limited access to freshwater, solar desalination can be employed to provide irrigation water. This supports crop cultivation in areas where traditional water sources are insufficient. Industries that require large quantities of freshwater for various processes, such as manufacturing or cooling, can benefit from solar desalination. This decentralized water supply can contribute to the sustainability of industrial operations.

Solar desalination systems utilize renewable solar energy, reducing reliance on nonrenewable energy sources. This aligns with global efforts to transition towards sustainable and eco-friendly technologies. Compared to traditional desalination methods that often rely on fossil fuels or grid electricity, solar desalination has a lower environmental impact. It reduces greenhouse gas emissions and minimizes the ecological footprint associated with freshwater production. While the initial installation cost may be higher, solar desalination systems offer long-term cost savings. The abundance of solar energy reduces operational costs, making it an economically viable solution over time. Desalination processes can be energy-intensive, and the efficiency of solar desalination may be influenced by factors such as solar radiation variability and the chosen desalination technology. Advances in technology and energy storage can help address these challenges. Solar desalination is highly dependent on sunlight, and its feasibility may be limited in regions with extended periods of cloud cover or reduced solar radiation. Solar desalination represents a promising solution to address water scarcity challenges globally. As technology continues to advance and the demand for sustainable water treatment solutions grows, the integration of solar desalination systems is likely to play a significant role in providing reliable freshwater sources for communities, agriculture, and industries.

3.4.7 Solar Powered LED Lighting

Solar-powered LED lighting represents a revolutionary approach to outdoor and off-grid illumination (Katzin et al. 2021). Harnessing the energy of the sun, these systems combine solar photovoltaic technology with efficient LED lighting, offering a sustainable and cost-effective alternative to traditional lighting solutions. This chapter explores the components, working principles, applications, advantages, and challenges associated with solar-powered LED lighting. Solar panels, composed of photovoltaic cells, convert sunlight into electricity. In solar-powered LED lighting systems, these panels are crucial for generating the energy required to power the LEDs. The efficiency of the solar panels directly impacts the overall performance of the system (Sundararajan et al. 2021). Light emitting diode (LED) lamps are the primary light source in solar-powered LED lighting systems. These lamps are highly energy-efficient and durable and can produce a range of colors and intensities, making them ideal for various outdoor lighting applications. To ensure continuous illumination during periods of low sunlight or at night, solar-powered LED lighting systems incorporate battery storage. The excess energy generated during the day is stored in these batteries, providing a reliable power source when sunlight is unavailable. Modern solar-powered LED lighting systems often include intelligent lighting control systems. These systems can be programmed to adjust the brightness of the LEDs based on factors such as ambient light levels, motion sensors, or specific time schedules (Sundararajan et al. 2021). During daylight hours, solar panels harvest sunlight and convert it into electrical energy. This energy is either directly used to power the LEDs or stored in batteries for later use. Excess energy generated during the day is stored in batteries, ensuring a continuous power supply during periods of low sunlight or at night. As darkness falls, the stored energy is utilized to power the LED lamps. The lighting control system may adjust the brightness of the LEDs based on preset parameters or external factors such as motion detection. Solar-powered LED lighting is widely used for outdoor applications, including street lighting, pathway illumination, garden lighting, and landscape features. These systems provide reliable and energy-efficient lighting without the need for an external power source. Solar-powered LED lighting systems are suitable for emergency lighting applications, providing a reliable source of illumination during power outages or in remote locations where access to the grid is limited. In off-grid or remote areas without access to traditional electricity sources, solar-powered LED lighting offers an independent and sustainable lighting solution, enhancing safety and visibility.

Solar-powered LED lighting systems are highly energy-efficient, utilizing sunlight to generate electricity and LED technology to convert that energy into visible light (Ling et al. 2021). This combination results in significant energy savings compared to traditional lighting systems. While the initial investment may be higher, solar-powered LED lighting systems offer long-term cost savings by eliminating electricity bills and reducing maintenance costs. Many regions also offer incentives and rebates to encourage the adoption of solar technology. Solar-powered LED lighting has a lower environmental impact compared to conventional lighting systems. It reduces reliance on fossil fuels and decreases greenhouse gas emissions, contributing to sustainable and eco-friendly practices. The upfront cost of installing solar-powered LED lighting systems can be a barrier for some users. However, as technology advances and prices decrease, the return on investment improves. The efficiency of solar panels is weather-dependent, and extended periods of cloud cover can impact the performance of the system. This challenge can be mitigated by incorporating larger battery storage or supplementary charging methods. Solar-powered LED lighting systems represent a cutting-edge solution for sustainable outdoor illumination (Sundararajan et al. 2021). As technology continues to advance and awareness of environmental sustainability grows, the adoption of solar-powered LED lighting is expected to increase, contributing to energy-efficient and eco-friendly lighting practices globally.

3.4.8 Solar Cooking System

A solar cooking system is a setup that utilizes solar energy to cook food. It typically involves the use of solar cookers, which are devices designed to harness sunlight and convert it into heat for cooking. Solar cookers offer an eco-friendly and sustainable alternative to conventional cooking methods that rely on fossil fuels. Solar cookers use reflective surfaces, such as mirrors or reflective materials, to concentrate sunlight onto a specific area. The reflectors focus sunlight onto the cooking chamber, enhancing the intensity of the solar radiation. Solar cookers have a cooking chamber where the food is placed (Khatri et al. 2021). This chamber is designed to absorb and retain the concentrated solar energy, converting it into heat. The materials used in the cooking chamber are selected for their ability to absorb and hold heat effectively. The working principles of a solar cooker revolve around efficiently harnessing sunlight, concentrating it onto a cooking area, and using that energy to heat and cook food (O. K. Singh 2021). While the effectiveness of solar cookers depends on factors such as sunlight availability and the specific design of the cooker, they offer a sustainable and eco-friendly alternative to traditional cooking methods that rely on nonrenewable energy sources.

3.5 Indian Scenarios of Solar Energy and Future Environmental Impact

The Indian states of Andhra Pradesh, Gujarat, Madhya Pradesh, Rajasthan, Punjab, Haryana, and Maharashtra have a highly favorable geographical disposition to harness solar energy, shown in Table 3.3. Among these states, Rajasthan's Thar region hosts some of the most remarkable solar projects in the country, generating nearly 2100 GW of power. Gujarat, on the other hand, boasts one of India's largest solar power plants, consolidating its position as a leader in the solar energy sector. Moreover, the Indian government has sanctioned a master plan to transform 50 cities into solar cities, a significant step toward accomplishing the country's renewable energy objectives. The implementation of large-scale solar photovoltaic (PV) installations in India since 2010 has led researchers to predict that by 2040, solar PV waste due to regular loss scenarios will reach a critical point from a waste generation perspective (Farrell et al. 2020). To estimate solar PV waste in India, researchers have projected the advanced PV installed capacity under two distinct scenarios—business as usual (BAU) and the ambitious scenario. Managing solar waste effectively is crucial and necessitates a proactive approach (Sharma et al. 2024). Employing a mathematical probability distribution function based on projected growth, researchers can accurately estimate solar waste without having to await the end of life of solar cells. Typically, solar cells have a lifespan of 25–30 years. The study is also utilizing mathematical modeling with Weibull distribution to

Table. 3.3 Indian scenarios of solar energy generation capacity (state wise)

State	Installed capacity in 2023
Rajasthan	17055.7
Gujarat	9254.57
Karnataka	8241.41
Tamil Nadu	6736.43
Maharashtra	4722.9
Telangana	4666.03
Andhra Pradesh	4534.19
Madhya Pradesh	2802.14
Uttar Pradesh	2515.22
Punjab	1167.26
Haryana	1029.16

forecast waste volumes by projecting the weight of waste generated using tonnes/MW. This approach provides a robust estimation of the magnitude of solar PV waste generation in India, enabling effective planning and management of this critical issue.

The guiding principle for waste management is crystal clear: prioritize reduction, reuse, and recycling. To significantly reduce the volume of solar waste, we must improve product design and manufacturing processes. This involves conducting extensive research and development to limit the use of precarious substance like lead, cadmium, and selenium. Fortunately, advancements in inkjet and screen-printing technologies have made it possible to avoid using silver. By implementing innovative printing and pasting techniques, we can confidently reduce silver consumption from 0.16 g/Wp in 2005 to less than 0.1 g/Wp in 2020. Therefore, it's crucial to design products with fewer hazardous materials and handle modules with the utmost care during transportation and operation stages to minimize waste.

Both reuse and recycling are critical strategies for reducing waste, and it's important to note that they have distinct differences. Reuse involves the custom of an item for its new purpose or repurposing it for a different utility, thus preventing it from ending up in landfills. This approach not only provides useful products for those in needs but also increases the occupational well-being of citizens (Kralj and Markič 2008). Recycling, on the other hand, involves collecting, segregating, processing, and manufacturing waste materials into new products, which helps to divert them from landfills. In the early stages of a photovoltaic (PV) panel's life, flaws and imperfections can be effectively managed through specialized training and skills for PV panel repairs. This approach not only extends the duration of PV panels but also creates a market for mended panels that can be traded at a lesser price compared to original or used panels. Furthermore, a second-hand business can be established for mended panels and those that have reached the end of their nominal lifespan after necessary repairs. By implementing these strategies, businesses and individuals can make a significant contribution to reducing waste and promoting sustainability. This is especially important given the pressing need to address environmental concerns and reduce our impact on the planet.

The production of solar panels is an energy-intensive process that involves the use of hazardous materials such as lead and cadmium. These materials can pose a significant threat to the environment and human health if not disposed of properly. However, recycling used solar panels can significantly reduce energy waste, mitigate pollution, and prevent health risks. Recent studies have shown that recycled solar panels are just as efficient as new ones, making recycling a viable and eco-friendly option. The recyclability rates of constituent materials of solar panels are high, and recycling them can significantly reduce end-of-life solar panel waste. Governments can create dedicated solar panel recycling plants by incentivizing waste recovery units. While recycling solar panels may not be economically feasible at the moment, it is nonetheless crucial for the environment and the economy. Recovered materials can be reused as raw materials in various phases of solar panel assembly, making the solar panel waste reprocessing industry a secondary mining industry. Therefore, treating solar panel waste is of great importance.

Improper disposal of electronic waste in landfills poses severe environmental and health risks. The leakage of toxic substances like lead and cadmium into groundwater can result in further pollution and lead to reduced plant and animal growth, habitat destruction, and adverse health effects on the human body's nervous, immune, lungs, heart, kidney and reproductive systems. Cadmium, for instance, is a highly toxic substance that can cause lung, prostate, and renal cancer. While landfill disposal of solar waste may appear an easy option, it is ill-advised as it results in the loss of valuable materials and economic losses. Therefore, it should be prioritized last when designing solar waste disposal strategies. If photovoltaic (PV) waste is not treated appropriately, it can outcome in the loss of reusable substance such as aluminum and glass, which account for 88% of the weight (Latunussa et al. 2016). Moreover, landfill disposal can lead to the loss of rare and critical metals such as silicon metal and precious metals found in small quantities. The waste electrical and electronic equipment (WEEE) directive, based on the principle of extended producer responsibility (EPR), inspires resource efficiency through reuse, recycling, and minimizing waste discarding in landfills. Hence, it is crucial to prioritize these practices to prevent severe environmental damage and protect human health. The consequences of improper disposal of electronic waste are far-reaching and can result in long-term environmental damage and adverse health effects. As such, it is necessary to adopt sustainable waste management practices that prioritize resource efficiency, reuse, recycling, and minimize waste dumping in landfills (Farrell et al. 2020).

3.6 Conclusion

The integration of solar energy into our energy landscape heralds a promising era for environmental sustainability. The environmental applications of solar energy represent a beacon of hope in the face of climate change and environmental degradation. As we strive for a cleaner, greener future, solar power stands out as a versatile and eco-friendly solution. By significantly reducing greenhouse gas emissions, mitigating air and water pollution, and fostering the conservation of natural resources, solar energy emerges as a key player in promoting a healthier planet. Its land use efficiency, contribution to sustainable development, and role in enhancing energy independence underscore its multifaceted benefits. Moreover, as the solar industry advances, innovations in recycling and the use of sustainable materials are paving the way for a more circular and responsible approach to solar technology. This progress not only extends the life cycle of solar products but also addresses concerns related to waste management. In embracing solar energy and its environmental applications, we are not just adopting a clean energy source; we are investing in the resilience of our ecosystems, the health of our communities, and the stability of our global climate. As we continue to harness the power of the sun, collaboration between governments, industries, and individuals will be crucial in accelerating the transition to a sustainable energy future. The journey toward a cleaner and more sustainable world is ongoing, and solar energy stands as a testament to our collective commitment to preserving the environment for generations to come. Through continued research, innovation, and widespread adoption, solar energy will undoubtedly play a pivotal role in shaping a more sustainable and resilient planet.

References

Alami AH, Olabi AG, Mdallal A, Rezk A, Radwan A, Rahman SMA et al (2023) Concentrating solar power (CSP) technologies: status and analysis. International Journal of Thermofluids 18:100340
Aliyu M, Hassan G, Said SA, Siddiqui MU, Alawami AT, Elamin IM (2018) A review of solar-powered water pumping systems. Renew Sust Energ Rev 87:61–76
Al-Mamun MR, Roy H, Islam MS, Ali MR, Hossain MI, Aly MAS et al (2023) State-of-the-art in solar water heating (SWH) systems for sustainable solar energy utilization: a comprehensive review. Sol Energy:111998
Alshuhail LA, Shaik F, Sundar LS (2023) Thermal efficiency enhancement of mono and hybrid nanofluids in solar thermal applications–a review. Alex Eng J 68:365–404
Amanowicz Ł, Ratajczak K, Dudkiewicz E (2023) Recent advancements in ventilation systems used to decrease energy consumption in buildings—literature review. Energies 16(4):1853
Andharia JK, Solanki JB, Maiti S (2023) Performance evaluation of a mixed-mode solar thermal dryer with black pebble-based sensible heat storage for drying marine products. J Energy Storage 57:106186
Bagher AM, Vahid MMA, Mohsen M (2015) Types of solar cells and application. Am J Opt Photonics 3(5):94–113
Balaras CA, Grossman G, Henning H-M, Ferreira CAI, Podesser E, Wang L, Wiemken E (2007) Solar air conditioning in Europe—an overview. Renew Sust Energ Rev 11(2):299–314
Bean J, Diver R (1993) Performance of the CPG 7. 5-kW [sub e] Dish-Stirling system. Retrieved from conference: 28. Intersociety energy conversion engineering conference, Atlanta, GA, USA
Benghanem M, Daffallah KO, Alamri SN, Joraid AA (2014) Effect of pumping head on solar water pumping system. Energy Convers Manag 77:334–339
Dawn S, Tiwari PK, Goswami AK, Mishra MK (2016) Recent developments of solar energy in India: perspectives, strategies and future goals. Renew Sust Energ Rev 62:215–235
Dixit S (2020) Solar technologies and their implementations: a review. Materials Today: Proceedings 28:2137–2148
Farrell C, Osman A, Doherty R, Saad M, Zhang X, Murphy A et al (2020) Technical challenges and opportunities in realising a circular economy for waste photovoltaic modules. Renew Sust Energ Rev 128:109911
Fernández-García A, Zarza E, Valenzuela L, Pérez M (2010) Parabolic-trough solar collectors and their applications. Renew Sust Energ Rev 14(7):1695–1721
Fernández EF, Almonacid F, Rodrigo PM, Pérez-Higueras PJ (2018) Chapter II-4-A – CPV systems. In: Kalogirou SA (ed) McEvoy's handbook of photovoltaics (Third Edition). Academic Press, pp 931–985
Gebreyohannes Y, Bayray M, Lauwaert J (2020) A review on solar thermal utilization for industrial heating and cooling processes: global and Ethiopian perspective. Momona Ethiop J Sci 12(2):232–256
Ghavami M, Al-Zaili J, Sayma AI (2022) A methodology for techno-economic and operation strategy optimisation of micro gas turbine-based solar powered dish-engine systems. Energy 251:123873

Ghosh T, Prelas M (2011) Energy resources and systems, volume 2: renewable resources, chapter 3 hydropower. Springer Heidelberg, London, New York
Goetzberger A, Hoffmann VU (2005) Photovoltaic solar energy generation, vol 112. Springer
Gorjian S, Shukla A (2020) Photovoltaic solar energy conversion: technologies, applications and environmental impacts. Academic Press
Green MA, Ho-Baillie A, Snaith HJ (2014) The emergence of perovskite solar cells. Nat Photonics 8(7):506–514
Green MA, Hishikawa Y, Dunlop ED, Levi DH, Hohl-Ebinger J, Ho-Baillie AW (2018) Solar cell efficiency tables (version 51). Prog Photovolt Res Appl 26(1):3–12
Green MA, Dunlop ED, Hohl-Ebinger J, Yoshita M, Kopidakis N, Bothe K et al (2022) Solar cell efficiency tables (version 60). Prog Photovolt Res Appl 30(7):687–701
Guerrero-Lemus RVTKAKLSR, Vega R, Kim T, Kimm A, Shephard LE (2016) Bifacial solar photovoltaics–a technology review. Renew Sust Energ Rev 60:1533–1549
Hassanien RHE, Li M, Dong Lin W (2016) Advanced applications of solar energy in agricultural greenhouses. Renew Sust Energ Rev 54:989–1001. https://doi.org/10.1016/j.rser.2015.10.095
Hilarydoss S (2023) Suitability, sizing, economics, environmental impacts and limitations of solar photovoltaic water pumping system for groundwater irrigation – a brief review. Environ Sci Pollut Res 30(28):71491–71510
Islam MT, Huda N, Abdullah A, Saidur R (2018) A comprehensive review of state-of-the-art concentrating solar power (CSP) technologies: current status and research trends. Renew Sust Energ Rev 91:987–1018
Ismaeel HH, Yumrutaş R (2020) Investigation of a solar assisted heat pump wheat drying system with underground thermal energy storage tank. Sol Energy 199:538–551
Ismail MI, Yunus NA, Kaassim AZM, Hashim H (2022) Pathways and challenges of solar thermal utilisation in the industry: ASEAN and Malaysia scenarios. Sustain Energy Technol Assess 52:102046
Jacobson MZ (2009) Review of solutions to global warming, air pollution, and energy security. Energy Environ Sci 2(2):148–173
Katzin D, Marcelis LFM, van Mourik S (2021) Energy savings in greenhouses by transition from high-pressure sodium to LED lighting. Appl Energy 281:116019
Khatri R, Goyal R, Sharma RK (2021) Advances in the developments of solar cooker for sustainable development: a comprehensive review. Renew Sust Energ Rev 145:111166
Kong D, Wang Y, Li M, Keovisar V, Huang M, Yu Q (2020) Experimental study of solar photovoltaic/thermal (PV/T) air collector drying performance. Sol Energy 208:978–989
Kralj D, Markič M (2008) Building materials reuse and recycle. WSEAS Trans Environ Dev 4(5):409–418
Kumar KR, Chaitanya NVVK, Kumar NS (2021) Solar thermal energy technologies and its applications for process heating and power generation–a review. J Clean Prod 282:125296
Kurtz S, Newmiller J, Kimber A, Flottemesch R, Riley E, Dierauf T, … Krishnani P (2013) Analysis of photovoltaic system energy performance evaluation method. Technical Report of National Renewable Energy Laboratory
Laine HS, Salpakari J, Looney EE, Savin H, Peters IM, Buonassisi T (2019) Meeting global cooling demand with photovoltaics during the 21st century. Energy Environ Sci 12(9):2706–2716
Latunussa C, Mancini L, Blengini G, Ardente F, Pennington D (2016) Analysis of material recovery from silicon photovoltaic panels. Luxembourg (Luxembourg): Publications Office of the European Union
Lei W, Lu H, Qi X, Tai C, Fan X, Zhang L (2023) Field measurement of environmental parameters in solar greenhouses and analysis of the application of passive ventilation. Sol Energy 263:111851
Ling H, Wu J, Su F, Tian Y, Liu YJ (2021) Automatic light-adjusting electrochromic device powered by perovskite solar cell. Nat Commun 12(1):1010
Lingayat A, Balijepalli R, Chandramohan VP (2021) Applications of solar energy based drying technologies in various industries–a review. Sol Energy 229:52–68

Lingayat A, Zachariah R, Modi A (2022) Current status and prospect of integrating solar air heating systems for drying in various sectors and industries. Sustain Energy Technol Assess 52:102274
Liu M, Tay NS, Bell S, Belusko M, Jacob R, Will G et al (2016) Review on concentrating solar power plants and new developments in high temperature thermal energy storage technologies. Renew Sust Energ Rev 53:1411–1432
Lovegrove K, Csiro WS (2012) Introduction to concentrating solar power (CSP) technology. In: Concentrating solar power technology. Elsevier, pp 3–15
Ma T, Yang H, Lu L (2014) Solar photovoltaic system modeling and performance prediction. Renew Sust Energ Rev 36:304–315
Malik MA, Tiwari GN, Kumar A, Sodha M (1982) Solar distillation (a practical study of a wide range of stills and their optimum design, construction, and performance). Pergamon Press, Oxford
Maxmut O'g'li XF (2023) Renewable energy sources: advancements, challenges, and prospects. International Journal of Advance Scientific Research 3(08):14–25
McVeigh JC (2013) Sun power: an introduction to the applications of solar energy. Elsevier
Menon A, Stojceska V, Tassou SA (2020) A systematic review on the recent advances of the energy efficiency improvements in non-conventional food drying technologies. Trends Food Sci Technol 100:67–76
Nasri F (2023) Numerical Simulation of a efficient solar-powered ventilation system. Eng Technol Appl Sci Res 13(4):11459–11465
Nowak S (2005) The IEA PVPS programme: towards sustainable global deployment of PVâ€ oe. Paper presented at the Proceedings 20th European Photovoltaic Solar Energy Conference
Østergaard PA, Duic N, Noorollahi Y, Kalogirou S (2023) Advances in renewable energy for sustainable development. Renew Energy 219:119377. https://doi.org/10.1016/j.renene.2023.119377
Patel AK (2023) Comparative analysis of solar heaters and heat exchangers in residential water heating. Int J Sci Res Arch 9(2):830–843
Pietruschka D, Eicker U, Schumacher J, Hanby V (2006) Simulation based design methods and economical analysis for solar driven absorption cooling systems. Paper presented at the Proceedings of the Eurosun
Price H, Lüpfert E, Kearney D, Zarza E, Cohen G, Gee R, Mahoney R (2002) Advances in parabolic trough solar power technology. J Sol Energy Eng 124(2):109–125
Rathore N, Panwar NL, Yettou F, Gama A (2021) A comprehensive review of different types of solar photovoltaic cells and their applications. Int J Ambient Energy 42(10):1200–1217
Reinders A, Verlinden P, Van Sark W, Freundlich A (2017) Photovoltaic solar energy: from fundamentals to applications. Wiley
Rodriguez-Pastor DA, Becerra JA, Chacartegui R (2023) Adaptation of residential solar systems for domestic hot water (DHW) to hybrid organic Rankine Cycle (ORC) distributed generation. Energy 263:125901
Ruelas J, Velázquez N, Cerezo J (2013) A mathematical model to develop a Scheffler-type solar concentrator coupled with a Stirling engine. Appl Energy 101:253–260
Saliba M, Correa-Baena JP, Grätzel M, Hagfeldt A, Abate A (2018) Perovskite solar cells: from the atomic level to film quality and device performance. Angew Chem Int Ed 57(10):2554–2569
Sampaio PGV, González MOA (2017) Photovoltaic solar energy: conceptual framework. Renew Sust Energ Rev 74:590–601
Sayigh AAM (2012) Solar energy engineering. Elsevier
Sehrawat R, Sahdev RK, Tiwari S (2023) Heat storage material: a hope in solar thermal. Environ Sci Pollut Res 30(5):11175–11198
Sharma A, Mahajan P, Garg R (2024) End-of-life solar photovoltaic panel waste management in India: forecasting and environmental impact assessment. Int J Environ Sci Technol 21(2):1961–1980
Sharon H (2023) A detailed review on sole and hybrid solar chimney based sustainable ventilation, power generation, and potable water production systems. Energy Nexus:100184

Singh OK (2021) Development of a solar cooking system suitable for indoor cooking and its exergy and enviroeconomic analyses. Sol Energy 217:223–234

Singh S, Anand A, Shukla A, Sharma A (2021) Environmental, technical and financial feasibility study of domestic solar water heating system in India. Sustain Energy Technol Assess 43:100965

Soumya C, Deepanraj B, Ranjitha J (2021) A review on solar photovoltaic systems and its application in electricity generation. Paper presented at the AIP Conference Proceedings

Stanojevic AB (2021) Application of photovoltaic technology in the use of solar energy. Ann Environ Sci Toxicol 5(1):094–098

Stevovic I, Mirjanic D, Petrovic N (2021) Integration of solar energy by nature-inspired optimization in the context of circular economy. Energy 235:121297

Sukhatme SP, Sukhatme SP (1996) Solar energy: principles of thermal collection and storage. Publisher, Tata McGraw-Hill

Sundararajan R, D'Couto H, Mugerwa J, Tayebwa M, Lam NL, Wallach ES et al (2021) Use, cost-effectiveness, and end user perspectives of a home solar lighting intervention in rural Uganda: a mixed methods, randomized controlled trial. Environ Res Lett 17(1):015002

Tiwari GN, Mishra RK (2012) Advanced renewable energy sources. Royal Society of Chemistry. RSC Publishing, Cambridge

Tiwari G, Dimri V, Chel A (2009) Parametric study of an active and passive solar distillation system: energy and exergy analysis. Desalination 242(1–3):1–18

Udomkun P, Romuli S, Schock S, Mahayothee B, Sartas M, Wossen T et al (2020) Review of solar dryers for agricultural products in Asia and Africa: an innovation landscape approach. J Environ Manag 268:110730

Vick B, Clark RN (2005) Effect of panel temperature on a solar-PV AC water pumping system. Proceedings of the International Solar Energy Society (ISES):159–164

Yu K, Jia M, Yang Y, Liu Y (2023) A clean strategy of concrete curing in cold climate: solar thermal energy storage based on phase change material. Appl Energy 331:120375

Zahedi A (2007) Technical analysis of an electric power system consisting of solar PV energy, wind power, and hydrogen fuel cell. Paper presented at the 2007 Australasian Universities Power Engineering Conference

Zhang Y, Sivakumar M, Yang S, Enever K, Ramezanianpour M (2018) Application of solar energy in water treatment processes: a review. Desalination 428:116–145

Chapter 4
Role of Solar Energy for Sustainable Environment

Vandana Kaushik, S. K. Sharma, and Manoj Kumar

Abstract Two essential challenges for contemporary society are energy and the environment. Fossil fuels, such as coal, oil, and natural gas, are nonrenewable and cannot provide an endless source of energy. Moreover, there are significant environmental problems associated with the use of these conventional energy sources, such as air pollution and global warming. The main cause of global warming, which raises the average global temperature and sea level, is CO_2 emissions. It is clear that in order to build a sustainable and eco-friendly future, both energy resources and energy-consuming systems need to be reevaluated given the existing energy backdrop. Among the best energy supply choices, renewable energy sources including hydropower, wind power, and solar energy are a few of them. In comparison to conventional energy sources, renewable energy sources are more abundant and environmentally friendly. One of the most appealing forms of renewable energy, which is now accessible, is solar energy. Solar energy has emerged as a crucial element of the renewable energy revolution in providing a sustainable response to our expanding energy requirements. For the given extraordinary rate at which scientific discoveries are being made, solar technology offers a bright future full of prospects for innovation and advancement. This chapter aims to emphasize the critical role that solar energy plays in sustainable development concerning environmental issues, problems, and solar energy applications. The thorough analysis covers the commercial and environmental effects of sustainable solar energy. It illustrates how the use of solar energy can help people to meet their energy needs, encourage the creation

---

V. Kaushik
Department of Physics, Sanskaram University, Jhajjar, Haryana, India

S. K. Sharma · M. Kumar (✉)
Department of Physics, Harcourt Butler Technical University (HBTU), Kanpur, Uttar Pradesh, India
e-mail: manoj.k@hbtu.ac.in

S. Kumar, V. K. Singh (eds.), *Renewable Energy Development: Technology, Material and Sustainability*, Clean Energy Production Technologies,
https://doi.org/10.1007/978-981-97-9626-7_4

of new jobs, and enhance environmental protection. Utilizing solar energy, which is often a steady and efficient energy source, is essential for a sustainable future. As such, its development and deployment must be given top priority in global energy solutions.

Keywords Solar energy · Environment · Photovoltaic (PV) system · Concentrating solar power (CSP) system · Solar heating and cooling (SHC) system

4.1 Introduction

Energy is important in every aspect for society. To maintain economic growth and living standards in our highly interdependent society, we must have access to reliable and sufficient energy resources. But the pace at which energy is produced and consumed now cannot continue. It has become more difficult to defend fossil fuels coal, oil, and natural gas as a reliable long-term energy source. In particular, greenhouse gas emissions, nitrous oxides, and carbon dioxide are considered to be the primary causes of global warming. Furthermore, the rate, at which the world's population is increasing, makes the need of energy as an essential resource. Energy is at the heart of the climate challenge and key to the solution. By burning fossil fuels to produce heat and electricity, energy production produces a significant portion of the greenhouse gases that cover the earth and trap solar radiation. Fossil fuels, such as coal, oil, and gas, are responsible for over 90% of carbon dioxide emissions and about 75% of greenhouse gas emissions globally. These fuels are by far the biggest cause of climate change. Science has made it clear that carbon emissions must be halved by 2030 and nearly eliminated by 2050 to prevent the worst effects of climate change.

The entire amount of energy utilized worldwide is shown in Fig. 4.1a. Oil makes up most of the world's consumption of 600 quads, or one quadrillion, with all fossil fuels accounting for 83% of the total until 2011 and 78% by 2040 due to increased usage of renewable energy resources (Abdalla et al. 2018). The world's energy consumption is still growing, especially in developing countries (Fig. 4.1b) (World Energy Use Physics n.d.). The world's energy demand has increased threefold over the last 50 years, and it may climb three times more over the next 30 years. Most of this expansion will come from the fast-expanding economies of China, India, and many established countries, especially to those in Europe, which expect to meet their energy needs by increasing their use of renewable sources.

As per projected energy needs globally, it is imperative to project and predict the amount of energy that will be consumed. Over the last 30 years, there has been a noticeable increase in the global yearly consumption of renewable energy. Figure 4.1c illustrates the average annual growth rate from 1990 to the present, along with projections for growth through 2040 (Ier 2012).

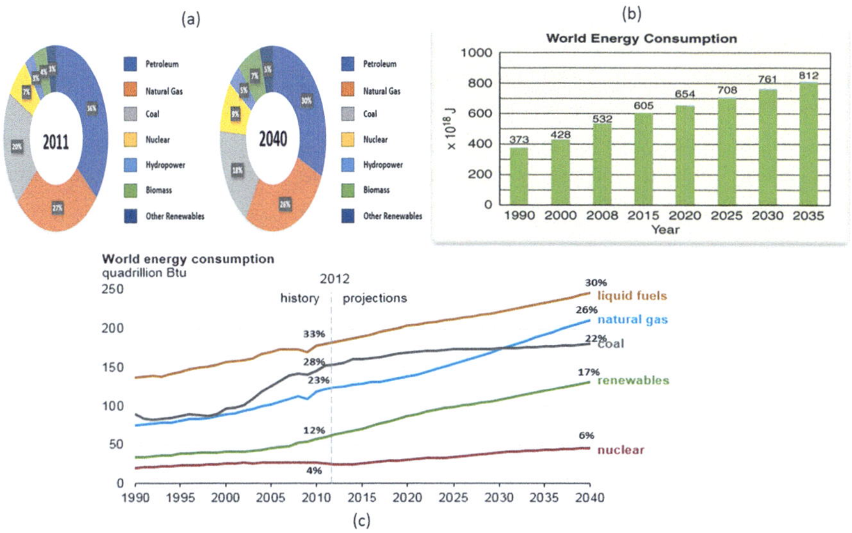

Fig. 4.1 (**a**) Worldwide multiple sources of energy consumption by 2040 "reproduced with permission from Creative Commons an international nonprofit organization (Abdalla et al. 2018)," (**b**) Global energy use in the past and the future "reproduced with permission from Creative Commons an international nonprofit organization (World Energy Use Physics n.d.)," and (**c**) The annual global energy consumption over the preceding three decades and the projected quadruples by 2040 from a variety of sources, including fossil fuels and renewable energy "reproduced with permission from, U.S. Energy Information Administration (Pitatzis 2016)"

Through its ability to contribute to a dispersed and diversified energy infrastructure, renewable energy has the potential to meet demand with a considerably smaller environmental imprint and ease other urgent issues, such as energy security. The values of fossil fuels are increasing swiftly (Canakci et al. 2013); thus, the foremost challenge for agronomic greenhouses is to search the ways that improve energy efficiency in addition to an absolute decrease in the overall energy consumption and related CO_2 releases of the greenhouse industry.

It is essential to take into account the utilization of renewable and sustainable alternative sources. Additionally, this technology may be broadly applied at different scales to meet human needs and help the nation to move away from fossil fuels (coal, oil, and natural gas). Renewable energy sources offer numerous socioeconomic and environmental benefits, making them highly promising for promoting sustainable development. Presently, renewable energy sources are used erratically and insufficiently. Despite the fact that many renewable energy sources are readily available and have great real economic potential, they deliver an unsatisfactory amount of power (Malik 1982). Renewable energy has several positive environmental effects beyond lowering greenhouse gas emissions and other forms of air pollution.

Since renewable energy is more inexpensive in many places, it is anticipated that its use would rise sharply. It is now the most economically viable option for new power generation, and by 2030, it is expected to account for 45–50% of global generation. This contribution is anticipated to increase to 65–85% by 2050, regardless of the circumstances (Ruelas et al. 2013; Stevovic et al. 2021). It is anticipated that solar energy would surpass wind energy as the most common renewable energy source.

4.2 Solar Energy: Mother of All the Energies

Solar energy is one of the renewable energy sources, which is nothing but simply defined as the conversion of solar energy from the sun. As sunlight enters into Earth's atmosphere, most of its part is transformed into visible light and infrared radiation. Solar energy is a powerful source of energy that may use for both domestic and commercial purposes. If the earth is our home, then solar energy is our primary source of energy. To put things into perspective, the majority of Earth's creatures produce their building components for life through chlorophyll photosynthesis, which depends on solar energy. Fossil fuels, winds, and tides are all products of solar radiation, which is the primary source of energy on our planet. Solar energy has become a key element of the worldwide transition toward sustainable and clean energy solutions because to technological breakthroughs, growing worries about climate change, and the depletion of fossil fuels.

Throughout history, people have taken advantage of the close relationship between plants and the sun, particularly in learning how to control plant life cycles to turn them into food crops. With continued technical advancement, it became possible to convert solar energy into electrical energy by the end of the nineteenth century. A gadget that takes advantage of the photovoltaic effect in physics is the solar cell. In 2019, the IRENA research stated that solar energy is the renewable energy source with the fastest annual growth rate (+24% annually), having developed over approximately 150 years since the first photovoltaic cell was invented (Solar energy n.d.). As a result, the technological efficiency of solar power plants is increasing rapidly. The global growth trend is being driven by Asia, and the benchmark of 1.000 GW installed capacity is about to be fulfilled and exceeded. So long as the sun shines, green energy will continue to rise unabatedly. In the light of the growing worries over climate change and the pressing need for sustainable energy solutions, solar energy use has surfaced as a promising avenue for combating environmental degradation. In the midst of the global push to eliminate greenhouse gas emissions and move away from fossil fuels, solar energy has emerged as a cheap, clean, and sustainable substitute.

4.3 Benefits of Solar Energy Toward Environment and Society

In order to lowering energy costs for years to come, solar energy solutions help us become less dependent on fossil fuels by lowering greenhouse gas emissions, conserving our valuable water, and improving air quality. Following are the benefits of solar energy for the environment:

4.4 Reduce Water Scarcity

Two common energy sources, hydropower, and nuclear energy, both require a lot of water to produce electricity. The traditional process uses thousands of liters of water to produce energy for use in appliances. Water is also utilized for generator cooling fuel treatment and refinement. However, the greatest advantage of solar energy and its uses is that no water is needed when we make use of solar panels or other solar power sources to generate electricity. Since solar photovoltaic cells require no water to generate energy. But all that's needed for solar electricity is rainfall for the solar panels to naturally clean themselves when they become a little filthy.

4.5 Curb Pollution and Mitigate Climate Change

Solar energy helps to reduce air pollution since it doesn't emit any greenhouse gases, including carbon dioxide and methane, when it generates electricity. However, producing power from fossil fuels can release these dangerous chemicals into the atmosphere, which degrades the quality of air. Breathing air of poor quality on a daily basis contributes to a host of health problems, including cancer, heart attacks, bronchitis, pneumonia, and asthma. Burning fossil fuels produces more damaging emissions than using solar power or sunshine to generate electricity. Numerous residences and commercial establishments are shifting to solar power because carbon footprint of solar energy systems is substantially lower than that of fossil fuels because they generate little to no greenhouse emissions when in use. Because of this, solar energy is an essential weapon in the fight against global warming. Use of this technique will help reduce airborne emissions of harmful substances from fossil fuels. A major reduction in carbon dioxide emissions, which raise global temperatures and fuel climate change, may result from moving away from fossil fuels and toward solar power for energy generation. Rising temperatures cause an array of extreme weather phenomena, such as melting glaciers, rising sea levels, and altered ecosystem processes. The use of solar energy reduces the carbon footprint by reducing the need for fossil fuels and greenhouse gas emissions. For example, utility-scale solar electricity is estimated to produce between 394 and 447 MWh per acre annually by the Lawrence Berkeley National Laboratory. Therefore, 121 to 138 metric tons of carbon dioxide are kept out of the atmosphere every year by one acre of solar panels with zero emissions (Global Renewable Energy Market Report and Forecast 2024–2032 n.d.).

4.6 Long-Term Human Health

The use of solar energy does enable us to lessen our water shortage. Furthermore, it is helpful in cleaning the air we breathe, since better air quality is beneficial to human health. According to a survey (solar energy advantages statistics), we might prevent over thousands of deaths by utilizing solar energy and improving the quality of the air every ear. Cleaner lungs result from cleaner air for this reason. Reducing pollutants can reduce unnecessary healthcare costs and save lives.

4.7 Location as per Choices

Solar power-producing units can be installed exactly where customers want them to be, saving money on production and transmission while meeting their needs as they occur. This is made possible by technological developments and increased flexibility.

4.8 Scalability

The issue of size and scalability is mitigated by the greater adaptability of solar technologies compared to large-scale infrastructure, which is required for other forms of electricity generation such as power plants.

4.9 Satisfy the Customer's Peak-Time Needs

Improvements in solar technology have made it possible to supply and gather electricity when customers need it most during peak hours.

4.10 Solar Energy Market: Overview

The worldwide solar energy market is divided into various segments based on factors, such as technology, solar module, application, end-use, and geography. Based on technology, concentrated solar power systems and photovoltaic systems comprise the global market. It is projected that the photovoltaic systems industry is expected to hold the biggest market share over the projection period. PV systems are widely used in distributed generation to allow individual homes and businesses to produce their own electricity. By doing this, dependency on centralized power infrastructure is decreased and energy resilience is increased. These devices are usually mounted on the rooftops of housing and business buildings in order to produce renewable energy and reduce power costs. Customers benefit monetarily from net metering's capacity to feed excess power into the grid.

The solar energy market is divided into two segments based on the solar module: amorphous silicon cells and monocrystalline and polycrystalline solar cells, cadmium telluride solar cells, and others. The polycrystalline category is anticipated to lead the market over the projected timeframe. For polycrystalline solar plates, melted raw silicon is initially poured into square or rectangular molds. This method produces a panel with a distinctive blue or speckled appearance because it contains several silicon crystals of varying sizes. Polycrystalline panels represent an economical option for a wide range of applications since their production is typically less expensive than that of monocrystalline panels. The expansion of this market is driven by the affordability of polycrystalline solar modules.

Based on the applications, the worldwide solar energy market is split into four segments: power generation, charging, lighting, and heating. As per the applications, the solar energy sector is split into three segments: domestic housing, business places, and industrial. For duration of forecast period, the industrial section is likely to lead the market. Global industrialization and population growth have increased demand for a wider variety of commercial goods. As output expanded, the usage of solar concentrators in the electricity-producing business leads to increase. In countries with scarce water supplies, especially those in the Middle East, solar concentrator energy is also used to power desalination plants.

It is predicted that the Asia Pacific area will dominate the global solar energy industry throughout the duration of the projection. Many nations in the Asia-Pacific area, including China, India, and Japan, have set ambitious targets for renewable energy and implemented laws that promotes use of solar energy. Together with incentives for solar connection, these policies also include feed-in tariffs and regulations for the renewable portfolio.

The scale of the global market for renewable energy is shown in Fig. 4.2a by type, including solar, wind, hydro, biomass, and geothermal (ZM. Research and Z. M. Research 2024). According to estimates, the global market for solar energy would be worth USD 285.9 billion by 2033, having grown from USD 92.5 billion

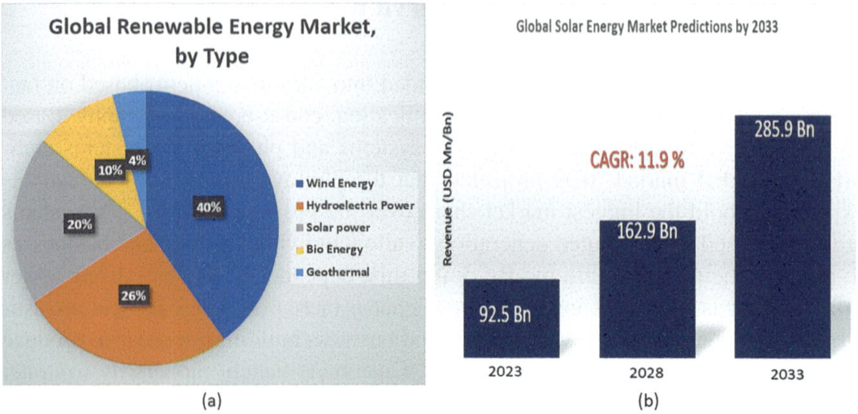

Fig. 4.2 (**a**) Global renewable energy market size, by type: solar, wind, hydro, biomass, geothermal; (**b**) Global solar energy market "reproduced with permission from (Global Solar Energy Market Size, Share, Forecast 2023–2033 n.d.)"

in 2022 to USD 162.9 billion in 2023, with a cumulative annual growth rate (CAGR) of roughly 11.5% (Global Solar Energy Market Size, Share, Forecast 2023–2033 n.d.) as is depicted in Fig. 4.2b.

4.11 Indian Solar Energy Market

The market for solar energy in India was estimated to be worth USD 38 billion in 2021, and between 2022 and 2030, it is expected to grow at a compound annual growth rate (CAGR) of almost 40%, reaching USD 238 billion (as projected in Fig. 4.3). The study report offers a thorough analysis of the variables influencing demand over the course of the projection period as well as those propelling and obstructing the Indian solar energy market (Discover thousands of collaborative articles on 2500+ skills n.d.).

4.12 Market for Solar Energy: Challenges

The expansion of commerce is hampered by environmental concerns. Excessive energy consumption and the use of rare minerals throughout the production process might have negative effects on the environment during and after use. Recycling and appropriate disposal should always be through the proper norms, but they are also essential. Thus, one of the main challenges to the market's growth over the projected time would be the environmental effect of solar panel disposal.

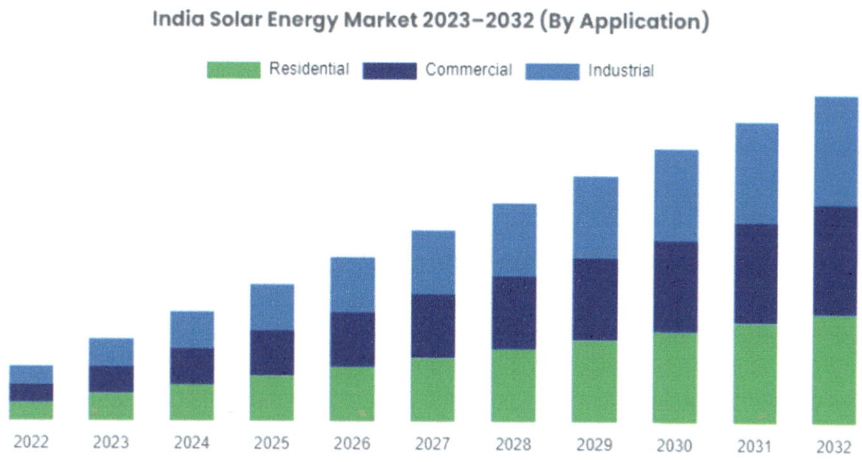

Fig. 4.3 India solar energy market report "reproduced with permission from (Discover thousands of collaborative articles on 2500+ skills n.d.)"

4.13 Solar Energy Harvesting Techniques

The solar energy that the earth receives in the 60 min that follow the dispersion and absorption of solar radiation by the earth's surrounds may be sufficient to provide the world's energy demands for a year. Therefore, the real question is not whether solar energy is available or not, but how effectively it can be converted into the many forms of energy that human requires. Solar energy is conveyed by electromagnetic waves, which allow for unrestricted travel in a vacuum. It is possible to transform solar heat energy into both useable heat and electrical energy by using devices like solar photovoltaics and solar thermal collectors.

The two primary categories that result from the technology utilized for both collecting and delivery of solar energy are active and passive systems. The active technique involves utilizing solar water heaters, photovoltaic cells, and other devices to directly harvest this radiation energy. Passive techniques complement active techniques by arranging the solar cells for optimal solar exposure (Tiwari et al. 2009). Solar energy can be used in a variety of ways, such as standalone and subsidiary systems that enable the conversion of solar radiation into useful energy like electricity.

To extract energy from solar radiation, three main technologies are used, including solar water heating, concentrated solar power, and photovoltaic systems. All these technologies aim to capture solar light and convert it into a practical energy source. In photovoltaic systems, sunlight is directly converted into power through the use of solar panels. Conversely, concentrated solar power plants use mirrors or lenses to concentrate sunlight onto a small area. In the process, a fluid heats up and steam is created, which drives a turbine to produce energy. Solar water heating systems employ solar collectors to heat water for use in homes and commercial buildings. Using solar energy, which is a promising form of renewable energy overall, may help us to become less dependent on nonrenewable energy sources. However, a number of variables, such as climate, location, and energy needs, will affect the solar energy technology.

4.14 Photovoltaic (PV) System

The vast array of uses for solar cells has led to the growth of the photovoltaic technology sector. When photoelectric experiments began in the nineteenth century, it was unanticipated that these optoelectronic devices would play a crucial role in battling the ecological damage caused by nonrenewable sources of energy ever since the Industrial Revolution. Photovoltaic cells can be categorized into several generations depending on the materials and production methods used. For a considerable amount of time, the market was controlled by the first-generation crystalline silicon cells (Luo et al. 2023). The second generation, which comprises thin-film cells, uses materials such amorphous silicon, cadmium telluride, and copper indium gallium selenide (CIGS) as examples (Lloyd and Forest 2010). Third-generation cells are currently being produced using constituents such as dye-sensitized solar cells (DSSC), perovskite solar cells, and organic semiconductors (Romeo and Artegiani 2021). Fourth-generation solar cells, utilizing technologies including tandem solar cells, hot-carrier solar cells, and quantum dot solar cells, are being researched and developed in an effort to increase the efficiency and robustness of the earlier generations of solar cells (Solar Cell Central n.d.).

First-generation cells are quite efficient, but producing them is costly and lengthy process. Even though second-generation cells exhibited inferior efficiency and lower production costs, they are nevertheless suitable for large-scale installations (Zhang et al. 2020). Although they are still in the experimental stage, third-generation cells have the capability to enhance the efficiency and decrease expenses. Fourth-generation cells not yet commercially available and are still in the development stage, but by using state-of-the-art materials and techniques, they should perform better than previous generations (India's Solar Mission n.d.).

Photovoltaic efficiency is the ability of a photovoltaic device, such as a solar cell, to convert sunlight into electrical energy. It indicates the proportion of electrical power output to solar energy input and is expressed as a percentage. PV efficiency is calculated by measuring the amount of electricity produced by a solar cell or panel under standard test circumstances, such as a specific temperature and light intensity. The performance of photovoltaic devices is influenced by various parameters such as the kind of material, cell design, temperature, sun spectrum, inverter efficiency, reflection and absorption losses, and dust (Fazal and Rubaiee 2023).

There are benefits and drawbacks to every generation of solar cells, and efforts are being made to improve their cost-effectiveness, durability, and efficiency (Wu and Hou 2011). The problems associated with the adoption of PV technology

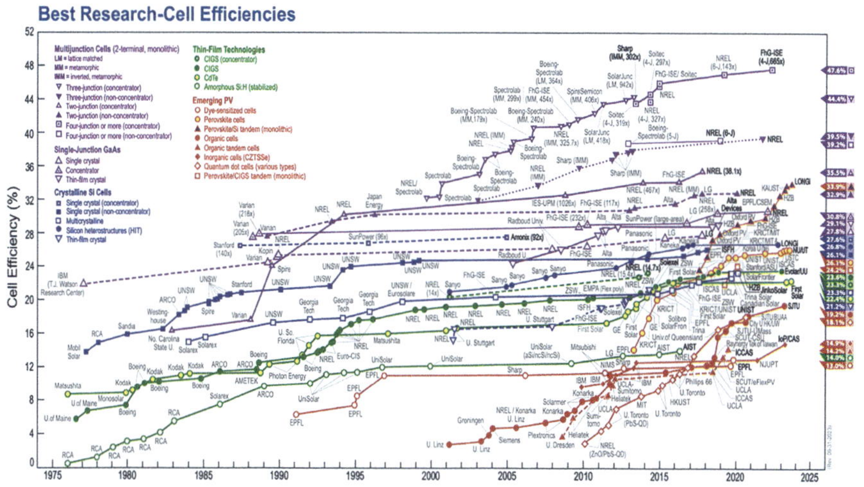

Fig. 4.4 The best research-cell efficiency graph "reproduced with permission from NREL Transforming Energy (Best Research-Cell Efficiency Chart| Photovoltaic Research |NREL n.d.)"

include its intermittency, the high cost of energy storage, disputes over land use, resource constraints, and competition from alternative energy sources, early cost barriers, infrastructural integration, and environmental concerns. To ensure that PV technology continues to expand and is widely accepted, it is imperative that these problems be addressed through research, innovation, and legislation that is favorable (Akhmedov et al. 2023).

The most notable advancements in photovoltaic systems as of 2022 include roll-to-roll printing, flexible and lightweight solar panels based on thin-film technologies for multipurpose fitting on unusual surfaces, cheaper and highly efficient perovskite solar cells, bifacial solar panels that capture sunlight from both sides, transparent solar cells integrated into surfaces, and better light-harvesting methods using inventive structures, manufacturing techniques, and nanomaterials like artificial intelligence and machine learning (PV Status Report 2011). The graph (Fig. 4.4) held by the National Renewable Energy Laboratory (NREL) shows the highest independently verified efficiency figures for several solar cell types. The Best Research-Cell Efficiency graph is this one, and it is updated on a regular basis to reflect the most recent research findings (Best Research-Cell Efficiency Chart| Photovoltaic Research |NREL n.d.).

4.14.1 Sustainability Impact

Because photovoltaic technologies can address large sustainability challenges, there has been a lot of interest in them. This section looks at the overall sustainability impact of solar systems, taking into account social, economic, and environmental aspects.

(a) *Environmental Benefits.*

Reduced air pollution, less water used for power production, and less dependency on the extraction of environmentally hazardous resources are just a few of the environmental advantages that PV systems offer. Solar panels don't need water to run, unlike fossil fuel-based power plants (Green 2000). PV systems undergo a thorough lifecycle assessment that looks at the environmental impact of the product from the point of manufacture to disposal. Emissions of greenhouse gases from fossil fuels are continuously higher than those from greenhouse gas emissions, according to lifecycle assessments. Clean electricity produced by PV systems significantly reduces climate change (Pelle et al. 2020).

Environmental issues arise from photovoltaic material production and disposal. The most urgent issues are those related to energy use, resource extraction, hazardous materials, insufficient reusing infrastructure, water use, carbon footprint, and land-use impact. Initiatives being pursued include sustainable practices, greener manufacturing, efficient recycling, energy efficiency, alternative materials, and responsible land-use planning. To overcome these obstacles and improve solar technology's environmental sustainability, ongoing research and innovation are needed (Skoplaki and Palyvos 2009).

(b) *Economic Benefits.*

The PV systems have major financial advantages. Over time, they reduce consumers' energy expenditures by producing cheaper electricity bills for them. In addition, the solar industry fosters energy independence, local economies, and job creation (Kim et al. 2021). The economic viability of PV technology is influenced by a wide range of factors, including continual cost reductions in the production of PV modules, government incentives, and subsidies. A greater range of homes and businesses may now afford PV systems because of the falling cost of solar panels. Due to the sharp decline in PV module costs, solar PV systems are now more financially feasible than power generation based on fossil fuels.

Some of the factors that make this competitive are lower module costs, lower levelized cost of electricity, economies of scale in large-scale projects, monetary encouragements like tax credits and feed-in tariffs, technological improvements that boost efficiency, innovative manufacturing, financing options like power purchase agreements, and the achievement of grid parity in some areas. These attributes, along with environmental constraints and the cost of carbon, make solar PV more modest and appealing in the energy sector (Solar Cell Central n.d.).

(c) *Social Consideration.*

PV technologies are essential for lowering disparities in energy availability, especially in underdeveloped or remote areas. They improve quality of life and stimulate economic growth by making it possible to provide energy to locations lacking a stable grid connection (Romeo and Artegiani 2021). The PV initiatives with a community focus have become a powerful force for good in society.

Involvement of community participation in the design and ownership of solar arrays, these projects strengthen social ties, encourage energy independence, and generate employment opportunities.

(d) *Technology Inventions for Sustainability.*

Energy storage systems enable the best possible use of solar energy even in low-light conditions, which enhances the sustainability of photovoltaic installations. The reliability and adaptability of PV systems could be greatly enhanced by developments in battery technology (Singh 2013). Through improved grid management, demand response, and energy flow optimization, smart grids increase the sustainability and effectiveness of PV installations. Sustainable building practices are being supported by new solar-integrated building materials like solar windows and roof tiles. The seamless integration of energy production and aesthetics in these solutions promotes the increased usage of PV systems in urban environments (Lameirinhas et al. 2022).

(e) *Long Term Viability and Challenges.*

The longevity of photovoltaic systems is contingent upon their robustness and dependability. This section discusses how to ensure the long-term operation of PV installations by regular maintenance, quality assurance, and material improvements (Adeeb et al. 2019). Examining the availability of solar technologies raw materials is essential to ensuring sustainability. Long-term sustainability planning must include looking at alternative materials and addressing resource limitations in the future. It is essential to conduct ongoing studies and find methods for reducing any possible negative environmental effects of solar technology development. This section examines strategies to reduce the environmental impact of PV manufacturing, such as recycling initiatives and sustainable material sourcing (Vinod et al. 2018).

4.14.2 Future Research on PV Technology

Recent advancements in PV technology have made it a more important and feasible source of renewable energy. Still, there is a great deal of room for development and study in this field (Green 1984). These are some potential areas for PV technology study in the future.

(a) *Energy Storage*

The fact that solar energy is only usable when the sun is shining is one of its biggest drawbacks. Research into more advanced energy storage techniques is needed to increase the dependability of solar energy (Sima et al. 2020). This could entail developing new kinds of energy storage technologies, such as thermal or hydrogen storage, as well as improvements in battery technology.

(b) *Efficiency*
Solar panels have improved over time; however, there is still an opportunity for their development. Researchers can look into novel materials and technologies that can improve the efficiency of solar cells, such as perovskites or tandem cells (Sima et al. 2020).

(c) *Durability*
Even though solar panels are meant to last for a number of years, inclement weather and other environmental factors might damage them. Subsequent investigations may focus on creating solar materials with greater durability and improving the design of solar panels to make them more resistant to harm (Passari and Susi 1983).

(d) *Artificial Intelligence and Machine Learning*
Artificial intelligence and machine learning can be used to maximize the efficiency of solar energy systems. Subsequent investigations might concentrate on creating algorithms that forecast solar irradiance, maximize energy storage, and enhance microgrid management to enhance the efficiency of solar energy systems (Al-Dahidi et al. 2018).

(e) *Integration with Other Technologies*
Smart homes, microgrids, and electric vehicles can all be powered by solar energy. Research in this field may help to lower the cost of these systems while boosting their reliability and efficiency (Restuccia et al. 2016).

(f) *Large-Scale Deployment*
Our energy needs can be greatly satisfied by solar energy, but this will require much wider use of the technology. The development of novel manufacturing techniques that might speed up and improve the efficiency of production as well as strategies to lower the price of solar panels are potential areas for research (Khan et al. 2023).

(g) *Solar-Driven Water Treatment*
The long-term provision of clean drinking water could be facilitated by the integration of solar energy into water treatment facilities. Research in this area could focus on developing reasonably priced and efficient solar-powered water treatment devices (Haffaf and Lakdja 2022).

4.15 Concentrated Solar Power (CSP)

The CSP is a technology that concentrates sunlight to generate heat, which is then used to produce electricity. CSP system uses quite different technology than PV systems. They employ the sun as a heat source rather than the photon energy used by PV systems. CSP plants employ mirrors or lenses to concentrate a significant amount of sunlight, or solar thermal energy, on a small area. A heat engine, usually a steam turbine cycle, is powered by concentrated light and generates electricity when it is coupled to an electrical power generator. When needed, electricity can be produced using the concentrated thermal energy that can be stored. The majority of power plants use sun concentrators in addition to fossil fuel-fueled burners of some kind to boost thermal heat output.

CSP systems are based on the concentration of photons that come directly from the sun and are not scattered by the Earth's atmosphere. In this technology, the major energy source is direct normal irradiance, which is often available in subtropical locations and/or high elevations. Reflective surface is capable of tracking the Sun in two or three dimensions, which are used to concentrate direct sunlight. The redirected photons then heat up a fluid, which is utilized to give power to a heat engine that generates electricity. As a result, CSP is sometimes called solar thermal electricity. It is easily linked to thermal energy storage, allowing energy gathered during periods of high solar irradiation to be used to improve dispatchability (Van Sark and Corona 2020).

The four primary kinds of CSP schemes are parabolic trough, linear Fresnel, power tower, and dish. They all use the sun's thermal energy to power a heat engine, which turns mechanical energy into electrical energy to produce electricity.

Thermal energy, which is frequently used in Stirling engine systems, is concentrated in the dish. The parabolic trough, power tower, and linear Fresnel are Brayton or Rankine cycle-based CSP technologies. Figure 4.5 depicts the fundamental and all-encompassing process in CSP systems for transforming solar thermal energy into electrical energy.

There are two types of solar collectors: non-concentrating and concentrated. The shape employs the same region to intercept and absorb solar light. In order to enhance radiation flow, concave reflecting surfaces of concentrating solar collectors used in CSP plants are usually employed to intercept and concentrate solar energy into a small receiving region.

The major technologies utilized in concentrating solar thermal power plants include parabolic troughs, linear Fresnel reflectors (LFRs), solar towers in heliostat fields, and parabolic dishes (PDs), which produce the heat of high temperature.

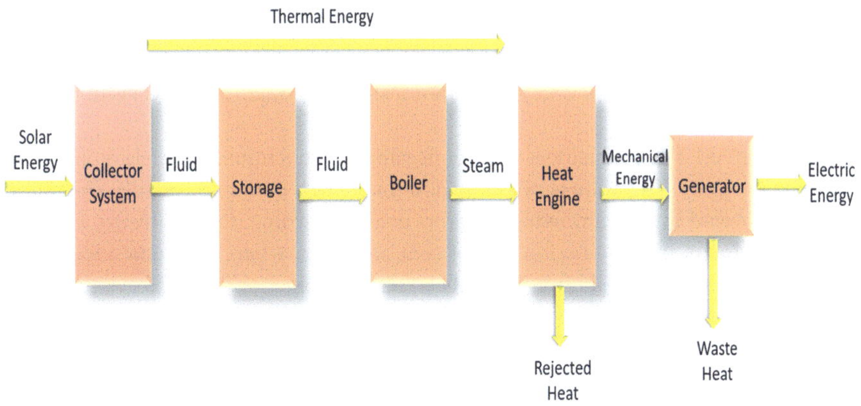

Fig. 4.5 General method of converting solar thermal energy to electric energy in CSP systems

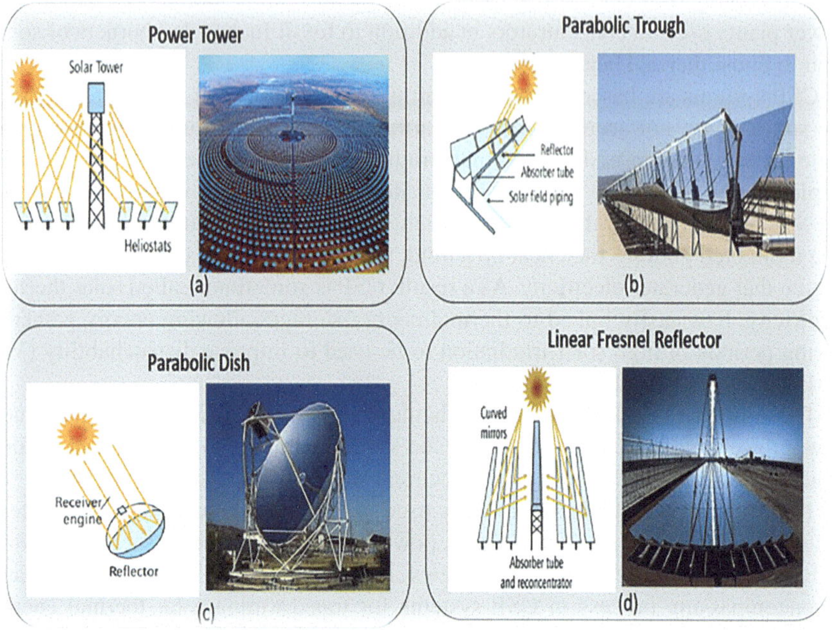

Fig. 4.6 Solar concentrators: (**a**) power tower, (**b**) parabolic trough, (**c**) PD, and (**d**) linear Fresnel reflector, "reproduced with permission from Creative Commons an international nonprofit organization (Van Sark and Corona 2020)"

Figure 4.6 depicts the concepts used in the solar concentrator: (a) power tower, (b) parabolic trough, (c) PD, and (d) linear Fresnel reflector. Heliostats are enormous, flat mirrors used in power towers to reflect sunlight onto a solar receiver at the top of a central tower (Fig. 4.6a). A direct steam power tower heats water up to 540 °C after pumping it up to the receiver. The superheated steam then drives a standard steam turbine. In this case, heat is delivered through steam from the receiver to the power block. Melted salt is used in certain power towers in place of steam and water. Melted salt that is too hot to handle can be saved for later use or used immediately to create steam and energy through a steam turbine that powers an electric generator.

Long, curved mirrors inside a tube focus sunlight on a liquid to create a parabolic trough (Fig. 4.6b). This tube is positioned in line with the mirror. Usually, the liquid heat is converted into steam, which powers a steam turbine. Power towers use mirror fields to focus sunlight on the tops of central towers, where the tremendous heat, typically conveyed by molten salts, boils water and spins turbines.

The PD is another form of CSP plant (Fig. 4.6c), although it is not widely used at present. This system focuses sunlight on a single location. Dish concentrating technique generates power via a Stirling engine. Mirrors on the outside of a Stirling engine reflect sunlight, generating power. Concentrating linear Fresnel reflector devices (Fig. 4.6d), like parabolic troughs, use many flat mirrors to focus solar energy onto the receiving tube. The same idea is at work when boiling water and powering a steam turbine with the liquid passing through the heated tube. Linear solar concentrations are the foundation of both of these technologies.

4.15.1 Future Potentials of CSP

The evidence of global warming caused by carbon dioxide releases, the hazardous materials associated with nuclear energy, and the scarcity of resources necessitate the transition to a sustainable energy system. Solar energy may be one of the most effective ways to solve these issues because of its unlimited capacity for worldwide radiation. The amount of Sahara Desert needed for CSP systems would only need to be 1% in order to supply the electricity worldwide. Post-2030, CSP is expected to have a major impact in Sunbelt countries due to its dispatchable nature and low electricity cost.

Figure 4.7 depicts the global electrical generation breakdown projections for 2050. It is predicted that renewable energy will account for around 85% of worldwide energy output. Furthermore, there is a high predicted reliance on wind energy followed by solar PV and a modest reliance on CSP (4%). According to the IEA 2DS scenario (Energy Technology Perspectives 2012), in order to keep global temperature increases to 2 °C, CSP should produce 8–10% of the world's electricity supply by 2050, or 800 GW of installed capacity. While increasing energy efficiency has the greatest promise in the short term, utilizing a variety of renewable energy sources is also essential. While wind and photovoltaic technologies are progressing at a rapid rate, other technologies including bioenergy, ocean energy, geothermal energy, and CSP are still developing too slowly to reach this goal.

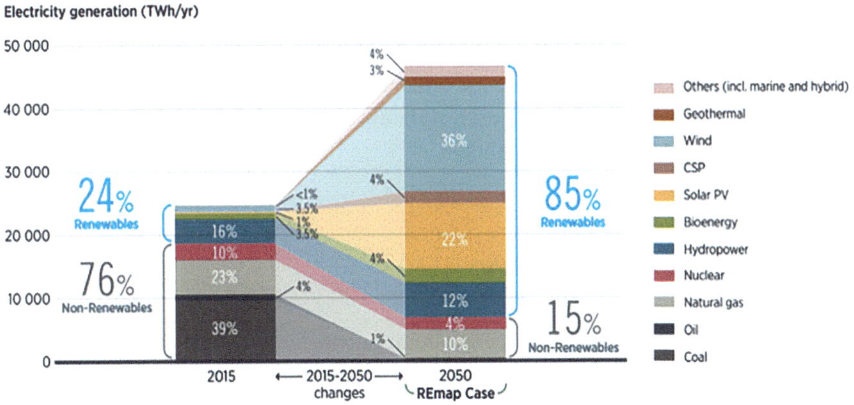

Fig. 4.7 Global electrical generation breakdown projections for 2050 "reproduced with permission from (Santos et al. 2018)"

4.16 Solar Heating and Cooling (SHC) Systems

SHC systems use solar thermal energy to heat pools and provide hot water, cool rooms, and power appliances in commercial, industrial, and residential environments. These techniques fulfill the requirement for natural gas or electricity. Solar heating and cooling systems, which harness the sun's plentiful and sustainable energy, can be advantageous for industrial, residential, and commercial structures (Pitz-Paal 2020). These technologies considerably reduce dependence on fossil fuels and electricity, hence promoting energy efficiency and environmental sustainability.

4.16.1 Solar Heating Systems

Solar heating systems collect solar radiation, transform it into heat energy, and use that heat energy for water and room heating and other things. The primary part of these systems is the solar collector. Different types of solar collectors are available, and they are all appropriate for diverse climates and uses.

(a) *Flat-Plate Collectors*

The most common version has a glazed box covering a dark absorber plate. In moderate areas, they work well for room heating and domestic water heating.

(b) *Evacuated Tube Collectors*

These consist of transparent glass tubes arranged in parallel rows, each containing an absorber tube. Even in colder climates, these collectors are quite effective because of the excellent insulation provided by the vacuum inside the tubes.

(c) *Solar Thermal Panels*

These panels can be incorporated into façades and roofs of buildings. They have multiple uses and can be employed for heating as well as construction. The heat storage device, typically a water tank or a phase-change substance, receives the thermal energy gathered by the collectors. Even in times of poor solar irradiation, this stored heat can be used right away or retained for later, providing a consistent source of thermal energy.

4.16.2 Solar Cooling Systems

Thermally powered chillers are the main components by which solar energy is used in solar cooling systems to power cooling processes. Technologies for solar cooling fall into two main categories:

(a) *Absorption Chillers*
Heat is used in absorption chillers to power the chilling cycle. The method makes use of a mixture of absorbent and refrigerant, like lithium bromide and water. The absorbent is separated from the refrigerant by solar heat, and later on, it is then condensed and evaporated to provide cooling.
(b) *Adsorption Chillers*
The function of adsorption chillers is similar to absorption chillers; except they use a solid material to attract and retain refrigerant. Similar to absorption chillers, the adsorption-desorption cycle generates cooling using solar heat. In hotter areas, where there is an abundance of solar energy to meet cooling needs, solar cooling is highly advantageous. These gadgets have the potential to significantly lower grid peak demand and electricity consumption, which will benefit businesses and the environment.

4.16.3 Integration and Efficiency

Numerous variables, including as building design, system configuration, solar collector type, and geographic location, affect how well SHC systems work. Overall efficiency and dependability can be increased by integrating these systems with conventional HVAC (heating, ventilation, and air conditioning) systems (Solar Energy Industries Association (SEIA) n.d.). Hybrid systems ensure constant thermal comfort by having the flexibility to switch automatically between solar and conventional energy sources based on availability and demand.

4.16.4 Advances and Future Prospects

The efficiency and practicality of SHC systems have grown due to recent advancements in solar technology. Innovations include changed thermal storage materials, better collector designs with increased absorption and reduced heat losses, and cultured control systems that maximize energy consumption. These developments have increased the attraction of mass-deployment solar heating and cooling systems (U.S. Department of Energy n.d.). SHC systems have significant environmental advantages. These technologies are less needed for fossil fuels, which helps to mitigate climate change and reduce greenhouse gas emissions. They also reduce the cost of energy for customers and promote energy independence.

4.17 Solar Energy Over the Next Decade

The production and uses of solar energy has grown exponentially in recent years, which took placed due to lower costs, government incentives, and more environmental consciousness. Solar power has progressed from an expert technology to a widely used energy source as it has become more affordable and accessible. However, the route to a fully sustainable energy future is far away from completion. The future of solar technology holds the key to achieving its full potential and overcoming the challenges that lie ahead. Some projections for the future of solar energy comprise the following:

- Solar panel prices will continue to drop, making solar installations more affordable for consumers and companies.
- Solar power will become a vital source of electricity, especially in poor nations where it may assist to decrease dependency on fossil fuels and can use electricity at lower cost.
- The usage of solar energy in transportation will continue to increase as solar-powered automobiles and solar panels for electric vehicles become more common.
- The usage of solar energy in buildings will grow as emerging new technologies, such as solar shingles and transparent solar panels that may be integrated into household areas.
- New technologies like solar steam production and solar thermal storage are developed, which will promote the use of solar energy in the industrial sector.

The precise evolution of solar energy over the next 10 years is hard to forecast because it will depend on a number of variables, including economic trends, regulatory decisions, and technology breakthroughs. Nonetheless, it's probable that solar energy will keep becoming more and more important, contributing significantly to the world's energy balance. Over the following 10 years, a few particular trends and advancements that are anticipated to influence the direction of solar energy include continued cost reductions, increasing adoption in developing countries, growing role in the transportation sector, continued innovation in technology, and increasing policy and regulatory support (Vr 2024).

4.18 Conclusions

A bright future for environmental sustainability is being heralded by the incorporation of solar energy into our energy landscape. In light of climate change and environmental damage, solar energy applications play crucial role. Solar power sticks out as a flexible and environmentally beneficial option as we work toward a cleaner, greener future. Solar energy is emerging as a major force in promoting a healthy planet through its large reduction of greenhouse gas emissions, mitigation of air and water pollution, and promotion of natural resource conservation. Its many advantages are highlighted by its efficient use of land, support of sustainable development, and role in boosting energy independence. Furthermore, developments in recycling and the use of sustainable materials in the solar industry are opening the door to a more moral and circular approach to solar technology. In addition to extending the lifespan of solar products, this innovation addresses waste management problems. When we embrace solar energy and its environmental applications, we are not only putting into practice a clean energy source but we are also making an investment in the resilience of our ecosystems, the health of our communities, and the stability of our global climate. As we continue to harness the power of the sun, cooperation between governments, corporations, and individuals will be crucial to accelerating the transition to a sustainable energy in future. As we move closer to being a cleaner and more sustainable society, solar energy is a symbol of our common dedication to preserving the environment for coming generations. Through continued research, innovation, and widespread use, solar energy will undoubtedly be essential in building a more resilient and sustainable planet.

References

Abdalla AM et al (2018) Nanomaterials for solid oxide fuel cells: a review. Renew Sust Energ Rev 82:353–368. https://doi.org/10.1016/j.rser.2017.09.046

Adeeb J, Farhan A, Al-Salaymeh A (2019) Temperature effect on performance of different solar cell technologies. J Ecol Eng 20(5):249–254. https://doi.org/10.12911/22998993/105543

Akhmedov AK, Abduev AK, Murliev EK, Belyaev VV, Asvarov AS (2023) Transparent conducting amorphous IZO thin films: an approach to improve the transparent electrode quality. Materials 16(10):3740. https://doi.org/10.3390/ma16103740

Al-Dahidi S, Ayadi O, Adeeb J, Alrbai M, Qawasmeh BR (2018) Extreme learning machines for solar photovoltaic power predictions. Energies 11(10):2725. https://doi.org/10.3390/en11102725

Best Research-Cell Efficiency Chart| Photovoltaic Research |NREL (n.d.). https://www.nrel.gov/pv/cell-efficiency.html

Canakci M, Yasemin Emekli N, Bilgin S, Caglayan N (2013) Heating requirement and its costs in greenhouse structures: a case study for Mediterranean region of Turkey. Renew Sust Energ Rev 24:483–490. https://doi.org/10.1016/j.rser.2013.03.026

Discover thousands of collaborative articles on 2500+ skills. (n.d.). https://www.linkedin.com/pulse/india-solar-energy-market-report-2023-industry-size-share-

Energy Technology Perspectives (2012) https://doi.org/10.1787/energy_tech-2012-en

Fazal MA, Rubaiee S (2023) Progress of PV cell technology: feasibility of building materials, cost, performance, and stability. Sol Energy 258:203–219. https://doi.org/10.1016/j.solener.2023.04.066

Global Renewable Energy Market Report and Forecast 2024–2032. (n.d.) Expert Market Research. https://www.expertmarketresearch.com/reports/renewable-energy-market

Global Solar Energy Market Size, Share, Forecast 2023–2033 (n.d.) Spherical Insights. https://www.sphericalinsights.com/reports/solar-energy-market

Green MA (1984) Limits on the open-circuit voltage and efficiency of silicon solar cells imposed by intrinsic Auger processes. IEEE Trans Electron Devices 31(5):671–678. https://doi.org/10.1109/t-ed.1984.21588

Green MA (2000) Photovoltaics: technology overview. Energy Policy 28(14):989–998. https://doi.org/10.1016/s0301-4215(00)00086-0

Haffaf A, Lakdja F (2022) Feasibility and performance analysis of using solar water heating system in Algeria. UPB Sci Bull Ser C Elect Eng Comput Sci 84:271–286
Ier (2012) Fossil Fuels Still King in EIA's Annual Energy Outlook 2013 – IER. IER, Dec. 07, 2012. https://www.instituteforenergyresearch.org/uncategorized/fossil-fuels-still-king-in-eias-annual-energy-outlook-2013/
India's Solar Mission (n.d.). http://india.gov.in/allimpfrms/alldocs/15657.pdf
Khan HR, Khan FS, Ahmed S, Akhtar J (2023) Smart nanomaterials and three-dimensional printing for flexible solar cell applications. Elsevier eBooks, pp 389–411. https://doi.org/10.1016/b978-0-323-91145-0.00016-5
Kim J, Rabelo M, Padi SP, Yousuf H, Cho E-C, Yi J (2021) A review of the degradation of photovoltaic modules for life expectancy. Energies 14(14):4278. https://doi.org/10.3390/en14144278
Lameirinhas RAM, Torres JPN, De Melo Cunha JP (2022) A photovoltaic technology review: history, fundamentals and applications. Energies 15(5):1823. https://doi.org/10.3390/en15051823
Lloyd B, Forest AS (2010) The transition to renewables: can PV provide an answer to the peak oil and climate change challenges? Energy Policy 38(11):7378–7394. https://doi.org/10.1016/j.enpol.2010.08.014
Luo Y et al (2023) Carrier transport enhancement mechanism in highly efficient antimony selenide thin-film solar cell. Adv Funct Mater 33(14). https://doi.org/10.1002/adfm.202213941
Malik MAS (1982) Solar distillation. Pergamon. [Online]. http://books.google.ie/books?id=9bITAAAAMAAJ&dq=9780080286792,+0080286798&hl=&cd=1&source=gbs_api
Passari L, Susi E (1983) Recombination mechanisms and doping density in silicon. J Appl Phys 54(7):3935–3937. https://doi.org/10.1063/1.332568
Pelle M, Lucchi E, Maturi L, Astigarraga A, Causone F (2020) Coloured BIPV technologies: methodological and experimental assessment for architecturally sensitive areas. Energies 13(17):4506. https://doi.org/10.3390/en13174506
Pitatzis A (2016) Global future energy mix, fossil fuels vs alternative sources of energy-energy routes, June. 08, 2016. https://energyroutes.eu/2016/06/08/global-future-energy-mix-fossil-fuels-vs-alternative-sources-of-energy/
Pitz-Paal R (2020) Concentrating solar power. Elsevier eBooks, pp 413–430. https://doi.org/10.1016/b978-0-08-102886-5.00019-0
PV Status Report (2011). http://www.ncpre.iitb.ac.in/userfiles/files/PV_Status_Report_2011.pdf
Restuccia D, Spizzirri UG, De Luca M, Parisi OI, Picci N (2016) Biogenic amines as quality marker in organic and fair-trade cocoa-based products. Sustain For 8(9):856. https://doi.org/10.3390/su8090856
Romeo A, Artegiani E (2021) CdTe-based thin film solar cells: past, present and future. Energies 14(6):1684. https://doi.org/10.3390/en14061684
Ruelas J, Velázquez N, Cerezo J (2013) A mathematical model to develop a Scheffler-type solar concentrator coupled with a Stirling engine. Appl Energy 101:253–260. https://doi.org/10.1016/j.apenergy.2012.05.040
Santos JJCS, Palacio JCE, Reyes AMM, Carvalho M, Freire AJR, Barone MA (2018) Concentrating solar power. Elsevier eBooks, pp 373–402. https://doi.org/10.1016/b978-0-12-812959-3.00012-5
Sima CA, Popescu MO, Popescu CL (2020) Sensitivity analysis of optimal economic dispatch. UPB Sci Bull Ser C Electr Eng Comput Sci 82:223–236
Singh GK (2013) Solar power generation by PV (photovoltaic) technology: a review. Energy 53:1–13. https://doi.org/10.1016/j.energy.2013.02.057
Skoplaki E, Palyvos JA (2009) On the temperature dependence of photovoltaic module electrical performance: a review of efficiency/power correlations. Sol Energy 83(5):614–624. https://doi.org/10.1016/j.solener.2008.10.008
Solar Cell Central (n.d.). http://solarcellcentral.com/
Solar energy (n.d.) Enel green power. https://www.enelgreenpower.com/learning-hub/renewable-energies/solar-energy

Solar Energy Industries Association (SEIA) (n.d.) Solar Heating & Cooling (SHC). Retrieved from SEIA

Stevovic I, Mirjanic D, Petrovic N (2021) Integration of solar energy by nature-inspired optimization in the context of circular economy. Energy 235:121297. https://doi.org/10.1016/j.energy.2021.121297

Tiwari GN, Dimri V, Chel A (2009) Parametric study of an active and passive solar distillation system: energy and exergy analysis. Desalination 242(1–3):1–18. https://doi.org/10.1016/j.desal.2008.03.027

U.S. Department of Energy (n.d.) Solar Thermal Technologies. Retrieved from Energy.gov.

Van Sark W, Corona B (2020) Concentrating solar power. Elsevier eBooks, pp 221–231. https://doi.org/10.1016/b978-0-12-818762-3.00012-1

Vinod N, Kumar R, Singh SK (2018) Solar photovoltaic modeling and simulation: as a renewable energy solution. Energy Rep 4:701–712. https://doi.org/10.1016/j.egyr.2018.09.008

Vr A (2024) Future of solar energy: predictions for 2024 and beyond. Republic of Solar, Mar 06, 2024. https://arka360.com/ros/future-of-solar-energy-predictions/

World Energy Use Physics (n.d.). https://courses.lumenlearning.com/suny-physics/chapter/7-9-world-energy-use/

Wu H, Hou Y (2011) Recent development of grid-connected PV systems in China. Energy Procedia 12:462–470. https://doi.org/10.1016/j.egypro.2011.10.062

Zhang H, Lu Y, Han W, Zhu J, Zhang Y, Huang W (2020) Solar energy conversion and utilization: towards the emerging photo-electrochemical devices based on perovskite photovoltaics. Chem Eng J 393:124766. https://doi.org/10.1016/j.cej.2020.124766

ZM. Research and Z. M. Research (2024) Solar energy market size report, share, growth, forecast by 2030. Zion Market Research, May 06, 2024. https://www.zionmarketresearch.com/report/solar-energy-market

Chapter 5
Solar Energy and Smart Cities: A Holistic Approach for Sustainable Development

Uplabdhi Tyagi and Neeru Anand

Abstract Megacities' energy supply is regarded as a hot research area in the emerging field of urban management, particularly in developing nations. The smart city concept, which begins with an overview of the sustainable development goals, offers a novel approach to energy provision in a city through the use of artificial intelligence (AI), renewable energy sources like photovoltaic (PV) technologies, and transformational participation (TP) based on citizen-motivating programs. A variety of energy sources are needed to meet the need for dependable and reasonably priced electricity, with production costs frequently surpassing environmental considerations. The idea behind smart cities is an urban environment that uses data and technology to optimize energy use and improve quality of life. In this perspective, solar energy is essential since it offers a dependable, clean power supply that adheres to sustainability ideals. Smart cities use solar energy to offer a seamless and connected urban experience, from solar-powered electric vehicle charging stations to intelligent street lighting powered by solar panels. Furthermore, cutting-edge energy management technologies guarantee efficient energy use, cutting expenses, and waste.

Keywords Smart city · Photovoltaic · Transformational participation · Urban environment · Optimize energy · Solar energy · Electric vehicle

5.1 Introduction

The population of people living in cities has increased dramatically over the twentieth century as a result of fast-increasing urbanization and population growth (55% of the global population as of 2018). Major cities make up only 5% of the planet's

U. Tyagi (✉) · N. Anand
University School of Chemical Technology, Guru Gobind Singh Indraprastha University, New Delhi, New Delhi, India

S. Kumar, V. K. Singh (eds.), *Renewable Energy Development: Technology, Material and Sustainability*, Clean Energy Production Technologies,
https://doi.org/10.1007/978-981-97-9626-7_5

land area, yet they consume 75% of its fossil fuels; therefore, the world's growing urbanization and skyrocketing energy demand in cities are concerning. Furthermore, since it is predicted that by 2050, 68.4% of the world's population will live in cities, action must be taken to address both the growing amount of fossil fuel consumed in cities and the excessive load placed on local power plants as a result of the imbalance in energy consumption. Specifically, 40% of the world's energy is consumed by buildings and infrastructure, with the residential sector making up the remaining 7% (Ghadami et al. 2021). Therefore, there is a great deal of interest in the development of new technologies to efficiently save energy in single-family or multi-family homes.

Numerous elements and components are needed in the development of energy systems to support cities and modernize them to become smart cities. These energy sources have to be abundant, clean, renewable, and safe for the environment. In a same vein, these cities' energy systems that provide them with electricity, heat, and air conditioning must be dependable, functional, safe, and efficient. Moreover, energy storage technologies, such as thermal energy storage (TES), pumped hydro, battery storage, and compressed air energy storage (CAES), are valuable resources since they improve the robustness and integrity of the energy infrastructure (Abu-Rayash and Dincer 2021). Because of the world's population growth, which is increasing at an exponential rate, and the potential risks associated with using fossil fuels extensively for the environment, global health, and climate, there is a pressing need to develop these systems. The higher economic performance for harvesting energy from various sources can be attributed to the proactive advancement and acceptance of renewable energy solutions, particularly solar energy. Utilizing a variety of technologies, solar energy is the radiant light and heat from the sun that can be used to create electricity or thermal energy for heating and cooling. It is among the most plentiful and sustainable energy sources on the planet. Here are some key aspects of solar energy:

5.2 Solar Photovoltaic Technology

Solar photovoltaic (PV) technology uses semiconductors, usually silicon, to directly convert sunlight into power. Electrons in the material of the photovoltaic cell are excited by sunlight, which makes them mobile and produces an electric current (commonly known as photovoltaic effect). A typical PV cell consists of multiple layers of semiconductor materials. Crystalline silicon, which comes in monocrystalline and polycrystalline forms, is the most prevalent kind. Producing monocrystalline cells is more expensive due to their single crystal structure, which makes them more efficient. Multiple silicon crystals are used to create polycrystalline cells, which are generally less expensive but marginally less efficient. PV cells are commonly assembled into bigger pieces known as modules or panels. After that, these modules are put together to form arrays, which range in size from modest residential setups to substantial utility-scale solar farms. An inverter is used to change the DC (direct current) electricity produced by the solar panels into AC (alternating current) electricity. After that, this AC electricity can be supplied into the grid for distribution or used to power electrical appliances. In order to collect sunlight, solar panels are typically installed on rooftops or on ground-level structures. The angle and orientation of the panels are optimized based on factors such as latitude, local climate, and shading to maximize energy production (Menniti et al. 2017). Grid-connected PV systems allow users to obtain credits for the electricity they contribute, and excess power produced by the solar panels can be transferred back into the grid. This process, known as net metering, enables homeowners and businesses to offset their electricity bills and further encourages the adoption of solar energy.

5.2.1 Advantages of Solar PV Technology

With so many benefits to offer, solar photovoltaic (PV) technology is one of the most widely used and quickly expanding renewable energy sources. Some key advantages are as follows:

Clean and Renewable: Solar PV technology uses the sun's infinite and plentiful energy to create electricity, which makes it a clean and renewable energy source. When in use, it emits no greenhouse gases or air pollution, which helps to lower carbon emissions and slow down global warming.

Low Operating Costs: When compared to traditional power generation technologies, solar PV systems have comparatively low operating and maintenance expenses after installation. There are no fuel costs since sunlight is free, and maintenance requirements are minimal, primarily limited to occasional cleaning and inspections.

Energy Independence: Solar PV systems can be deployed on a wide range of scales, from residential rooftops to large utility-scale installations (Telang et al. 2021). This decentralized nature of solar power generation promotes energy independence by reducing reliance on centralized power plants and imported fossil fuels.

Scalability and Modularity: Solar PV systems are highly scalable and can be tailored to meet the specific energy needs of any application, whether it's a small off-grid system or a large utility-scale solar farm. Additionally, they are modular, allowing for easy expansion or integration with other energy systems as demand grows.

Reduced Electricity Bills: By generating electricity on-site, solar PV systems can help reduce electricity bills for homeowners, businesses, and communities. Depending on factors such as local solar resource availability and electricity prices, solar PV installations can provide significant long-term savings on energy costs.

Job Creation and Economic Benefits: Numerous jobs in manufacturing, installation, maintenance, and associated industries have been created as a result of the solar industry's rapid expansion. Solar PV deployment also stimulates local economies through investments in infrastructure and technology.

Versatility and Compatibility: Solar PV technology can be integrated into various environments and applications, including residential, commercial, industrial, and agricultural settings. It can also be combined with energy storage systems, smart grid technologies, and other renewable energy sources to enhance reliability and efficiency.

Environmental Benefits: In addition to reducing greenhouse gas emissions, solar PV technology helps conserve water resources by not requiring water for electricity generation, unlike conventional power plants that rely on water for cooling purposes.

Longevity and Durability: Solar PV panels typically come with long warranties and have a lifespan of 25 years or more. They are designed to withstand various environmental conditions, including wind, hail, and snow, ensuring reliable performance over their operational lifetime.

Grid Stability and Resilience: Distributed solar PV installations can enhance grid stability and resilience by reducing transmission losses, minimizing the need for costly grid upgrades, and providing power during peak demand periods, thereby contributing to a more reliable and resilient electricity supply.

5.2.2 Challenges and Considerations

While solar photovoltaic (PV) technology offers numerous advantages, there are also several challenges and considerations associated with its deployment:

Intermittency and Variability: Sunlight availability, which changes with the time of day, season, weather, and geographic location, is a prerequisite for solar photovoltaic power. In order to maintain a steady supply of electricity, this erratic and unpredictable behavior might make grid integration difficult and need the use of backup power or complementing energy storage technologies.

Land Use and Environmental Impact: Large-scale solar PV installations need a lot of land, which can cause conflicts between land uses, damage to habitats, and the disappearance of natural or agricultural landscapes (Kuznetsov et al. 2021). Additionally, the production and disposal of PV panels involve some environmental impacts, such as energy-intensive manufacturing processes, use of rare materials, and end-of-life recycling challenges.

Energy Storage and Grid Integration: In order to balance supply and demand, stabilize grid frequency, and control voltage swings, large-scale solar PV integration into the electrical grid needs sophisticated grid management strategies and energy storage technologies. Energy storage technologies, such as batteries, pumped hydro storage, or thermal storage, can help store excess solar energy for use during periods of low sunlight or high demand.

Cost and Financial Considerations: While the cost of solar PV has declined significantly in recent years, upfront capital costs for installation and equipment procurement can still be a barrier to adoption, especially for residential and commercial customers. Financing options, incentives, and policies, such as tax credits, rebates, and net metering, play a crucial role in making solar PV more economically viable.

Performance and Maintenance: To guarantee maximum performance and efficiency, solar PV systems need to have regular maintenance, which includes cleaning the panels, checking the electrical connections, and replacing any broken parts. Factors such as dust, shading, soiling, and degradation over time can affect the output of PV modules and may require corrective actions to maintain system reliability.

Grid Infrastructure and Distribution Challenges: Integrating distributed solar PV installations into existing grid infrastructure can present technical challenges, such as voltage regulation, power quality issues, and reverse power flow management. Upgrading and modernizing grid infrastructure, including distribution networks and inverters, may be necessary to accommodate high penetrations of solar PV generation.

Policy and Regulatory Uncertainty: Policy and regulatory frameworks, including incentives, tariffs, and permitting processes, can significantly influence the deployment and economics of solar PV technology. Uncertainty or changes in government policies, such as subsidy reductions or tariff revisions, can affect investor confidence and project viability, leading to market fluctuations and project delays.

Geographical and Resource Constraints: The feasibility and effectiveness of solar PV deployment vary depending on geographical factors, such as solar irradiance levels, climate conditions, land availability, and proximity to load centers. Remote or off-grid locations may face logistical challenges and higher costs associated with transportation, infrastructure, and maintenance.

In order to address these issues and maximize the advantages of solar PV technology while limiting any potential disadvantages, a comprehensive strategy integrating technological innovation, legislative support, financial incentives, and stakeholder engagement is needed. Collaboration among governments, utilities, industry stakeholders, and research institutions is essential to overcoming barriers and accelerating the transition to a sustainable energy future. All things considered, solar photovoltaic technology has had tremendous breakthroughs and cost reductions in the last few years, positioning it as one of the energy sources with the fastest global growth rates. Continued innovation, supportive policies, and economies of scale are expected to further drive the adoption of solar PV as a key component of the global energy transition.

5.3 Solar Thermal Technology

Using solar thermal technology, heat energy is produced by the sun and used directly for a variety of purposes, including space and water heating, industrial activities, and the production of electricity. In contrast to solar photovoltaic (PV) technology, which uses solar radiation to create electricity directly, solar thermal systems use solar radiation to generate thermal energy. The essential parts of solar thermal systems are solar thermal collectors, sometimes referred to as sun collectors. Heat is produced by these collectors when sunlight is absorbed. Solar collectors come in various varieties, such as concentrating collectors, evacuated tube collectors, and flat-plate collectors. A dark-colored absorber plate is located inside a flat, insulated box that has a clear cover (sometimes made of glass) on top. This is known as a flat-plate collector. Heat from the sun is captured by the absorber plate and transferred to a fluid (such as water or antifreeze) that is circulated through pipes or channels in the collector. After that, the heated fluid is either pumped straight into the building's heating system or to a storage tank. The glass tubes in evacuated tube collectors are arranged in rows in a parallel fashion, with an absorber tube within each tube to capture sunlight and transform it into heat. Reduced heat loss from conduction and convection is achieved by creating a vacuum in the area between the inner and outer tubes. Particularly in colder areas, evacuated tube collectors are more efficient than flat-plate collectors. Concentrating collectors focus sunlight onto a narrow focal point with the use of mirrors or lenses. A receiver then collects the concentrated solar energy and raises the temperature there. In concentrating solar power (CSP) systems, which use steam turbines to generate electricity, concentrating collectors are frequently employed (Razmjoo et al. 2022). Thermal storage systems are frequently a part of solar thermal systems, which allow the excess heat produced during sunny spells to be stored for use during cloudy or nighttime hours. Phase-change materials, water tanks, and other thermal storage media can all be used to achieve thermal storage.

5.4 Applications

Solar thermal technology harnesses sunlight to generate heat energy, which can be utilized in various applications across different sectors. Some common applications of solar thermal technology include:

Water Heating: Water for industrial, commercial, and household consumption can be heated using solar thermal collectors. Flat-plate or evacuated tube collectors, which collect solar radiation and transfer the heat to a fluid (such as water or antifreeze solution) flowing through the collector, are the standard components of solar water heating systems (Ha and Phung 2019). After heating, the fluid is kept in a tank for use in industrial operations, space heating systems, or hot water systems for homes.

Space Heating and Cooling: Systems like solar air heaters, solar space heating panels, or solar thermal absorption chillers can be utilized to exploit solar thermal energy for space heating and cooling purposes in buildings. Solar air heaters, solar space heating panels, and solar thermal absorption chillers are a few examples of systems that employ solar thermal energy for building space heating and cooling purposes.

Solar Cooking: Solar cookers and solar ovens use solar thermal energy to cook food without the need for traditional cooking fuels, such as gas or electricity. These devices typically consist of reflective surfaces or concentrators that focus sunlight onto a cooking chamber, where the heat is absorbed and used for cooking or baking. Solar cooking is especially beneficial in off-grid or rural areas where access to cooking fuels may be limited or expensive.

Industrial Process Heat: Solar thermal energy can provide process heat for a wide range of industrial applications, such as food processing, textile manufacturing, chemical production, and desalination. Solar thermal collectors can deliver high-temperature heat directly to industrial processes or be integrated with existing heating systems to supplement conventional fuel sources, reducing operating costs and greenhouse gas emissions.

Solar Thermal Power Generation: Utilizing mirrors or lenses, concentrated solar power (CSP) systems focus sunlight onto a receiver, where it is transformed into heat to create steam and power a turbine engine to generate electricity. In order to generate grid-connected power, CSP plants can be installed at a utility scale. They provide a dependable and dispatchable renewable energy source that can be stored for use in times of high demand or low sunshine.

Solar Desalination: Solar thermal desalination systems use solar energy to evaporate water and separate it from salt or other contaminants, producing freshwater for drinking, irrigation, or industrial use. Solar stills, solar ponds, and solar desalination plants are examples of solar thermal desalination technologies that can provide clean water in arid or remote regions where freshwater resources are scarce.

Solar Drying: Solar thermal energy can be used for drying agricultural crops, food products, and biomass materials using solar dryers or solar drying systems. These systems capture sunlight to heat air or fluid, which is then circulated through drying chambers to remove moisture from the materials, preserving their quality and extending their shelf life without the need for fossil fuels or electricity.

Solar Thermal Energy Storage: Solar thermal systems can incorporate thermal energy storage technologies, such as molten salt, phase change materials, or sensible heat storage, to store excess heat for later use, enabling continuous operation or dispatchable power generation. Energy storage enhances the flexibility and reliability of solar thermal applications, allowing for load shifting and better integration with grid systems.

5.4.1 Advantages

Solar thermal technology offers several advantages, making it a promising renewable energy solution for various applications:

High Efficiency: Solar thermal systems can achieve high conversion efficiencies, especially in concentrating solar power (CSP) plants where sunlight is concentrated to generate heat at elevated temperatures. This high efficiency enables solar thermal technology to produce significant amounts of heat energy from sunlight, making it a cost-effective solution for heating and power generation.

Versatility: Numerous uses for solar thermal technology exist, such as desalination, power production, industrial process heat, space heating and cooling, water heating, and solar drying. Its adaptability enables customized solutions to satisfy various energy requirements in various industries and geographical areas.

Energy Storage Integration: Thermal energy storage technologies, including phase change materials or molten salt, are easily integrated into solar thermal systems to store excess heat for later use. In times of low sunshine or heavy demand, energy storage allows for continuous operation or power generation, improving the dispatchability and reliability of solar thermal applications.

Cost-Effectiveness: Solar thermal technology can offer cost-effective solutions for heating and power generation, especially in regions with abundant sunlight. With decreasing costs of solar thermal collectors and increasing efficiencies, solar thermal systems are becoming more competitive with conventional energy sources, providing long-term cost savings and reducing reliance on fossil fuels.

Low Environmental Impact: Solar thermal technology is an environmentally benign energy source since it uses sunlight to produce heat energy without releasing greenhouse gases or air pollution. Solar thermal systems support sustainable growth and environmental preservation by lowering dependency on fossil fuels and enhancing air quality.

Local Energy Production: There are several sizes at which solar thermal systems can be used, ranging from modest home setups to massive utility-scale power plants. This decentralized approach to energy production promotes energy independence and resilience by reducing reliance on centralized power generation and imported fossil fuels.

Job Creation and Economic Benefits: The deployment of solar thermal technology creates jobs and stimulates economic growth in manufacturing, installation, maintenance, and related industries. Local sourcing of materials and labor further enhances the economic benefits of solar thermal projects, contributing to job creation and community development.

Reliability and Durability: Solar thermal systems are known for their reliability and durability, with minimal moving parts and long operational lifetimes. Properly designed and maintained solar thermal installations can provide reliable heat energy and power generation for decades, with low maintenance requirements and operational costs.

Scalability and Modularity: Solar thermal technology is scalable and modular, allowing for flexible deployment and expansion according to energy demand. Systems can be easily scaled up or down to meet changing requirements, making solar thermal technology suitable for both small-scale and large-scale applications.

Resilience to Fuel Price Volatility: Solar thermal technology is independent of fossil fuels, which are prone to fluctuations in price and interruptions in supply. By harnessing sunlight as an abundant and free energy source, solar thermal systems offer stability and predictability in energy costs, reducing vulnerability to fluctuations in fuel prices and geopolitical risks.

5.4.2 Challenges

While solar thermal technology offers numerous advantages, it also faces several challenges that can affect its widespread adoption and deployment. Some of these challenges include:

High Initial Costs: The upfront capital costs of solar thermal systems, especially concentrating solar power (CSP) plants, can be significant compared to conventional energy sources. This includes the costs of solar collectors, heat storage systems, power generation equipment, and associated infrastructure. High initial costs can hinder investment in solar thermal projects, particularly in regions with lower solar resource availability or competing energy sources with lower upfront costs.

Land and Space Requirements: Large-scale solar thermal installations, such as CSP plants, require significant land area for solar collectors, heat storage facilities, and other infrastructure (Koch et al. 2023). Securing suitable land for solar projects can be challenging, especially in densely populated areas or regions with competing land uses, environmental sensitivities, or regulatory restrictions. Land use considerations can also impact ecosystem conservation, biodiversity, and land access rights.

Intermittency and Variability: The availability of sunlight, which changes depending on the time of day, season, weather, and location, is a prerequisite for the creation of solar thermal energy. This intermittency and variability can pose challenges to grid integration and energy dispatchability, especially for CSP plants without energy storage (Pandiyan et al. 2023). Energy storage technologies, such as molten salt or thermal storage, can mitigate intermittency issues but add complexity and costs to solar thermal systems.

Complexity of Technology and Design: Solar thermal systems, particularly CSP plants, involve complex technology and engineering designs, including solar collectors, heat transfer fluids, heat exchangers, and power generation cycles. Designing, constructing, and operating solar thermal projects require specialized expertise and skills in solar energy, thermodynamics, materials science, and system integration. Technical challenges related to system efficiency, reliability, and performance optimization may arise during project development and operation.

Water Use and Environmental Impact: Some solar thermal technologies, such as CSP plants with wet cooling systems, require significant amounts of water for steam generation and power plant cooling. Water use can pose environmental challenges, particularly in arid regions or areas with water scarcity, competing water demands, or ecological sensitivities. Efforts to minimize water consumption, improve water efficiency, and adopt dry cooling or alternative cooling technologies can help mitigate water-related impacts of solar thermal projects.

Transmission and Distribution Infrastructure: Large-scale solar thermal projects, such as CSP plants, are often located in remote or desert areas with abundant sunlight but limited grid infrastructure. Connecting solar thermal power plants to existing transmission and distribution networks may require significant investments in new infrastructure, transmission lines, substations, and grid upgrades. Integration challenges, such as grid stability, voltage regulation, and power dispatchability, must be addressed to ensure the reliable and efficient operation of solar thermal plants.

Permitting and Regulatory Hurdles: Solar thermal projects are subject to various permitting, regulatory, and licensing requirements at the local, regional, and national levels. Obtaining permits for land use, environmental impact assessments, water rights, air quality, and wildlife protection can involve lengthy and complex approval processes, leading to project delays and cost overruns. Inconsistent or uncertain regulatory frameworks, land tenure issues, and permitting bottlenecks can pose barriers to solar thermal development and investment.

Competitive Market Dynamics: Conventional energy sources like nuclear power and fossil fuels, as well as alternative renewable energy sources like geothermal, wind, and solar photovoltaics (PV), compete with solar thermal technology. Technological advancements, declining costs, policy support, and market dynamics influence the competitiveness of solar thermal projects relative to alternative energy options. Economic factors, such as electricity prices, fuel costs, subsidies, and carbon pricing, also impact the business case for solar thermal investments and deployment.

In spite of these obstacles, solar thermal technology has a great deal of promise to improve energy security, cut greenhouse gas emissions, and reduce energy consumption—especially in areas with plenty of sunshine. The objective of ongoing research and development is to enhance the scalability, affordability, and efficiency of solar thermal systems in order to hasten their acceptance as a clean energy source.

5.5 Smart Cities: A Holistic Approach to Urban Development

Smart cities are metropolitan regions that effectively handle urban concerns, improve sustainability, and improve the quality of life for its citizens by utilizing technology and data-driven solutions (Su and Fan 2023). These cities use a variety of smart technology, infrastructure, and services to improve connectivity, optimize

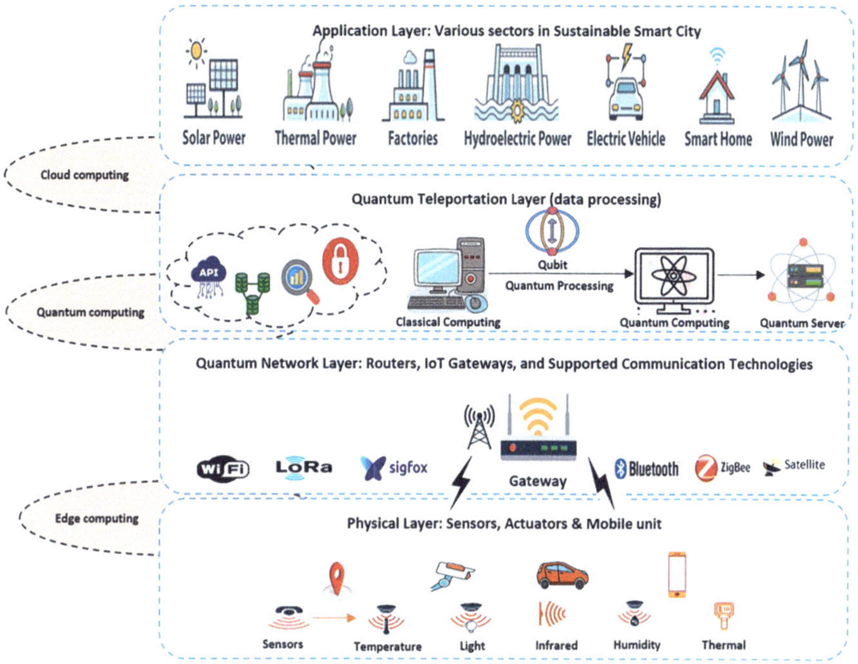

Fig. 5.1 Quantum IoT network architecture driven by energy for sustainable smart cities

resource use, stimulate economic growth, and encourage innovation. Figure 5.1 shows the quantum IoT network architecture driven by energy for sustainable smart cities. Some key components and characteristics of smart cities are as follows:

5.5.1 Information and Communication Technology (ICT) Infrastructure

Robust ICT infrastructure is implemented in smart cities, such as wireless networks, sensor networks, and high-speed internet access, to facilitate data sharing and smooth communication amongst stakeholders, systems, and devices.

5.5.2 Internet of Things (IoT) and Sensor Networks

In order to gather real-time data on a variety of aspects, including traffic flow, air quality, energy consumption, trash management, and public safety, IoT devices and sensor networks are placed throughout the city (Hatem 2023). Decision-making, resource allocation, and service delivery are all improved with the utilization of this data.

5.5.3 Data Analytics and Urban Intelligence

Smart cities employ sophisticated data analytics methods, such as artificial intelligence, machine learning, and predictive modeling, to examine vast amounts of information gathered from Internet of Things (IoT) sensors and other sources. Insights regarding urban trends, patterns, and dynamics are thereby gained by city planners and policymakers, resulting in better decision-making and policy formulation.

5.5.4 Smart Mobility and Transportation

Intelligent transportation systems (ITS) are used in smart cities to optimize traffic flow, lessen congestion, and increase resident mobility. This includes solutions such as real-time traffic monitoring, smart parking systems, electric vehicle (EV) charging infrastructure, and integrated public transportation systems.

5.5.5 Sustainable Energy and Infrastructure

Smart cities prioritize sustainability and environmental stewardship by integrating renewable energy sources, energy-efficient technologies, and green infrastructure into urban planning and development (Gracias et al. 2023). This includes initiatives such as solar power generation, energy-efficient buildings, smart grid systems, and sustainable water management practices.

5.5.6 Urban Planning and Governance

Smart cities adopt innovative urban planning strategies and governance models to enhance efficiency, transparency, and citizen engagement. This includes participatory decision-making processes, open data initiatives, and digital platforms for civic engagement and service delivery.

5.5.7 Quality of Life and Livability

By giving citizens access to high-quality housing, healthcare, education, and recreational opportunities, as well as cultural attractions, smart cities aim to improve their quality of life and livability. This includes initiatives to promote social inclusion, public safety, health and well-being, and equitable access to services and opportunities.

5.5.8 *Economic Development and Innovation*

Smart cities foster a culture of innovation, entrepreneurship, and economic development by supporting research and development initiatives, technology incubators, and startup ecosystems. This includes initiatives to attract investment, stimulate job creation, and promote economic resilience and diversification.

5.6 Solar Power for a Smart City's Sustainable Development

By supplying clean, renewable power to meet the expanding energy demands of metropolitan areas, solar energy may play a critical role in the sustainable development of smart cities. Solar energy contributes to the sustainable development of smart cities in the following ways;

I. *Reducing Carbon Footprint*: Since solar energy produces power without releasing any pollutants or greenhouse gases into the atmosphere, it helps to slow down global warming and enhance air quality. Smart cities may drastically lower their carbon footprint and dependency on fossil fuels for electricity generation by switching to solar power (Hashim et al. 2023).

II. *Energy Efficiency and Grid Stability*: Smart grid technology can be integrated with solar PV installations to manage and distribute electricity more efficiently. By incorporating renewable energy sources like solar electricity, smart networks may improve grid stability, optimize energy use, and balance supply and demand in real time.

III. *Resilience and Energy Security*: In smart cities, solar energy improves the energy infrastructure's robustness and dependability. When paired with energy storage options like batteries, distributed solar photovoltaic systems can deliver backup power in the event of an emergency or grid failure, guaranteeing a steady flow of electricity to vital buildings and services.

IV. *Cost Savings*: Long-term cost savings are possible with solar energy when compared to traditional fossil fuel-based electricity generating. While the initial investment in solar infrastructure may be higher, solar power has minimal operating and maintenance costs once installed. Over time, smart cities can realize significant economic benefits from reduced energy bills and avoided costs associated with the environmental and health impacts of fossil fuels.

V. *Promoting Sustainable Urban Development*: Solar energy facilitates sustainable urban development by supporting energy-efficient buildings, green infrastructure, and low-carbon transportation systems (Li and Tan 2023). Solar panels can be integrated into building designs, public spaces, and transportation infrastructure to provide clean energy, enhance urban aesthetics, and promote sustainable living practices among residents.

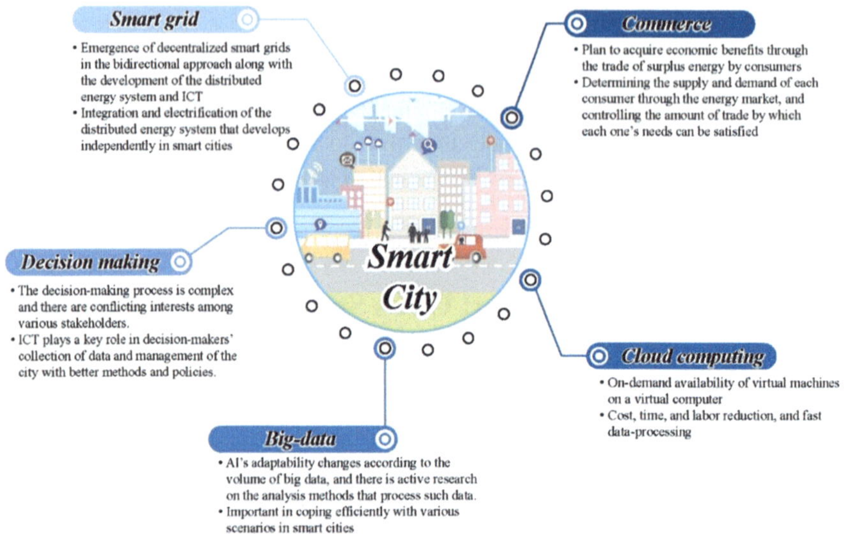

Fig. 5.2 Conceptual diagram of energy conservation systems in smart cities

VI. *Community Engagement and Education*: Solar energy projects in smart cities offer opportunities for community engagement, education, and participation in sustainable development initiatives. Public outreach programs, educational campaigns, and demonstration projects can raise awareness about the benefits of solar energy, empower residents to adopt renewable energy solutions, and foster a culture of environmental stewardship.

VII. *Job Creation and Economic Growth*: The implementation of solar energy infrastructure in smart cities promotes economic growth and employment creation across a range of industries, including solar system production, building, installation, and maintenance. By investing in the solar industry, smart cities can create employment opportunities, attract investments, and support local businesses, contributing to overall socio-economic development. Figure 5.2 showed the conceptual diagram of energy conservation systems in smart cities.

5.7 Developing and Analyzing an Integrated Solar Energy System for Smart Cities

Developing and analyzing an integrated solar energy system for smart cities is important for the following several reasons including renewable energy integration, energy efficiency and cost savings, grid stability and resilience, peak load management, resilience to climate change, urban planning and development, technological innovation and economic growth, and community engagement and empowerment. It involves several key steps and considerations. Some comprehensive approaches to the process are as follows:

I. *Assessment of Energy Needs*: Begin by assessing the energy needs and requirements of the smart city, including electricity demand for residential, commercial, and industrial sectors. Conduct thorough energy audits and analysis to understand consumption patterns, peak demand periods, and potential areas for energy efficiency improvements.

II. *Solar Resource Assessment*: Evaluate the solar potential of the city by conducting a solar resource assessment. This entails examining past data on solar irradiance, mapping the city's solar availability, and choosing appropriate sites for solar systems based on elements including land availability, shade, and sunlight exposure.

III. *System Design and Integration*: Based on the evaluation of solar resources and energy consumption, determine the ideal size and arrangement for the integrated solar energy system. Depending on the unique needs and applications of the smart city, choose the right solar technologies, such as solar thermal collectors, solar photovoltaic (PV) systems, and concentrating solar power (CSP) systems. In order to maximize energy generation, distribution, and consumption, integrate solar energy systems with the current infrastructure and smart grid technology. Consider factors such as grid compatibility, energy storage options, and demand response capabilities.

IV. *Financial Analysis and Funding Mechanisms*: Conduct a comprehensive financial analysis to evaluate the costs and benefits of the integrated solar energy system, including upfront capital costs, operating expenses, and long-term savings. Explore various financing options and funding mechanisms, such as public-private partnerships, government incentives, grants, and subsidies, to support the deployment of solar energy infrastructure in the smart city (Hashim et al. 2023).

V. *Environmental Impact Assessment*: Assess the environmental impact of the integrated solar energy system, including considerations such as land use, habitat disruption, and potential effects on biodiversity. Implement measures to minimize environmental impacts, such as site selection criteria, habitat restoration initiatives, and sustainable land management practices.

VI. *Implementation and Deployment*: Develop a detailed implementation plan outlining the phased deployment of the integrated solar energy system, including timelines, milestones, and resource requirements. Coordinate with relevant stakeholders, including government agencies, utilities, technology providers, and local communities, to facilitate the smooth implementation of the solar energy projects. Monitor and evaluate the performance of the solar energy system throughout the deployment process, making adjustments as needed to optimize efficiency and effectiveness.

VII. *Monitoring and Performance Evaluation*: Install control and monitoring devices to keep an eye on the integrated solar energy system's generation, consumption, and efficiency in real time. Conduct regular performance evaluations and data analysis to identify opportunities for optimization, troubleshoot issues, and ensure that the solar energy system meets its objectives.

VIII. *Public Outreach and Education*: Engage with the community through public outreach initiatives, educational programs, and awareness campaigns to promote the benefits of solar energy and encourage participation in sustainability efforts. Provide training and capacity-building opportunities for local stakeholders, including residents, businesses, and policymakers, to enhance their understanding of solar technologies and their role in smart city development.

5.8 Parametric Optimization of Solar Power Plants for the Construction of Sustainable Smart Cities

It involves fine-tuning various design parameters to maximize energy output, efficiency, and economic viability while minimizing environmental impact. Figure 5.3 shows the new paradigm of the sustainable smart city. Here's a systematic approach to perform parametric optimization:

(i) *Define Objectives and Constraints*

- Clearly define the objectives of the optimization process, such as maximizing energy generation, minimizing costs, or reducing environmental footprint.
- Identify constraints related to technical, economic, regulatory, and environmental factors that must be considered during optimization.

(ii) *Selection of Optimization Parameters*

Identify key design parameters that can be optimized to improve the performance of solar energy power plants. These parameters may include:

- Solar panel orientation and tilt angle.
- Selection of solar PV technology (e.g., monocrystalline, polycrystalline, thin-film).
- Tracking systems (e.g., single-axis, dual-axis tracking).
- Energy storage capacity and type (e.g., batteries, molten salt).
- Integration with other renewable energy sources (e.g., wind, biomass).
- Land use optimization and site selection.
- Grid integration strategies (e.g., smart grid technologies, demand response).

(iii) *Development of Optimization Model*

- Formulate a mathematical optimization model that represents the relationship between the optimization parameters, objective function, and constraints (Mishra and Singh 2023).
- Depending on the complexity of the optimization problem, use analytical models, numerical simulations, or optimization algorithms (e.g., genetic algorithms, particle swarm optimization) to find the optimal solution.

Fig. 5.3 New paradigm of the sustainable smart city

(iv) *Data Collection and Analysis*

- Gather relevant data on solar resource availability, site characteristics, energy demand profiles, equipment specifications, cost parameters, and environmental considerations.
- Analyze the data to identify patterns, trends, and correlations that can inform the optimization process and support decision-making.

(v) *Simulation and Sensitivity Analysis*

- Conduct simulation studies to evaluate the performance of different design configurations under varying operating conditions.
- Perform sensitivity analysis to assess the impact of changes in input parameters on the overall system performance and identify critical factors influencing the optimization outcomes.

(vi) *Optimization Algorithm Implementation*

- Implement optimization algorithms or optimization software tools to search for the optimal solution within the defined parameter space.
- Iteratively adjust the optimization parameters and evaluate the objective function to converge towards the optimal solution.

(vii) *Validation and Verification*

- Validate the optimized design solutions using real-world data, field tests, or validation studies to ensure that they meet performance requirements and constraints.
- Verify the accuracy and reliability of the optimization model by comparing the predicted outcomes with observed results and historical data.

(viii) *Decision Making and Implementation*

- Evaluate the optimized design solutions based on their technical feasibility, economic viability, and environmental sustainability.
- Make informed decisions regarding the selection and implementation of optimized solar energy power plant configurations in the context of smart city development.
- Develop a phased implementation plan and monitor the performance of deployed systems to assess their effectiveness and make continuous improvements over time.

5.9 Conclusion

To meet the increasing demand for energy due to urbanization, methodological approaches based on alternative urban models are needed to facilitate the necessary shift towards more energy-efficient cities. This study addresses the built environment's potential for solar energy and looks at how urban planning might help turn current cities into solar smart cities. An existing urban area was subjected to a performance-based analysis in order to support this study. This analysis considered the solar potential of roofs as well as the optimization of power distribution networks across the city neighborhood. Urban morphological features and land use patterns have been identified and defined to support the urban planning process using GIS and dynamic simulation software. Peak to off-peak times were estimated using statistical data on energy consumption at the neighborhood level, accounting for the length, frequency, and temporal distribution of existing building uses throughout the day. This study demonstrates the benefits of solar energy at the building block level and the role that energy management and urban planning, when backed by a solar smart grid, may play in achieving a city's overall energy balance.

Competing Interest Declaration The authors have no conflicts of interest to declare that are relevant to the content of this book.

References

Abu-Rayash A, Dincer I (2021) Development and analysis of an integrated solar energy system for smart cities. Sustain Energy Technol Assess 46:101170

Ghadami N, Gheibi M, Kian Z, Faramarz MG, Naghedi R, Eftekhari M et al (2021) Implementation of solar energy in smart cities using an integration of artificial neural network, photovoltaic system and classical Delphi methods. Sustain Cities Soc 74:103149

Gracias JS, Parnell GS, Specking E, Pohl EA, Buchanan R (2023) Smart cities—a structured literature review. Smart Cities 6(4):1719–1743

Ha Q, Phung MD (2019) IoT-enabled dependable control for solar energy harvesting in smart buildings. IET Smart Cities 1(2):61–70

Hashim N, Neo TG, Mohammed MN, Sultan HS, Jameel Al-Tamimi AN, Alfiras M (2023) Toward sustainable smart cities: a new approach of solar and wind renewable energy in agriculture applications. In: Artificial intelligence and transforming digital marketing. Cham, Springer Nature, pp 555–563

Hatem FM (2023) Making cities smarter for an inclusive green transition towards a long-term sustainable development: a critical literature review. IET Smart Cities 5(4):243–253

Koch F, Beyer S, Chen CY (2023) Monitoring the sustainable development goals in cities: potentials and pitfalls of using smart city data. GAIA Ecol Perspect Sci Soc 32(1):47–53

Kuznetsov P, Rimar M, Yakimovich B, Kulikova O, Lopusniak M, Voronin D, Evstigneev V (2021) Parametric optimization of combined wind-solar energy power plants for sustainable smart city development. Appl Sci 11(21):10351

Li B, Tan W (2023) A novel framework for integrating solar renewable source into smart cities through digital twin simulations. Sol Energy 262:111869

Menniti D, Bayod-Rújula AA, Burgio A, García DA, Leonowicz Z (2017) Solar energy and PV systems in smart cities. Int J Photoenergy 2017:1–2

Mishra P, Singh G (2023) Energy management systems in sustainable smart cities based on the internet of energy: a technical review. Energies 16(19):6903

Pandiyan P, Saravanan S, Usha K, Kannadasan R, Alsharif MH, Kim MK (2023) Technological advancements toward smart energy management in smart cities. Energy Rep 10:648–677

Razmjoo A, Gandomi AH, Pazhoohesh M, Mirjalili S, Rezaei M (2022) The key role of clean energy and technology in smart cities development. Energ Strat Rev 44:100943

Su Y, Fan D (2023) Smart cities and sustainable development. Reg Stud 57(4):722–738

Telang S, Chel A, Nafdey R, Kaushik G (2021) Solar energy for sustainable development of a smart city. In: Security and privacy applications for smart city development. Springer, Cham, pp 155–169

Chapter 6
Solar Energy and Its Utilization in Smart Cities

Ambneesh Mishra, Mohammad Izhar Alam, Dan Bahadur Pal, and Shraddha Awasthi

Abstract The majority of people on the planet today reside in cities. Significant rural-to-urban migration is occurring, and it is anticipated that this trend will continue for the ensuing few decades (the World Migration Report 2015 estimates that 814 million people will live in urban India by 2050). Cities now face several difficulties as a result of the transition toward urbanization, including issues with waste management, few resources, air pollution, health risks, traffic congestion, and inadequate infrastructure. Cities all across the world are looking for creative ways to address these issues, which has given rise to the idea of "smart cities."

A smart city is an efficient, sustainable metropolitan area that strives to give its residents a more affordable, liveable, environmentally friendly, and stimulating environment. Utilizing renewable energy sources and optimizing urban energy use are essential for raising the standard of living for residents. The chapter delves deeply into the fascinating partnership between advances in smart city technology and solar electricity. It looks at how they may work together to create cities that are more hygienic, productive, and capable of overcoming obstacles. Reducing carbon emissions and building more resilient urban ecosystems are the main goals.

You will gain knowledge about a variety of solar technologies, including photovoltaics (PV), which turn light directly into electricity; concentrating solar power (CSP), which powers utility-scale electric turbines with solar heat (thermal energy); and solar heating and cooling (SHC) systems, which use thermal energy collected

A. Mishra
Faculty of Sciences and Technology, University of Algarve, Faro, Portugal

M. I. Alam
Department of Environmental Studies, Delhi University, New Delhi, New Delhi, India

D. B. Pal
Department of Chemical Engineering, Harcourt Butler Technical University, Kanpur, Uttar Pradesh, India

S. Awasthi (✉)
Department of Environmental Studies, Motilal Nehru College (E), Delhi University, New Delhi, New Delhi, India

S. Kumar, V. K. Singh (eds.), *Renewable Energy Development: Technology, Material and Sustainability*, Clean Energy Production Technologies,
https://doi.org/10.1007/978-981-97-9626-7_6

from the sun to heat water and air conditioning. Numerous facets of city life, including as infrastructure, services, buildings, and transportation, are powered by these technologies. This chapter sheds light on the obstacles that stand in the way of solar energy's potential to power sustainable, smart cities of the future. This chapter examines policy barriers, budgetary constraints, and technological limits.

To overcome these obstacles, the chapter suggests a course of action centred on creative policy (for instance, as of June 30, 2023, data from the Ministry of New and Renewable Energy indicates that the installed capacity for solar power under the Jawaharlal Nehru Solar Mission has reached roughly 70.10 GW), collaborations between public and private entities, and community involvement. We can develop dynamic urban environments and realize the full potential of solar energy by overcoming these obstacles. The chapter essentially emphasizes that solar energy is essential to creating the intelligent and sustainable cities of the future.

Keywords Urbanization · Eco-friendly · Sustainable city · Solar energy · Thermal energy

6.1 Introduction

The rapid development of information and communication technologies (ICTs) and their application in cities for a range of purposes have made smart cities a popular topic in recent years (Chourabi et al. 2012; Caragliu et al. 2009). Various terms, including "cyberville," "digital city," "electronic city," "flexicity," "information city," "telicity," "wired city," and "smart city," are used to characterize these cities. A smart city is characterized as a location where traditional networks and services are enhanced for the benefit of its residents through the use of digital and telecommunication technologies, making them more effective, adaptable, and sustainable (Hollands 2008; Caragliu et al. 2009). However, there is no universally accepted definition of the term among academics and practitioners (Anthopoulos and Vakali 2018). To put it another way, digital technologies improve public services for citizens and maximize resource efficiency while reducing their negative effects on the environment in smart cities (Mora et al. 2017).

A "smart city" is defined as one that "connects the physical infrastructure, the information-technology infrastructure, the social infrastructure, and the business infrastructure to leverage the collective intelligence of the city." A more thorough definition characterizes a smart sustainable city as an innovative city that makes use of ICTs and other tools to boost competitiveness, quality of life, and the effectiveness of urban operations and services while making sure that it satisfies the demands of both the current and future generations (Massaro et al. 2018) in terms of the social, economic, and environmental aspects.

A city can become smart in any combination of these components, but not all cities need them all. The quantity of smart components is determined by the cost and technology available.

6.2 Definition and Concept of Smart Cities

In recent decades, there has been a tremendous increase in the worldwide population and the expectation of higher living conditions. It is projected that over 70% of the world's population will reside in urban regions by 2050 (United Nations 2018). At the moment, cities use 75% of the energy and resources in the world, which results in the production of 80% of greenhouse gases. In the coming decades, the ecosystem may suffer greatly from this fast urbanization and population increase (IEA 2022). As such, the idea of smart cities has become essential. Smart city implementation can drastically cut carbon emissions, water and energy consumption, transportation needs, and trash generation in the city, even with related expenditures. Globally, there is variation in the traits, prerequisites, and elements of smart cities. Globally recognized specifications are provided by standards set by organizations, like the International Organization for Standardization (ISO), which promote expansion while guaranteeing efficiency, quality, and safety (Mora et al. 2017; Massaro et al. 2018). The creation and development of smart cities can benefit greatly from the application of standards. In addition to aiding in the fight against climate change, addressing security and transportation concerns, and guaranteeing the quality of water supplies, they can offer specifications for monitoring the technological and functional performance of smart cities. Standards monitor the smart city's performance and lessen its environmental impact by taking into account a number of elements, including resource management and business practices. For its various components—smart grids, IoT, eHealth, and intelligent transportation systems (ITS)—IEEE has been creating standards for smart cities. ISO 37120 is one such standard that outlines 100 city performance indicators, of which 46 are core and 54 are supporting indicators (IEEE Smart Cities). City civic bodies can use these indicators—which include those related to the economy, education, energy, and environment—to compare their city to others, learn from other cities' best practices, and assess how effectively their services are performing (MDPI).

6.3 Overview of the Relationship Between Smart Cities and Solar Energy

Solar energy is a plentiful and renewable form of energy that has gained attention due to global warming, greenhouse gas emissions, and the depletion of fossil resources. Both the need for solar energy and the integration of solar electricity into smart power systems are increasing. Demand-side management techniques will be used in this integration to enhance energy management systems (The International Telecommunication Union n.d). Liu et al. (2016) proved that to mitigate the negative impacts of large-scale interconnected power networks and enhance inter-area oscillations, solar photovoltaic systems can operate as actuators. In addition to its primary purpose, Wandhare and Garwal suggested a novel dispatch mechanism that attempts to alleviate the low-frequency power fluctuation issue in a central PV-grid arrangement Wandhare and Agarwal (2014a). The suggested strategy changed the stability of the system without requiring any more hardware or changes that could be used in SG applications. A different project Wandhare and Agarwal (2014b) that entails extending the PV grid's auxiliary circuit has been suggested. As a result, the system's reactive power consumption is increased by up to three times compared to the original PV inverter. The system consists of a miniature gas turbine, storage elements, and photovoltaic cells. To govern the system, a creative control scheme is implemented that was suggested by Kanchev et al. (2011). The primary energy management sector of the microgrid and the customer power supervision section made up the two components of the management scheme. The generation of PV electricity and the estimation of demands served as the foundation for this management plan, which allowed for two-way communication with the grid. An examination of the active photovoltaic system (APS) control approach, incorporating ultra-capacitors, batteries, and PV, was presented by Choudar et al. (2015). By employing a direct current (DC) link to govern power flow between the converter and the grid, the operator was able to maintain the power grid control petition. The outcomes of the simulation demonstrated that the system was enhanced by flattening power, quick dynamic adjustment, and efficient utilization of storage components. Through a management process, the study investigated a PV micro-inverter microgrid's capacity to function in both grid-connected and island modes by Rodríguez et al. (2013). They made use of grid-linked micro-inverters, which could operate in island mode while still utilizing the same control algorithms. They realized that they could use droop schemes to operate as a voltage source by rearranging the management of micro-inverters in the island mode. They also created a building-integrated photovoltaic (BIPV) technique that was integrated into an urban storage system by Sechilariu et al. (2013). Serving the buildings themselves was the system's main objective. A DC network distribution was used to remove several energy exchanges. The system showed promise for concurrently powering a tertiary building and producing electricity via a photovoltaic array. Figure 6.1 showed the solar water heater using sun.

Furthermore, two-way data interchange was used for classified management. Another project used a DC system to control how energy is distributed across a building by Byeon et al. (2013). The energy management system that was suggested took into account both the DC loads and the generation of DC systems like fuel cells and photovoltaics. It was shown that establishing a useful energy management system requires limiting the energy expenses associated with building. The work concentrated on online control and one-day advanced functional forecasting for an urban microgrid with active generators as its PV arrangement by Kanchev et al. (2011). The dynamic algorithm was employed to carry out multi-objective tasks, including the reduction of economic costs and CO2 emissions. In a different study, researchers looked into the cooperative control of interior expense-based demand response for combined heat and power (CHP) and photovoltaic (PV) prosumer

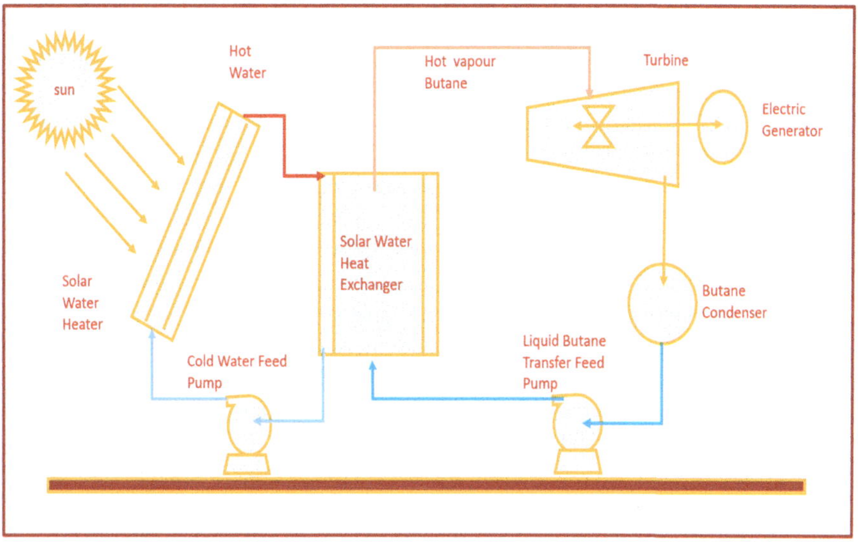

Fig. 6.1 Solar water heater using solar energy

systems by Ma et al. (2016), and it showed that the actual power curves and the anticipated ones differed significantly. For a two-way power demand/response configuration in a solar customer grid supply system (PCGSS), a novel algorithm was put forth by Guichi et al. (2018).

6.4 Energy

The environment and natural resources are being strained by the rising demand for energy worldwide. Because more fossil fuels are being used, carbon dioxide emissions are increasing. Despite being greener energy sources, solar and wind power are underutilized. The practices of globalization are unjust because they transfer adverse consequences to nations with inadequate governance. There is pressure on nations and areas to make the most of their natural resources to make money and lessen the effects of climate change. The primary problem is that the bulk of greenhouse gas emissions are caused by our current energy system's heavy reliance on fossil fuels. However, a secure and practical response to the problem of climate change is provided by renewable energy sources like wind, solar, biomass, hydro, wave, and tidal. These energy sources facilitate the shift to a low-carbon economy by supplying heat, cooling, power, and transportation fuels in a carbon-neutral manner (IEA 2022).

6.5 Renewable Energy

Because renewable energy sources cut greenhouse gas emissions like carbon and methane, they are essential to the fight against climate change. It is essential to the implementation of climate change policies and can significantly lower carbon dioxide emissions. Furthermore, by promoting energy efficiency and enhancing energy security, renewable energy lessens reliance on gas and oil. Technologies for renewable energy are affordable and make use of energy flows found in nature, such as biomass, solar, water, or wind. They can carry fuels and provide power, heating, and cooling. Of all energy sources, renewables have the least negative effects on the environment. They are also practical, safe, and reasonably priced energy-generating technologies that may boost resilience and aid in disaster response and recovery. Energy-intensive services including communications, water purification, refrigeration, water pumping, medical equipment, and lighting can be powered by wind and photovoltaic (PV) cell electricity. Renewable energy is the answer to climate change (IEA 2022, 2023).

6.6 Types of Renewable Energy Sources

6.6.1 Hydropower

Humans have been controlling water flow and capturing the energy of river currents through the construction of dams for many centuries. The world's largest renewable energy source at the moment is hydropower, which is mostly produced in China, Brazil, Canada, the United States, and Russia. Hydropower has several disadvantages that should be taken into account, even though it is a naturally replenishable energy source that is ostensibly pure and powered by rain and snow (Hannah 2022).

Big dams may destroy animals and force locals to relocate, among other detrimental effects on river ecosystems and the towns they affect. Silt accumulation can also affect hydropower output, compromising capacity and damaging equipment (Water Needed to Produce Various Types of Energy 2008). As demonstrated in the western United States, where utilities switched to coal and gas to replace lost hydropower during droughts, carbon dioxide emissions increased by 100 megatons over 15 years (Environmental Impacts of Dams n.d.). Droughts can also lead to other problems. Methane is released into the atmosphere by decaying organic material in reservoirs, so even hydropower at full capacity has emissions issues (McKeegan 2015). Water can also be used in different ways to generate electricity. Worldwide tidal and wave energy initiatives aim to capture the inherent rhythms of the ocean (Tidal Energy n.d.). Although the present output of marine energy projects is only expected to be 500 megawatts or less than 1% of total renewable energy, the potential is far higher. Initiatives such as Scotland's Saltire Prize are designed to promote creativity in this field (Renewable Energy Explained n.d.).

6.6.2 Wind

Energy from wind has been utilized for more than 7000 years. These days, electricity-producing wind turbines are growing more widespread worldwide, with Germany, the USA, and China being the leading wind energy producers. The total global wind capacity climbed from 23,900 megawatts to over 539,000 megawatts between 2001 and 2017, more than 22 times the baseline capacity (Renewable Energy Explained n.d.).

Notwithstanding some people's disapproval of their noise and look, wind turbines remain a useful source of electricity. Although onshore turbines generate the majority of wind electricity, offshore projects are increasingly prevalent. The USA is starting to catch up with the UK and Germany in terms of offshore wind projects (Renewable Energy Explained n.d.).

6.6.3 Biomass

Biofuels such as ethanol and biodiesel, wood, wood waste, biogas from landfills, and municipal solid waste are examples of biomass energy, which is a form of renewable energy. Similar to solar energy, this flexible energy source may heat buildings, power cars, and generate electricity (Trust 2020). However, there are several issues with biomass. Critics of corn-based ethanol, for example, claim that it encourages the same destructive farming techniques that have resulted in toxic algae blooms and other environmental problems, and it competes with the corn food market (Blair et al. 2021). Similar arguments have been made about the wisdom of transporting wood pellets from US forests to Europe to generate power. In the meantime, researchers and businesses are trying to figure out how to turn biomass sources like corn stover and wastewater sludge into electricity more effectively. Recovering value from materials that would otherwise be discarded is the goal (Wardle 2023).

6.6.4 Geothermal

In many nations, geothermal energy has been utilized for heating and cooking purposes for millennia. The Earth's interior heat is the source of this kind of energy (Use of Geothermal Energy – U.S Energy Information Administration (EIA) n.d.). By drilling wells down to a depth of 2 km or more, subterranean pools of hot water and steam can be tapped to produce large-scale geothermal energy. Electricity can be produced with this energy (Geothermal Heat Pumps n.d.). Geothermal heat pumps can be used to heat and cool buildings on a smaller scale by making use of temperature differences that exist several meters below the surface. Geothermal energy is constantly available, unlike solar and wind energy (All the Advantages of

Geothermal Energy 2021); however, it has some uncontrollable negative effects. The smell of rotten eggs, which is a result of hydrogen sulfide being emitted, is one of these adverse effects (Environmental Impacts of Geothermal Energy 2013).

6.6.5 *Solar Energy*

A great, safe, and pollution-free source of electricity is solar energy. It is crucial to combating climate change and advancing sustainability. Our solar energy resources are free to use and provide limitless, clean, and silent energy. They are made nationwide, on the ground and rooftops, with little harm done to the environment or living beings, without the need for water, and without causing any mishaps. The economy is stimulated and job possibilities are created by solar energy. It gives households and businesses a dependable source of heat and electricity while reducing the volatility of energy prices (Solar Energy, Wildlife, and the Environment n.d.). At the moment, solar technologies concentrate on using the sun to produce power and heat. Solar thermal energy heats water or air, whereas solar photovoltaic (PV) energy produces electricity.

Solar photovoltaic (PV) systems are mostly used to generate grid-connected electricity for air conditioning, lighting, business equipment, and home appliances in various kinds of buildings. It can also be utilized with stand-alone devices and battery power in isolated areas without access to an electrical source. Solar photovoltaic panels can be erected on the ground, mounted on building rooftops, or integrated into building materials in the process of manufacturing. The roads, car roofs, and clothing of the future will all be converted into solar collectors that can generate electricity (How Does Solar Work? n.d.).

To address the demand for energy-efficient hot water, heating, and cooling systems in residential, commercial, institutional, and industrial buildings, solar thermal technologies are growing in popularity (Stand Alone PV System for Off-grid PV Solar Power n.d.). These technologies will also be employed in the future to store solar energy, which will be useful on cloudy days. In the future, solar thermal technology could greatly improve industrial processes, spur the creation of novel products, and provide economies of scale that would be advantageous to all industries (What Is the Best Mounting System for a Solar Photovoltaic Project? 2024).

6.7 Solar Energy and Sustainability (Sustainable Energy)

Renewable energy technologies have seen a sharp increase in utilization in recent years. Natural energy sources are used by these technologies, and they replenish considerably more quickly than they are used up. Hydro, bioenergy, geothermal, aerothermal, solar, wind, and ocean energy are a few examples of these sources.

All of these technologies have special qualities that influence how they are implemented; therefore, before deploying any of them, their practicality must be taken into account. These distinctive characteristics—which are listed below—define the essential elements of energy sustainability (Energy.gov n.d.). The four basic pillars of sustainable energy development are energy storage, efficiency improvements in generation, transmission, distribution, and usage, conservation, and the development and use of renewable energy sources. The notion of sustainability is introduced, including five dimensions: technological, economic, institutional, environmental, and social, to attain sustainable energy development (IRENA – International Renewable Energy Agency 2024).

From its physical infrastructure, a technically sustainable system should generate an output that is both economical and accessible. The acceptability and accessibility of the energy supply for all members of society are measured by social sustainability. The goal of environmental sustainability is to lessen the harmful effects of energy use. To maintain institutional sustainability, local people must manage the energy system (IRENA – International Renewable Energy Agency 2024).

Two prerequisites must be satisfied for there to be a sustainable energy supply: all energy must originate from renewable resources that are managed responsibly, and energy must be distributed and used effectively. Any energy source is not sustainable if it violates these principles and interferes with the way humans and the natural world interact. The majority of renewable energy technologies are sustainable, including geothermal installations, hydropower, solar, wind, and ocean energy. As part of a comprehensive strategy, each pillar needs to be considered to create a sustainable framework. While the benefits of each renewable energy source are different, all renewable energy technologies have the potential to reduce environmental impact, boost efficiency, and promote sustainable resource management (THE 17 GOALS | Sustainable Development n.d.).

6.8 Solar Power Systems

Photovoltaic cells are used in solar power systems to harness the photovoltaic effect to produce electricity. They are employed in a variety of settings, including residences, businesses, and factories. More light striking the solar cells results in a greater amount of power being produced. One of the most affordable and quickly expanding energy sources in the world is solar electricity, and in the upcoming years, it will only grow more widely. The advantages of solar power for the environment and economy are growing stronger every year as solar panel technology advances. Figure 6.2 shows the overview of solar energy to electricity supply.

There are three types of solar power systems:

1. *On-grid solar power system.*
2. *Off-grid solar power system.*
3. *Hybrid solar power System.*

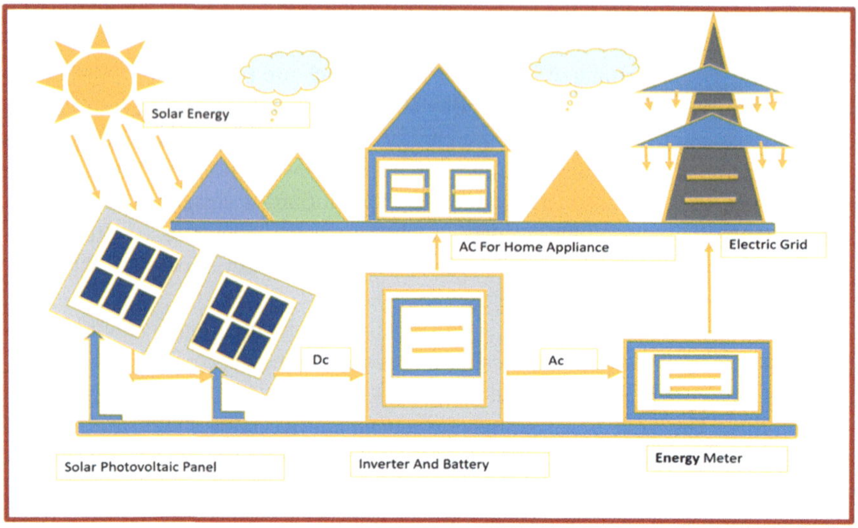

Fig. 6.2 Overview of solar energy to electricity supply

6.9 Advantages and Disadvantages of Solar Energy

6.9.1 Advantages

Using solar energy to power your home has several advantages. Here are some compelling arguments for converting to solar power for your house or place of business.

1. *Lowers Your Electricity Bill*: One of the most crucial elements is that using solar energy lowers the monthly cost of power. However, you can use the sun's power to make your energy. Plus, you have the option to make money. You can save money and choose a safer, more environmentally friendly electricity source by installing one of the state-approved solar energy systems (National Renewable Energy Laboratory (NREL) Home Page n.d.).
2. *Environmentally Friendly*: Compared to conventional non-green energy sources that mostly rely on fossil fuels, solar energy emits fewer greenhouse gases. The emission of greenhouse gases, like CO_2, and other hazardous pollutants, like sulfur oxides, nitrogen oxides, and particulate matter, is significantly decreased for every kilowatt-hour (kWh) of solar energy produced. Because solar power doesn't require burning fuel, greenhouse gas emissions are negligible, and we can feel the air getting cleaner the more renewable energy we consume (National Renewable Energy Laboratory (NREL) Home Page n.d.).

3. *More Energy Independence*: There are finite supplies of fossil fuels that cannot be replenished. On the other hand, solar energy never runs out and lets you become more self-sufficient as technology advances. There are methods to use renewable energy in daily life (National Renewable Energy Laboratory (NREL) Home Page n.d.).
4. *Low Maintainance*: Since solar energy has no moving parts, there is no chance of a malfunction or other issue. Solar energy systems require little upkeep and don't require costly repairs. The majority of solar energy systems have an extended guarantee, often good for 25 years. This is a crucial consideration for people who find solar energy systems appealing but are unable to maintain their home appliances (National Renewable Energy Laboratory (NREL) Home Page n.d.).
5. *Sustainable*: The most plentiful energy source on Earth is solar energy, which has the potential to remain steady for billions of years. More solar energy than the world uses in its entirety, or about 173,000 terawatts, strikes the Earth every day (National Renewable Energy Laboratory (NREL) Home Page n.d.).
6. *Benefits of the Electricity Grid*: The electric grid is less burdened by solar power, and net metering regulations are available in some nations, such as the USA and India. These regulations enable solar power producers to sell their excess electricity and get credit on their utility bills. As a result, you need to purchase less electricity from a utility (National Renewable Energy Laboratory (NREL) Home Page n.d.).

6.9.2 Disadvantages

Not everyone will benefit from solar energy. Here are a few explanations for why people choose not to use this environmentally beneficial technique.

1. *Energy Storage*: In certain situations, solar energy must be stored on batteries, which are expensive and need a lot of space, if it is not used right away. The batteries have to be charged during the day and used at night (National Renewable Energy Laboratory (NREL) Home Page n.d.).
2. *Substantial Initial Investment*: Solar power systems might have a high initial cost due to their various price options. Instead, the price varies based on your location, the type of installation you need, and the quantity of solar panels you require. It can be challenging for domestic use (National Renewable Energy Laboratory (NREL) Home Page n.d.).
3. *Needs More Space*: More solar panels translate into more room if you require more electricity. It might not be feasible for every roof to have enough room to install as many solar panels as clients desire or need (National Renewable Energy Laboratory (NREL) Home Page n.d.).

4. *Doesn't Work at Night*: When sunlight isn't present throughout the night, solar energy must rely on utility grids to produce electricity. A battery backup is an optional feature of some solar panel systems; however, it is not available in conjunction with other options (National Renewable Energy Laboratory (NREL) Home Page n.d.).
5. *Location Matters*: The sun is a plentiful energy source, yet its accessibility varies globally. It occasionally relies on factors like the climate, where you reside, etc. Relying only on solar energy sources may be problematic in some areas during the winter months because the sun doesn't always appear. Certain solar systems will need to supplement the drop in solar power energy during solar eclipses by drawing power from other electrical sources. To get you through the darkest moments, make sure you have enough battery backups (National Renewable Energy Laboratory (NREL) Home Page n.d.).

6.10 Case Studies

6.10.1 India

India is the second most populous country in the world, with a population of approximately 1.3 billion people, located in South Asia. New Delhi serves as its capital. The Arabian Sea is to the west, the Bay of Bengal is to the east, and the Indian Ocean is to the south. Its borders are shared by Bangladesh and Myanmar to the east; China, Nepal, and Bhutan to the northeast; and Pakistan to the west. The Maldives are to the southwest and Sri Lanka is to the southeast. The majority of the nation has tropical weather, but occasionally, monsoons and other atmospheric phenomena can produce unstable weather patterns that result in cyclones, floods, and droughts that have claimed millions of lives. Climate change, according to researchers, will cause more natural disasters in the area in the future that are more unpredictable, frequent, and intense (India Population (2024)—Worldometer n.d.).

With a current urbanization rate of 32%, India is seeing fast urbanization. Experts estimate that 843 million Indians, or 55% of the country's entire population, will reside in cities by 2050. The Indian government has launched a strategy to create 100 new smart cities and sustainably grow existing cities to address this tremendous urbanization. This strategy calls for the use of a paradigm of private-public partnerships and the funding of more than $1 billion in 2015. The government wants to install smart meters, set up test locations for smart grid projects, provide energy to every household for at least 8 h a day, and build a center of competence on the topic. By 2017, the government also intends to add 88,000 MW of rated power plants and produce 30,000 MW of electricity from renewable sources (India—the World Factbook n.d.).

The provision of essential services like roads, healthcare, and sanitation, which have an impact on millions of people's quality of life, is still a struggle for India. In addition, the Indian building sector is expected to grow at a rate never seen before, with 11.5 million new homes being produced a year by 2020. By 2016, the area of intelligent building management is projected to have grown to a value of $1891 billion. Thirty percent water savings, 40% energy savings, and a 10–30% decrease in building maintenance expenses can all result from effective building management (India | Latest News & Updates | BBC News n.d.).

6.10.2 Smart City Projects in India

Prime Minister Narendra Modi of India has promised to build 100 smart cities in retaliation for China including them in their urban development plan. With a $90 billion budget, seven new cities are currently being built along the Delhi-Mumbai Industrial Corridor (DMIC), which runs across six states. The project, which encompasses parts of Uttar Pradesh, Haryana, Rajasthan, Gujarat, Maharashtra, and Madhya Pradesh, is being built in partnership with Japan, a manufacturing and commerce hub. With its excellent infrastructure and the relationship between political and economic capital, India needs this initiative badly (Smart Cities: Ministry of Housing and Urban Affairs, Government of India n.d.).

However, there have been objections in the project's participating regions due to the Special Investment Region (SIR) Act, which permits property expropriation. The impacted areas are mostly agricultural, and the people living there are farmers who do not want to lose their land—which is prone to flooding during the rainy season and experiences annual sea level increases of a few centimeters. Despite the consequences, this natural hazard is frequently disregarded. Many smart city projects, especially those in the East, run the risk of being constructed without taking into account the fundamental tenets of urban life, like social inclusion, sustainable development, and smart security. Smart city projects are visually appealing, but they might not always follow the path of natural evolution as glossy ads would have us believe (Smart Cities: Ministry of Housing and Urban Affairs, Government of India n.d.).

The Indian government intends to construct 20 environmentally friendly cities with cutting-edge infrastructure, including drinking water, the electricity that is available around the clock, effective shared transportation systems, pathways for bicyclists and pedestrians, full recycling of waste and wastewater, remotely controlled smart electricity grids to regulate consumption, smart metering systems, and cultural events. Over the next 10 years, the government projects that these infrastructural projects will increase industrial output, quadruple exports, and create twice as much employment believe (Smart Cities: Ministry of Housing and Urban Affairs, Government of India n.d.).

Dholera, the first city, would be 110 km from Ahmedabad in the Gujarat region. The initial stage of land acquisition is under process, and the project is ready. Approximately 50,000 people live in the neighborhood, which is a part of the Delhi-Mumbai Industrial Corridor (DMIC) project. The other cities that will be comparable in size to Dholera are Dighi and Nasik-Igatpuri in Maharashtra,

But-Nesar-Bawal in Haryana, and Indore-Mhow in the Madhya Pradesh region. These cities will all be connected to other important Indian cities by newly built highways, trains, ports, and airports. The first seven cities will be established in the first phase of investment between 2018 and 2019 believe (Smart Cities: Ministry of Housing and Urban Affairs, Government of India n.d.).

Cities have been suggested to be developed using creative ideas, such as building roads and subterranean parking lots. Public transportation will be accessible within a minute's walk from residences or places of employment, and road design will emphasize reducing the use of private vehicles. A managing city (city-CEO) will guarantee that residents use services effectively believe (Smart Cities: Ministry of Housing and Urban Affairs, Government of India n.d.). Similar initiatives are currently underway in India, including the Hyderabad-Chennai and Chennai-Bangalore industrial corridors. These corridors' industrial locations will be developed according to their unique industrial strengths. For instance, Hyderabad would concentrate on the aerospace and pharmaceutical sectors, while Chennai and Bangalore would concentrate on the manufacturing of automobiles believe (Smart Cities: Ministry of Housing and Urban Affairs, Government of India n.d.).

The planning stages of other industrial corridors, such as Coimbatore-Salem and Chennai-Madurai-Tuticorin-Tirunelveli, are also underway. Cement manufacture, food processing, and subterranean extraction will all be more integrated by these corridors.

Redevelopment of already existing metropolitan areas is another priority. In the industrial city of Haldia, West Bengal, the European Business and Technology Centre is putting a smart city pilot project into action. The goal of this project is to lessen the urban community's ecological imprint.

The European Union launched the EBTC program to help European and Indian businesses transfer technologies for energy generation while reducing their negative environmental effects (Delegation of the European Union to India and Bhutan n.d.). In this instance, Danish businesses would spearhead the knowledge transfer to India.

The state of Gujarat is now developing the city known as Gujarat International Finance Tech City or GIFT. It is situated around 12 km from Ahmedabad International Airport. The primary goal of the 20-square-kilometre city is to provide high-quality infrastructure, including roads, broadband access, centralized cooling, water, and energy. The intention is to draw technology and financial firms from places like Bangalore, Gurgaon, Mumbai, and other places with expensive or insufficient infrastructure (GIFT | Welcome to the New Epicentre of Global Finance and IT. n.d.).

Special economic Zzones (SEZ) are a subsidized regime that will facilitate investment and benefit the project. Gujarat is going to have technological parks, shopping and business centres, lodging, recreational places, and an integrated and democratic governance framework. To lower road accidents, the transportation infrastructure will also be considerably safer. For regional and urban inter-displacements, such as Ahmedabad, Airport, Gandhinagar, and the city, an intermodal transport system like the mass rapid transport system (MRTS), light rapid transport system (LRTS), bus rapid transport (BRT), and more would make this feasible.

GIFT will integrate walk-to-work (soft mobility) in urban development, utilizing a 1:9 ratio between private and public transportation. Urban transportation in the city will also be electric-powered. Two towers, each with 29 floors, are now being built to house commercial businesses (GIFT | Welcome to the New Epicentre of Global Finance and IT. n.d.). The construction of the 33-story residential tower will soon get underway. The first of its kind in India, GIFT is envisioned to be a significant hub for finance, IT, and trade on par with other global endeavours like Singapore, Shinjuku-Tokyo, Lujiazui-Shanghai, La Defense-Paris, London-Dockyards, and La Defense-Paris (Ministry of External Affairs, Government of India n.d.).

6.11 Freiburg (Switzerland)

Fribourg is the capital of the Fribourg canton and a city in western Switzerland. The 26 cantons that makeup Switzerland's Confederation are the states that make up the country. Nestled between Lausanne, the headquarters of the Vaud canton, and Bern, the capital of Switzerland, Fribourg. is situated in the centre of the historic region of Üechtland. La Sarine is a river that flows through the town and contributes to its overall attractiveness. The Fribourg canton is home to lakes of various sizes, including the Murten and Neuchâtel lakes. The area, which lies between the Fribourg pre-Alps and the Swiss Plateau, is well-known for its abundant greenery (Switzerland – the World Factbook n.d.).

The warmest region of Switzerland is Fribourg, which has a moderate climate with plenty of sunshine. It receives almost 1800 hours of sunshine annually and 1117 kW/m^2 of radiation. With a population of slightly over 220,000, the town itself is home to numerous monuments of cultural, historical, and architectural significance. These include the city's ancient gateways, the green Schlossberg hill, the cathedral with its tall Gothic tower, and the Kaufhaus, an old artisan building (Switzerland – the World Factbook n.d.).

Freiburg has a long history of urban planning, which is exemplified by the renowned Bächle. They are called "streams" because they are little water-filled grooves found in the city center's pedestrian streets. They were initially constructed as a component of the sewerage system to handle fire situations. They are roughly 30 cm wide and 10–15 cm deep (Switzerland – the World Factbook n.d.).

Freiburg has made environmental protection a top priority since the 1990s. Even visitors can note the abundance of green places that have arisen from this, Freiburg, Germany, is renowned for being a "sunny city" that realized the possibilities of solar energy before other areas. In Freiburg, solar panels are found in churches, schools, private residences, the town hall, and the roof of the Badenova stadium. They can also be seen on towers and facades. Being the first football stadium with a solar system in the world, the local stadium has gained popularity (Switzerland – the World Factbook n.d.).

Numerous awards have been given to this trailblazing action, which has drawn attention to innovative energy projects like the Heliotrope, the solar village designed by Rolf Disch, the first solar building in the world to be energy self-sufficient, and the zero-energy houses of the Vauban district (Switzerland – the World Factbook n.d.).

Three years ago, the Vauban district was built and is the municipality's pride. More than 5000 people live in homes in Vauban that use solar thermal collectors, energy-efficient architecture, and solar panels to generate more energy than they require. Vauban's cutting-edge traffic control system is just one more incredible feature of its infrastructure. Owing to a thorough examination of the residents' transportation requirements, substitute options including carpooling and bike lanes have been presented. Consequently, Vauban has a car-to-population ratio of just 70 per thousand (Switzerland – the World Factbook n.d.).

Professionals in the fields of architecture, energy, and urban planning have been interested in this unusual reality, which has resulted in the growth of a niche tourism industry. These experts travel from all over the world to take part in what is known as the "Freiburg Tour" (Switzerland – the World Factbook n.d.).

6.12 Smart Mobility and Smart Policy

It is clear from Freiburg, Germany's mobility analysis that the city has been committed to environmental issues since the 1970s. Freiburg started creating an urban transport policy in 1969 to advance sustainable mobility and give the preservation of the urban environment and the natural world a top priority. The traffic and transport policy of the city has attracted global notice since it prioritizes ecologically friendly transportation options including walking, bicycling, and public transportation within the city (Germany – Freiburg – Green City | the EcoTipping Points Project n.d.).

The "European Local Public Transport Award" was given to the city in recognition of its dedication. A notable accomplishment of Freiburg's transportation strategy is the decrease in traffic volume within the city from 28% to 15% from 1982 to 1999. Riding a bike has helped make this feasible. In the same time frame, motor vehicle miles travelled fell from 38% to 30%, while public transportation climbed from 11% to 18%. With 423 automobiles per 1000 residents, Freiburg now has the lowest motor vehicle density of any German city (Germany – Freiburg – Green City | the EcoTipping Points Project n.d.).

The city has taken steps to accomplish these goals, including creating a compact, easily navigable city with robust neighborhood centres. Along major transportation routes, urban development has also been planned, with centralized device expansion given precedence. There are restrictions on where traffic can go, and most households don't have cars. One of the two garages in the neighborhood is used to park private automobiles (Germany—Freiburg—Green City).

The city's tram system has been expanded to residential areas since 2006, encouraging people to use bicycles or public transportation in place of their cars. The main issue is waste management. Due to the district's complete recycling of paper, plastic, and organic materials, the amount of garbage produced per person is far less than the national average. With almost 80% of its paper coming from recycling, Freiburg is a prime example of this approach. Since its introduction in 1991, recycling has received strong support—even from SC Freiburg, a football team. By offering discounts for group garbage disposal, rewards for using textile diapers, and incentives for producing compost for plants from waste disposal, the program encourages waste reduction. Waste that cannot be recycled has been burned at a plant in the Breisgau industrial park, which is 20 km south of Freiburg, since 2005 (Switzerland – the World Factbook n.d.).

The plant complies with strict environmental regulations and safety protocols while disposing of garbage. The energy produced by the fermentation of organic waste accounts for 1% of Freiburg's energy requirements.

Campaigns and events consistently increase consumer awareness of recycling. Waste and Sanitation (ASF), a disposal company that was partially privatized in Freiburg, has been working with schools since 1994 to arrange eco-station classes and tours, a garbage theater for elementary school students, and educational programs like "Ideas, Not Waste" and "Children and Agenda 21" (Switzerland – the World Factbook n.d.).

6.13 Sustainable Neighborhoods: Vauban, Siedlung Solar, and Rieselfeld

An intriguing illustration of a residential community where the locals actively participated in the design process is the Vauban neighborhood. Environmental, social, and cultural concerns were given priority in the project. The City of Freiburg, the nongovernmental organization Forum Vauban, the owners' committee, the building cooperative Genoa, the independent committee SUSI, and a few commercial construction businesses were the ones that started the project. The initiative involved collaboration with the Freiburg Eco-Institute for Applied Ecology and a few other businesses.

The district was constructed on ground that was formerly home to a French barracks that was abandoned in the early 1990s. It is located about 15 km from Freiburg's downtown. The district's reality was greatly aided by the 1994 founding of the nongovernmental organization Forum Vauban. The forum was formally designated by the city in 1995 as the community's representative for participatory processes and including locals in the planning process. The Forum aided in the development of neighborly ties and social networks. It established creative solutions that prioritized environmental and social care, participatory components, and communication while also offering a platform for the needs of the residents (StackPath n.d.).

Vauban Aktuell, a neighborhood newspaper, was also published by the NGO. Five thousand people are living in the district and 2000 sustainable homes (StackPath n.d.). A balance between residential and commercial sectors, as well as between social groupings, should be the goal of urban development. To lessen the use of cars and increase accessibility of public places, particularly for children, it should maintain the existing vegetation and give priority to walkers, cyclists, and public transportation. To guarantee convenience, services ought to be provided on-site (Germany – Freiburg – Green City I the EcoTipping Points Project n.d.). The Vauban neighborhood has used participatory planning to encourage the growth of green spaces and make technological and architectural decisions that will lower energy use.

Utilizing thermal solar and photovoltaic panels is one of the neighbourhood's main characteristics, making Vauban one of the European neighbourhoods with the highest concentrations of solar technology (Germany – Freiburg – Green City I the EcoTipping Points Project n.d.).

Designed by renowned Freiburg architect Rolf Disch, the Siedlung Solar is a residential development next to the Vauban neighborhood that was constructed in 2001. Because the homes in this solar district are equipped with excess energy production capabilities, it satisfies both environmental and financial criteria. This neighborhood, which is next to a 205-hectare natural reserve area that acts as a "green belt" for the people of Rieselfeld, values community involvement and cooperation (Switzerland – the World Factbook n.d.).

The neighborhood was constructed on 70 hectares of brownfield land that became accessible. The intervention's guiding planning principles include building multiple housing units up to five stories high, attaining a high population density, and implementing the idea of adaptive urban planning—which takes into account the needs of disabled people, families, and children. Integration of the district's industrial development zones has been made possible by the removal of the traditional "home-work division." Every home is constructed with minimal energy usage in mind, and many have solar thermal and photovoltaic systems installed (Switzerland – the World Factbook n.d.).

In addition to district heating and renewable energy sources, the district's overall energy supply is completed by a combined heat and power plant. Green areas, playgrounds, open areas, bike lanes, and traffic-light-controlled roadways enable kids to spend their leisure time with each other and are fundamentally important in metropolitan areas (Switzerland—the World Factbook n.d.).

6.14 Conclusions

Harvesting energy in megacities is a major difficulty, especially in developing nations. Producing solar energy can help energy systems manage electrical energy peaks and compensate for underutilized energy capacity in areas with abundant sunshine. Global interest in renewable energy sources has grown as a result of the growing demand for energy supplies and the depletion of fossil fuels. As a result, careful investigation and evaluation of different energy sources are needed. Other renewable energy sources, like wind power, biogas, and incineration, need more research based on the research findings. It is important to carefully consider each of these approaches and factors to create a framework for integrated energy resource management that may help smart cities realize their objectives. It is important to remember that additional machine learning and prediction algorithms could be the subject of future research to achieve smart city objectives and effectively manage energy.

References

All the advantages of geothermal energy (2021) September 6. https://www.enelgreenpower.com/learning-hub/renewable-energies/geothermal-energy/advantages
Anthopoulos L, Vakali A (2018) Security in smart cities literature: A review. Emerg Sci J 1(1):1–23
Blair MJ, Gagnon B, Klain A, Kulišić B (2021) Contribution of biomass supply chains for bioenergy to sustainable development goals. Land. https://doi.org/10.3390/land10020181
Byeon G, Yoon T, Oh S, Jang G (2013) Energy management strategy of the DC distribution system in buildings using the EV service model. IEEE Trans Power Electron 28:1544–1554
Caragliu A, Nijkamp P, van den Berg A (2009) Smart cities in Europe and the globe: today and tomorrow. J Urban Technol 16(4):307–323. https://www.tandfonline.com/journals/cjut20
Choudar A, Boukhetala D, Barkat S, Brucker J-M (2015) Local energy management of a hybrid PV-storage-based distributed generation for Microgrids. Energy Convers Manag 90:21–33
Chourabi H, Langemeyer J, Usman A, Harrington W (2012) Applications of ICT in smart cities: a survey. J Urban Technol 19(1):127–144. https://www.tandfonline.com/journals/cjut20
Delegation of the European Union to India and Bhutan (n.d.) EEAS. https://www.eeas.europa.eu/delegations/india_en?s=167
Energy.gov. (n.d.) The Department of Energy's Energy.gov. https://www.energy.gov/.
Environmental Impacts of Dams (n.d.) International rivers. https://archive.internationalrivers.org/environmental-impacts-of-dams
Environmental Impacts of Geothermal Energy (2013) Union of concerned scientists, March 5. https://www.ucsusa.org/resources/environmental-impacts-geothermal-energy
Geothermal Heat Pumps (n.d.) Energy.gov. https://www.energy.gov/eere/geothermal/geothermal-heat-pumps
Germany – Freiburg – Green City | The EcoTipping Points Project (n.d.). http://www.ecotipping-points.org/our-stories/indepth/germany-freiburg-sustainability-transportation-energy-green-economy.html
GIFT | Welcome to the new epicentre of Global Finance and IT (n.d.). https://www.giftgujarat.in/
Guichi A, Talha A, Berkouk EM, Mekhilef S (2018) Energy management and performance evaluation of grid-connected PV-battery hybrid system with inherent control scheme. Sustain Cities Soc 41:490–504
Hannah M (2022) Largest hydropower producer in the world. Gracon LLC, June 3. https://graconllc.com/largest-hydropower-producer-in-the-world/
Hollands R (2008) Edge city: planning for parts to a whole. Wiley
How Does Solar Work? (n.d.) Energy.gov. https://www.energy.gov/eere/solar/how-does-solar-work
Use of geothermal energy – U.S. Energy Information Administration (EIA) (n.d.). https://www.eia.gov/energyexplained/geothermal/use-of-geothermal-energy.php
India – The World Factbook (n.d.). https://www.cia.gov/the-world-factbook/countries/india/
India | Latest News & Updates | BBC News (n.d.). https://www.bbc.com/news/world/asia/india

India Population (2024) – Worldometer (n.d.). https://www.worldometers.info/world-population/india-population/
International Energy Agency (IEA) (2022) Global Energy Review 2022. https://www.iea.org/reports/world-energy-outlook-2022
IEA (2023) Renewables Report. https://www.iea.org/reports/renewables-2023
IRENA – International Renewable Energy Agency (2024), March 14. https://www.irena.org/
Kanchev H, Lu D, Colas F, Lazarov V, Francois B (2011) Energy management and operational planning of a microgrid with a PV-based active generator for smart grid applications. IEEE Trans Ind Electron 58:4583–4592
Liu Y, Zhu L, Zhan L, Gracia JR, King TJ, Liu Y (2016) Active power control of solar PV generation for large interconnection frequency regulation and oscillation damping. Int J Energy Res 40:353–361
Ma L, Liu N, Zhang J, Tushar W, Yuen C (2016) Energy management for joint operation of CHP and PV prosumers inside a grid-connected microgrid: a game theoretic approach. IEEE Trans Ind Inform 12:1930–1942
Massaro M, Rana MU, Jain S (2018) A review on urban water management practices: towards smart cities. Sustain Cities Soc 43:371–379
McKeegan N (2015) Scotland announces $20 million Saltire Prize for clean energy innovation. New Atlas, May 2. https://newatlas.com/saltire-prize-marine-renewable-energy/9231/
Ministry of External Affairs, Government of India (n.d.) Ministry of External Affairs, Government of India. https://www.mea.gov.in/
Mora L, Bolici R, Devault M (2017) The new era of smart cities: empirically assessing the functional and technological components of smart cities. Sustain For 9(8):1406
National Renewable Energy Laboratory (NREL) Home Page (n.d.) NREL. https://www.nrel.gov/
Renewable Energy Explained (n.d.). https://education.nationalgeographic.org/resource/renewable-energy-explained/
Rodríguez CLT, et al. (2013) Reconfigurable control scheme for a PV microinverter working in both grid-connected and Island modes. IEEE Trans Ind Electron 60:1582–1595
Sechilariu M, Wang B, Locment F (2013) Building integrated photovoltaic system with energy storage and smart grid communication. IEEE Trans Ind Electron 60:1607–1618
Smart Cities: Ministry of Housing and Urban Affairs, Government of India (n.d.). https://mohua.gov.in/cms/smart-cities.php
Solar Energy, Wildlife, and the Environment (n.d.). Energy.gov. https://www.energy.gov/eere/solar/solar-energy-wildlife-and-environment
StackPath (n.d.). https://use.metropolis.org/case-studies/sustainable-urban-district-vauban#:~:text=The%20building%20plan%20included%20a,and%20planning%20the%20new%20district
Stand-alone PV System for Off-grid PV Solar Power (n.d.) Alternative energy tutorials. https://www.alternative-energy-tutorials.com/solar-power/stand-alone-pv-system.html
Switzerland – The World Factbook (n.d.). https://www.cia.gov/the-world-factbook/countries/switzerland/
THE 17 GOALS | Sustainable Development (n.d.). https://sdgs.un.org/goals
The International Telecommunication Union (ITU). https://www.itu.int/en/about/Pages/default.aspx
Tidal Energy (n.d.) PNNL. https://www.pnnl.gov/explainer-articles/tidal-energy
Trust ES (2020) What role does biomass have to play in our energy supply? Energy Saving Trust, November 3. https://energysavingtrust.org.uk/what-role-does-biomass-have-play-our-energy-supply/#:~:text=A%20fuel%20for%20all%20needs,of%20wood%2Dfuelled%20heating%20systems.
United Nations, Department of Economic and Social Affairs, Population Division (2018) World urbanization prospects: the 2018 revision (ST/ESA/P/WP/241). https://population.un.org/wup/Publications/Files/WUP2018-Highlights.pdf

Wandhare RG, Agarwal V (2014a) Novel stability enhancing control strategy for centralized PV-grid systems for smart grid applications. IEEE Trans Smart Grid 5:1389–1396

Wandhare RG, Agarwal V (2014b) Reactive power capacity enhancement of a PV-grid system to increase PV penetration level in smart grid scenario. IEEE Trans Smart Grid 5:1845–1854

Wardle AR (2023) A review of the environmental effects of the renewable fuel standard's corn ethanol mandate. The CGO, August 11. https://www.thecgo.org/research/environmental-effects-of-renewable-fuel-standards/#:~:text=Both%20of%20these%20resources%20are,context%20of%20energy%20systems%20analysis

Water Needed To Produce Various Types Of Energy (2008) ScienceDaily, April 8. https://www.sciencedaily.com/releases/2008/04/080417173953.htm

What is the best mounting system for a solar photovoltaic project? (2024) January 17. www.linkedin.com. https://www.linkedin.com/advice/3/what-best-mounting-system-solar-photovoltaic-project-df5qe

Chapter 7
Integration of Renewable Energy in Buildings

Shivani Rani, Arijit Ghosh, and Anurag Bajpai

Abstract The integration of renewable energy technologies in buildings is a crucial strategy for achieving sustainable and energy-efficient built environments. This paper explores the various renewable energy technologies, including solar photovoltaic systems, solar thermal applications, biomass energy, and geothermal heat pumps, and examines their applicability across different building typologies. We discuss the evolution of buildings in response to climate change imperatives, emphasizing the shift toward energy efficiency and renewable energy adoption. Through a detailed renewable energy integration Matrix, we identify the optimal technologies for residential, commercial, institutional, industrial, healthcare, and assembly buildings. The paper also highlights the role of green building rating systems in enhancing building efficiency and presents case studies demonstrating the practical benefits of renewable energy integration. By providing comprehensive insights and actionable recommendations, this research underscores the potential of renewable energy to transform the built environment and contribute to global sustainability goals.

Keywords Sustainable · Energy-efficient · Building · Environments · Solar thermal · Biomass energy geothermal

7.1 Introduction

Like the rest of the world, India stands at a crossroads in the face of climate change. The consequences of global warming are increasingly tangible, manifesting in erratic weather patterns, rising sea levels, and unprecedented natural disasters (Ray 2023). In this climate emergency, the built environment, particularly buildings,

S. Rani (✉) · A. Bajpai
GreenTree Global, Noida, Uttar Pradesh, India

A. Ghosh
S.A.P. Automations (India) Private Limited, New Delhi, New Delhi, India

S. Kumar, V. K. Singh (eds.), *Renewable Energy Development: Technology, Material and Sustainability*, Clean Energy Production Technologies,
https://doi.org/10.1007/978-981-97-9626-7_7

assumes a central role both as a contributor to the problem and a potential solution. The urgency to adapt and adopt sustainable construction practices has never been more critical (MacGregor et al. 2018).

In recent years, India has witnessed a commendable surge in adopting environmentally conscious building practices. Architects, developers, and policymakers are steering the narrative toward sustainability, recognizing our built infrastructure's profound impact on the environment (Chen et al. 2024). From innovative architectural designs to the implementation of green building standards, the shift toward eco-friendly construction is gaining momentum.

The symbiotic relationship between renewable energy and the built environment unfolds as a key strategy in curbing carbon emissions and fostering resilience. This chapter explores the dynamic interplay between climate change, sustainable building practices, and the critical role of renewable energy integration in shaping a more sustainable and climate-resilient future for India's built landscape.

7.2 The Buildings

The structures are created for various purposes, providing shelter, workspace, living space, or storage come in diverse forms, sizes, and architectural styles, reflecting the diversity of human activities and the environments they inhabit. The realm of architecture is vast and varied, encompassing an infinite number of purposes, cultural influences, and design ideologies. Building typologies cover a wide range of constructions, each designed to meet certain needs and satisfy particular contexts (Energy Conservation Building Code 2017).

Residential A wide variety of dwelling types, representing different lifestyles, cultural preferences, and societal necessities, are represented in residential architecture. These are primarily to create a private, safe space where people may live, work, and go about their everyday lives. The following types of residential buildings can be broadly classified as:

Standalone Buildings: Independent structures intended to serve as separate dwelling quarters are known as standalone residential buildings. Every building serves as a distinct home, giving its residents privacy and independence.

Single-Family Homes: Typically, detached dwellings are occupied by a single household. These homes vary widely in architectural styles and sizes, ranging from cottages and bungalows to large estates.

Duplexes and Townhouses: Buildings are divided into two separate living units (duplex) or multiple attached units (townhouses) with shared walls. Each unit typically has its entrance and may be owned or rented independently.

Group Housing Complex *These structures are designed to accommodate multiple households in close proximity, fostering a sense of community and shared amenities. Key characteristics include apartments and condominiums: multiunit buildings*

Fig. 7.1 Different types of buildings

with multiple floors, each containing individual living units. Apartments are often rented, while condominiums involve individual ownership of units within the building, cooperative housing: residents collectively own and manage the entire residential complex, sharing responsibilities and decision-making, gated communities. Controlled-access residential developments with a collective infrastructure, often including shared amenities like parks, security, and recreational facilities, senior living communities: housing for older adults, providing various levels of care and communal services within a dedicated residential community (Architecture n.d.). Figure 7.1 showed the different types of buildings.

Commercial These structures are distinguished by their efficiency and usefulness, designed to meet a variety of business and institutional requirements. The goal of commercial architecture is to design environments that support customer interaction, productivity, and the smooth running of business operations (Types of Housing and Residential Blocks 2016).

24-Hrs running buildings: buildings or facilities that operate continuously without significant interruptions, running around the clock, 24 hours a day.

Hospitality: Hospitality buildings are designed to provide accommodation, entertainment, and other services for travellers and guests. It includes Hotels & Resorts etc.

Healthcare: Healthcare buildings are dedicated to providing medical services and support. It includes Hospitals & Clinics.

08-Hrs running buildings: buildings or facilities that operate for a standard workday of approximately 8 h.

Institutional: Institutional buildings serve public or organizational functions and are often associated with civic and community life. It includes government buildings and educational buildings.

Shopping complex: Shopping complexes are designed for retail activities, offering spaces for the sale of goods and services. It includes malls and strip malls.

Business: Business buildings are designed to accommodate office spaces and support professional services. It includes office buildings and co-working spaces.

Assembly: Assembly buildings are designed for gatherings, events, and communal activities. It includes theatres and auditoriums.

7.2.1 Building Performance

Green-rated buildings in India, fueled by initiatives like GRIHA and LEED, have steadily risen. These certifications, widely adopted in the construction sector, reflect a growing emphasis on sustainable and eco-friendly building practices among builders and organizations.

The efficacy and efficiency of a building in fulfilling its intended purposes and specifications is referred to as building performance. It involves evaluating a building's performance concerning several factors, such as operational functionality, occupant comfort, environmental impact, and energy efficiency (Geraldi and Ghisi 2020). Building performance evaluations usually cover a structure's whole lifecycle, from design and construction to use and upkeep. Building performance assessments are often conducted through energy audits, environmental impact assessments, post-occupancy evaluations, and other monitoring and measurement techniques. The goal is to continually improve and optimize a building's performance, considering both environmental sustainability and the well-being of its occupants. Certifications such as LEED (Leadership in Energy and Environmental Design) and BREEAM (Building Research Establishment Environmental Assessment Method) are two examples of frameworks that assess and identify high-performance buildings (WBDG n.d.).

7.2.2 Energy Performance Index (EPI)

It provides a quantitative measure of a building's energy performance by considering factors such as energy consumption, building size, and other relevant parameters. A building's annual energy use, expressed in kWh per square meter, is its Energy Performance Index (EPI). The goal of EPI is to assist stakeholders in understanding and comparing the energy efficiency of various buildings or a portfolio of buildings, including building owners, operators, and policymakers (Energy Conservation Building Code 2017). The Energy Performance Index is a tool for promoting energy efficiency, as it allows for the benchmarking and comparison of buildings, encourages best practices, and supports efforts to reduce the consumption of energy and greenhouse gas emissions. Various organizations and certification programs may use EPI as part of their assessment criteria for green building certifications (Guide for Using National Building Code of India 2016). As a building progresses along the efficiency field with measures well-versed by EPI assessments, it sets the path for the integration of renewable energy and, ultimately, the realization of Net Zero energy consumption.

7.3 Net Zero and Net Positive

"Net Zero" and "Net Positive" are terms used in the context of sustainability, particularly in the built environment. They describe approaches that balance or exceed resource consumption and environmental impact.

7.3.1 Net Zero

It is to balance the amount of greenhouse gas emissions produced and the amount removed from the atmosphere. In the context of buildings, it often relates to balancing the energy a building consumes and the renewable energy it generates on-site (Jaysawal et al. 2022).

Key Aspects

- Energy: A Net Zero Energy Building (NZEB) produces as much energy as it consumes over a specific period, typically a year.
- Carbon Emissions: Net Zero Carbon goes beyond energy and includes efforts to offset or mitigate all carbon emissions associated with the building's construction, operation, and embodied carbon.

Strategies

- Energy Efficiency: Implementing energy-efficient technologies and design principles.
- Renewable Energy: Incorporating on-site renewable energy sources such as solar panels.
- Carbon Offsetting: Investing in projects that reduce or capture an equivalent amount of carbon to offset emissions.

7.3.2 Net Positive

"Net Positive" takes the concept further than Net Zero. It means generating more resources than consumed and contributing positively to the environment and community. This approach goes beyond mitigating negative impacts and seeks to create a regenerative impact (Cole 2014).

Key Aspects

- Resource Generation: The building or system generates more energy, water, or other resources than it uses.
- Biodiversity: Consideration for enhancing biodiversity and ecosystem services.

Strategies

- Renewable Energy Surplus: Generating more renewable energy than the building consumes.
- Water Conservation and Generation: Efficient water use and possibly generating surplus clean water.
- Ecological Design: Incorporating green roofs, on-site agriculture, or other features that enhance biodiversity.

7.4 Approaches to Minimize Environmental Impact

By focusing on reducing energy consumption, through efficient design and technologies, and concurrently integrating renewable energy sources, buildings can significantly enhance their overall efficiency. The following approach not only minimizes environmental impact but also contributes to long-term sustainability and resilience in the face of evolving energy challenges (Chel and Kaushik 2018).

Reducing Energy Consumption

- Implementing Energy-Efficient Design: Optimize building orientation, incorporate passive design principles, and use advanced insulation to minimize heat transfer and maintain comfortable indoor temperatures.
- Upgrading HVAC Systems: Install energy-efficient heating, ventilation, and air conditioning (HVAC) systems. Implement smart controls to optimize energy use based on occupancy and environmental conditions.
- Adopting Efficient Lighting: Replace traditional lighting with energy-efficient LED fixtures and utilize controls like motion sensors and daylight harvesting to reduce unnecessary energy consumption.
- Implementing Water Conservation Measures: Install water-efficient fixtures and systems, such as low-flow toilets and rainwater harvesting, to minimize water consumption.

Integrating Renewable Energy

- *Solar Technologies*: Incorporate solar photovoltaic (PV) panels to harness clean and sustainable energy from the sun. Solar water heaters can also be employed to meet hot water demands efficiently.
- *Biomass Energy*: Utilize biomass heating systems and biogas production to generate heat and electricity from organic materials, contributing to a more sustainable energy mix.
- *Geothermal Energy*: Integrate geothermal heat pumps to harness the Earth's natural heat for heating and cooling applications, improving energy efficiency.
- *Wind Energy*: Explore the feasibility of wind turbines to generate on-site renewable power, especially in locations with sufficient wind resources.

7.5 Renewable Energy

As a source of sustainable energy, renewables provide a range of options for powering our planet with the least number of negative effects on the environment. Let's explore the technologies for renewable energy and how they can be used to create a more sustainable and greener future.

7.5.1 Solar Technologies

Solar technologies have demonstrated remarkable efficacy in the acquisition of sustainable energy. Photovoltaic systems are becoming more and more efficient at converting sunlight into electricity. Sustainable substitutes are offered by solar thermal applications, such as solar water heating and concentrated solar power. Research breakthroughs are constantly improving performance, which makes solar technologies essential for the global transition to clean energy (Hayter and Kandt 2011). Figure 7.2 *showed the classification of solar technologies.*

A. Solar Photovoltaic

Solar photovoltaic systems are at the forefront of the renewable energy revolution, converting sunlight directly into electricity. These systems, composed of solar cells, find applications across residential, commercial, and industrial sectors and even in large-scale solar farms contributing to grid power (ResearchGate n.d.).

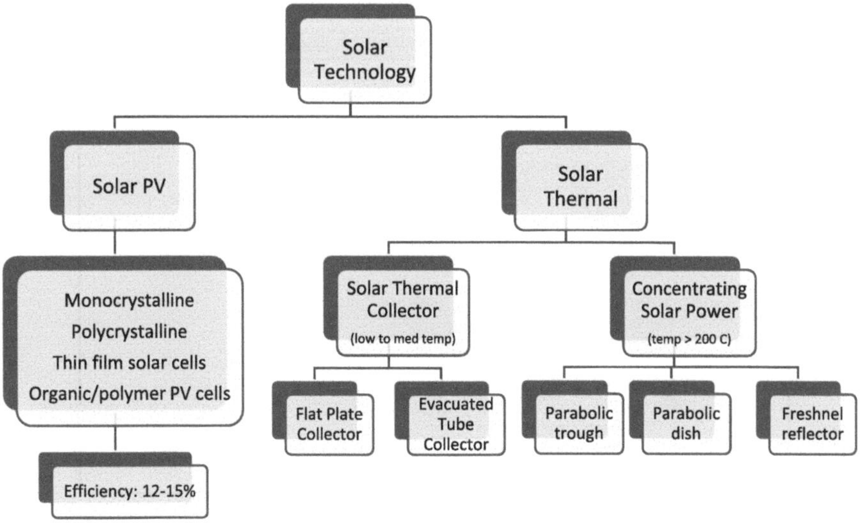

Fig. 7.2 Classification of solar technologies

The efficiency of solar panels is influenced by several factors, and the proper installation is crucial for maximizing their energy output (EIA,n.d.). Some key considerations for placing and installing solar panels to increase efficiency are:

Orientation

South-Facing (For Northern Hemisphere): In most parts of India, solar panels should ideally face south to capture maximum sunlight throughout the day. This orientation ensures optimal exposure to the sun, maximizing energy generation.

Tilt Angle

Latitude Adjustment: The tilt angle of solar panels should be adjusted based on the specific geographical latitude of the location within India. Aligning the tilt angle with the latitude helps optimize energy capture, ensuring the panels are positioned to receive sunlight effectively.

Maximizing Energy Production: An optimal tilt angle contributes to maximizing energy production over the entire year. By capturing more sunlight, solar panels can convert more solar energy into electricity, improving the overall efficiency of the solar energy system.

Avoiding Shading Effects: Shading can significantly reduce the efficiency of solar panels. The tilt angle should be chosen to minimize shading from nearby structures, vegetation, or other obstructions. This ensures that each solar panel in an array operates at its maximum efficiency without being shaded by others (Yvonne 2023). Figure 7.3 showed the efficient orientation of solar panels, and Fig. 7.4 showed the diagrammatic representation of tilt angle for maximum efficiency of solar panels.

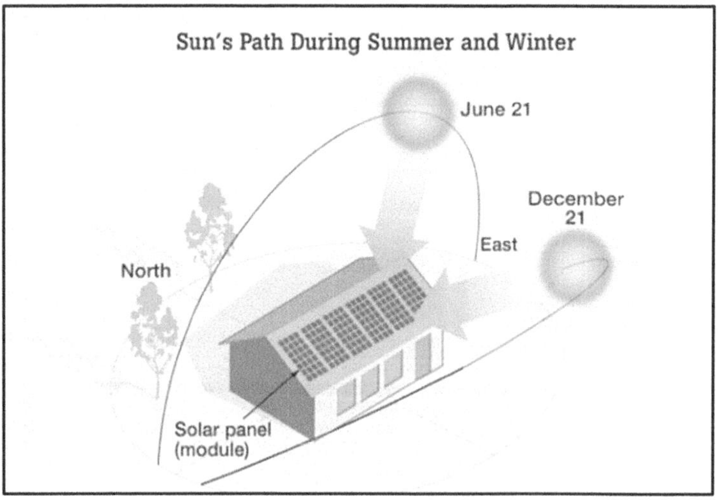

Fig. 7.3 Efficient orientation of solar panels. (Source: Solaretribe)

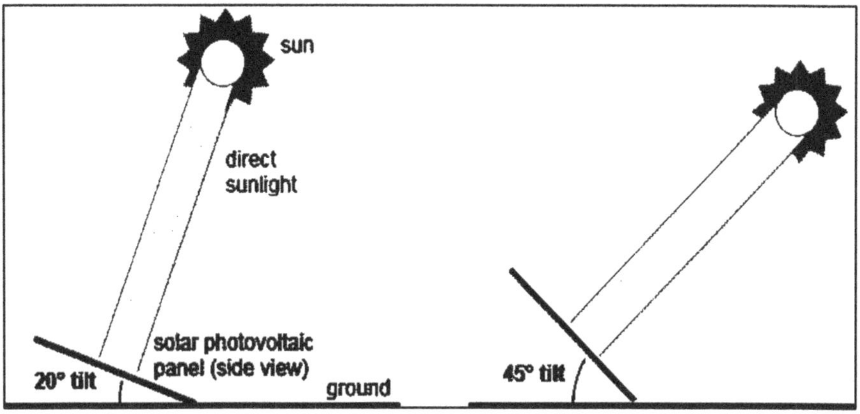

Fig. 7.4 Diagrammatic representation of tilt angle for maximum efficiency of solar panels. (Source: EelctronicsHub)

How to Calculate the Optimal Angle for Solar Panel (Sinovoltaics 2015)
Method 1:

The optimum tilt angle is calculated by adding 15 degrees to the latitude during winter and subtracting 15 degrees from the latitude during summer.
For instance, if the latitude is 34°, the optimum tilt angle for the solar panels during winter will be 34 + 15 = 49°. The summer optimum tilt angle on the other hand will be 34–15 = 19°.

Method 2:

In this method, the optimum tilt angle for solar panels during winter is calculated by multiplying the latitude by 0.9 and then adding 29°.
In the above case example of a latitude of 34°, the tilt angle will be (34 * 0.9) + 29 = 59.6°. This angle is 10° steeper than in the general method but very effective at tapping the midday sun which is the hottest in the short winter days.
For summer, the tilt angle is calculated by multiplying the latitude by 0.9 and subtracting 23.5°. In the above case example, this angle would be (34 * 0.9)—23.5 = 7.1°. For optimum tilt angles during spring and fall, 2.5° is subtracted from the latitude.

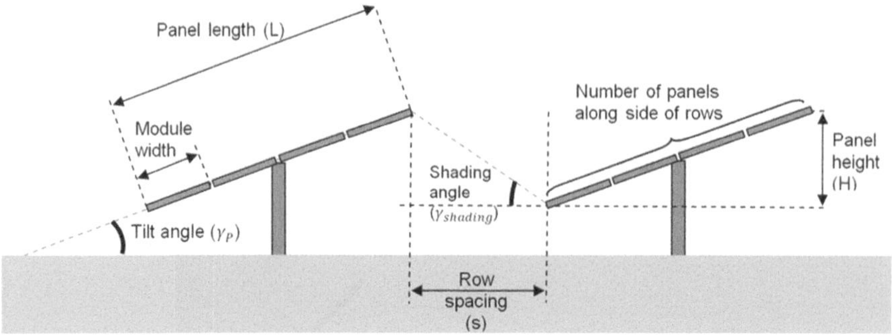

Fig. 7.5 Spacing between solar panels. (Source: Saint-Drenan and Barbier 2019)

Shading

Minimize Shade: To enhance solar panel efficiency in India, it is crucial to minimize shading from nearby structures, trees, or any potential obstructions. Even partial shading, common in urban and residential areas, can significantly impact solar energy generation. Regular trimming or removal of shading elements is advisable (SunWare n.d.).

Spacing Between Panels

Avoid Shadowing: Maintain adequate spacing between solar panels, particularly in larger installations. This prevents one panel from casting shadows on adjacent panels, ensuring each panel operates at its maximum efficiency. Proper spacing is essential for harnessing the collective energy potential of the entire array (Yvonne 2023; Saint-Drenan and Barbier 2019). Figure 7.5 showed the spacing between solar panels.

B. Solar Thermal

Solar Water Heater System: Solar thermal technologies extend beyond electricity generation, providing an efficient solution for water heating. Solar water heater systems absorb sunlight to heat water for domestic and industrial use, reducing the reliance on conventional water heating methods (Musunuri et al. 2007).

Solar Ventilation Preheating Systems: Innovative solar ventilation preheating systems leverage solar energy to preheat ventilation air, enhancing energy efficiency in buildings. By harnessing the sun's warmth, these systems contribute to a more sustainable and comfortable indoor environment.

7.5.2 Biomass Energy

Biomass, derived from organic materials, unfolds as a renewable energy source with diverse applications. From producing biogas to providing heating solutions, biomass energy exemplifies a circular and sustainable approach to power generation (Neoli 2015).

Biogas, a by-product of anaerobic digestion of organic matter, emerges as a versatile energy source. Biogas production finds application in waste treatment plants, agricultural facilities, and even at a household level, offering a clean and renewable alternative to traditional fuels (Vargas-Soplín et al. 2022).

7.5.3 Geothermal Energy

Geothermal energy, tapped from the Earth's internal heat, is a constant and reliable source of power. Heat pumps and other geothermal technologies have the potential to completely change the way we heat and cool our buildings.

The Technology

The geothermal loop exchange is done with the help of 32–40 mm borehole exchangers that are placed in the ground, in vertically drilled boreholes which are generally 100–140 m with a diameter of 6–7 inches. The BHE are HDPE RC pipes that are lowered into the borehole and filled with brine/fluid that is circulated as a medium for heat transfer between the soil and the heat pump. To transfer heat from the soil, a temperature difference is created from the geothermal heat pump.

The thermal conductivity of the soil is the deciding factor for the exchange parameters.

The heat (renewable thermal) transferred from the ground is then brought up to the geothermal heat pump which produces heat in the form of hot water, to be deposited in the buffer tank, at temperatures within 40–60 °C.

This hot water is then circulated through the secondary or distribution circuit that is used to heat or cool the building.

For heating purposes, the distribution uses PEX (polyethylene xylene) pipes of 16–20 mm diameter and are embedded in the floor, attempting to heat the floor with around 35/40 °C water and thereby raising the heat of the thermal mass. This radiant heating exactly matches the ideal heating curve, hence most energy efficient.

Cooling is done by fan coil units and from the ceiling and sidewalls as required. Heat is floor up and cooling is ceiling down. The underfloor pipes heat the slab/floor, and there is no distribution of hot air, which is very efficient in extremely cold climates.

Geothermal Exchange

- Is a thermal energy storage system.
- No generation of energy.
- Transfer between earth and building.
- The thermal energy is either sourced or rejected in the ground.
- It's the same energy that is used in different climates.

Understanding "Source" or "Sink"

- If the thermal energy is sourced or sunk from the ground beneath.
- It's very important to know how heat dissipation happens in the soil.

- If you do not know how much thermal energy can be sourced or sunk—how can you design the space exchange?
- You have to know how the ground will behave in the next 25/30 years.

The First Borewell

- Based on geological data, the first test bore well must be designed with a simulated thermal diffusion model.
- This test bore well results will be the basis of design post-TRT.
- After TRT is completed, this borewell shall be part of the GHX loop field and usable for exchange.

Understanding Grout

- Appropriate material selection, mixing, and positioning.
- Factors:
  - (i) Formation characteristics (diffusivity, permeability, and thermal conductivity).
  - (ii) Static groundwater level.
  - (iii) The chemistry of groundwater and its possible impacts on grout.
  - (iv) Vadose zone impact.
- Common Types: materials based on cement, graphite, bentonite, and combinations of bentonite and sand.

Thermal Response Test

- During a specific amount of time (36 hours), a closed loop borehole is subjected to a finite quantity of heat energy, and the rate at which heat drains into the surrounding ground is tracked as part of a thermal response test (TRT).
- Accurate measurements of ground thermal characteristics are made possible by appropriate analysis of the test data.

Why Is TRT Crucial to the Design of a Geothermal Collector?

The purpose of TRTs is to find the site-specific thermal characteristics of the earth surrounding a closed-loop borehole.

The test results enable the development of a cost-effective collector by minimizing capital costs and maximizing the number of boreholes that must be completed, preventing long-term operational issues, optimizing design performance, and lowering the system's overall operating costs while satisfying the necessary energy demands.

Once Analysis is Done

- Choose the GHX Pump ratings. Choosing the exact specifications for the exchange parameters.
- The Hydronic system must be calculated.

Flow and Volume
- These two factors solely determine the "thermal capacity" of the system.
- To balance the system, correct control-based circulator pumps are a must.
- The sensors must be connected properly for controls.

Components of Ground Source Heat Pump (GSHP) System
- Probes + weight.
- Collectors in the mechanical room.
- Ground source heat pump.
- Distribution: cooling.
- Distribution heating.
- Circulator pump.

Versatility
- Exchanges from the ground, hence no relation with outdoor ambient temperature.
- Ability to produce heating, cooling, and hot water from the same system.
- The ability to cascade, makes the entire system modular, with no need for standby.
- Ability to mix multiple exchange sources, e.g., ground and air.
- Ability to hybridize with solar through e-Manager.

ICAP Ready
- Low energy consumption: This can be achieved by exchanging in the correct exchange medium.
- The system should be consuming low electricity and replenished with direct solar hybridization.
- Low carbon cooling: Conventional. Electricity consumption produces carbon.
  - DeCarb the HVAC.
- Low refrigerant usage: 28 kgs for conventional and 9.1 kgs for geothermal/aerothermal.

Principle of Working
Geothermal exchanges the temperature from a stable source beneath the ground accessed at approximately 100–120 meters down the earth in a closed loop exchange called ground loop exchanger. This is done by an array of HDPE pipes. These figures depend upon the model, exchange, and host of other factors.

Features
- Utilizing the ground's energy storage capacity, an energy-efficient controlled HVAC-DHW system is achieved.

Advantages and Benefits
- The exchange provides the building with warm air in the winter and cooled air in the summer, which makes this system unique. Additional advantages consist of:

- Comfortably temperature-controlled using ground heat exchange.
- Ideal for hybridizing solar power systems with heat pumps.
- Low maintenance costs.
- Has no pitting and resilience to corrosion.
- Does not typically develop deposits.

Design

- Every geological location has different soil conditions, therefore even if geothermal energy is mostly site-specific, it still has to be constructed and provided according to the building's output requirements based on the soil test result or soil conditions.

7.6 Evolution of Buildings and the Climate Imperative

The evolution of buildings has been evidence of our ingenuity, in responding to changing needs, materials, and technologies. However, as we stand at the rise of a climate crisis, the focus on buildings has shifted dramatically, propelled by an urgent need to reevaluate and redesign in the face of environmental challenges (Aysha and Mani 2017).

The Climate Imperative

Buildings, once seen merely as static structures, are now recognized as dynamic entities with a substantial carbon footprint. The construction, operation, and demolition of buildings contribute significantly to greenhouse gas emissions, exacerbating climate change. The imperative to address this impact is rooted in the realization that buildings are not just contributors to the problem; they are key to the solution.

Buildings as Climate Change Mitigators

Buildings wield immense potential in the battle against climate change. Sustainable construction and retrofitting practices, coupled with energy-efficient technologies, can drastically reduce a building's carbon footprint. Moreover, buildings can become active participants in the generation of clean energy, transforming from mere consumers to contributors to the grid.

Integration of Renewable Energy Technologies

To achieve this transformative shift, the integration of renewable energy technologies in buildings has emerged as a key player strategy. By leveraging solar, wind, biomass, and geothermal technologies, buildings can not only reduce reliance on conventional energy sources but can also become resilient hubs of energy production (Renewable Energy: The Clean Facts n.d.). This integration represents a symbiotic relationship between the built environment and the natural resources, creating a paradigm where buildings not only adapt to a changing climate but also actively work towards its mitigation.

Table 7.1 Renewable energy integration matrix

Buildings		Technologies: Solar technologies: Solar PV	SWHS	Solar ventilation preheating system	Biomass energy	Geothermal energy
Residential	Standalone	✓	✓	✓		✓
	Group housing	✓	✓	✓		✓
Commercial	Office building	✓	✓	✓		✓
	Retail/shopping complex	✓	✓			
	Industrial buildings	✓				
	Institutional buildings	✓	✓			✓
	Hospitality	✓	✓	✓	✓	
	Healthcare	✓	✓			✓
	Assembly	✓	✓			✓

7.7 Reintegration Matrix: The Outcome

The renewable energy (RE) integration matrix provides a comprehensive overview of the most suitable renewable energy technologies for various types of buildings. This matrix serves as a valuable decision-making tool for architects, builders, and energy professionals, offering insights into the optimal adoption of RE technologies across distinct building categories.

The matrix, designed for clarity and ease of reference, facilitates strategic planning for integrating renewable energy into diverse building contexts, fostering a more resilient and eco-friendly built environment.

Table 7.1 illustrates the applicability of various renewable energy technologies across different building types. Solar PV and solar water heating systems (SWHS) are widely applicable, particularly in residential and commercial settings. Geothermal energy and biomass are effective for specific applications, highlighting their benefits for energy-efficient and sustainable building operations.

References

Architecture (n.d.) Ebook. Neufert, 3d edn. Free download, borrow, and streaming: Internet archive. Retrieved from https://archive.org/details/Architecture.Ebook.Neufert3dEdition/

Aysha S, Mani M (2017) Adaptation of buildings to climate change. In: Encyclopedia of sustainable technologies, pp 387–406. https://doi.org/10.1016/b978-0-323-90386-8.00165-0

Chel A, Kaushik G (2018) Renewable energy technologies for sustainable development of energy efficient building. Alex Eng J 57(2):655–669. https://doi.org/10.1016/j.aej.2017.02.027

Chen L, Hu Y, Wang R et al (2024) Green building practices to integrate renewable energy in the construction sector: a review. Environ Chem Lett 22:751–784. https://doi.org/10.1007/s10311-023-01675-2

Cole RJ (2014) Net-zero and net-positive design: a question of value. Build Res Inf 43(1):1–6. https://doi.org/10.1080/09613218.2015.961046

EIA (n.d.) Most utility-scale fixed-tilt solar photovoltaic systems are tilted 20 degrees-30 degrees – U.S. Energy Information Administration (EIA). Retrieved from https://www.eia.gov/todayinenergy/detail.php?id=37372

Energy Conservation Building Code (2017). Retrieved from https://beeindia.gov.in/sites/default/files/ECBC%20book%20final%20one%20%202017%20with%20Amendements.pdf

Geraldi MS, Ghisi E (2020) Building-level and stock-level in contrast: a literature review of the energy performance of buildings during the operational stage. Energ Build 211:109810. https://doi.org/10.1016/j.enbuild.2020.109810

Guide for Using National Building Code of India (2016). Retrieved from https://www.bis.gov.in/wp-content/uploads/2022/08/Booklet-Guide-for-Using-NBC-2016.pdf

Hayter SJ, Kandt A (2011) Renewable energy applications for existing buildings. National Renewable Energy Laboratory

Jaysawal RK, Chakraborty S, Elangovan D, Padmanaban S (2022) Concept of net zero energy buildings (NZEB) – a literature review. Clean Eng Technol 11:100582. https://doi.org/10.1016/j.clet.2022.100582

MacGregor C, Dowdell D, Jaques R, Bint L, Berg B (2018) The built environment and climate change: a review of research, challenges and the future. BRANZ Study Report SR403. Judgeford, New Zealand: BRANZ Ltd

Musunuri RK, Sánchez D, Rodriguez R (2007, October) Solar thermal energy. Retrieved from https://www.solarthermalworld.org/sites/default/files/story/2015-06-20/solarthermal.pdf

Neoli (2015, February 10) Biomass energy requirements to become part of green building standard. Retrieved from https://bee-inc.com/2015/02/10/biomass-energy-green-building-standard/

Ray K (2023, December 17) Weathering the storm: India grapples with climate change. Retrieved from https://www.deccanherald.com/india/weathering-the-storm-india-grapples-with-climate-change-2814503

Renewable Energy: The Clean Facts (n.d.). Retrieved from https://www.nrdc.org/stories/renewable-energy-clean-facts#sec-whatisSolarVenti (2021, November 22). Retrieved from https://alfasol.hu/solarventi/

ResearchGate (n.d.) Data-analysis and modelling of the effect of inter-row shading on the power production of photovoltaic plants – scientific figure on ResearchGate. Available from: https://www.researchgate.net/figure/Definition-of-module-row-interspacing-and-the-module-height_fig1_332265278

Saint-Drenan Y, Barbier T (2019) Data-analysis and modelling of the effect of inter-row shading on the power production of photovoltaic plants. Sol Energy 184:127–147. https://doi.org/10.1016/j.solener.2019.03.086

Sinovoltaics (2015, July 21) Solar panel angle: how to calculate solar panel tilt angle? Retrieved from https://sinovoltaics.com/learning-center/system-design/solar-panel-angle-tilt-calculation/

SunWare (n.d.) SunWare GmbH & Co KG – J.Schueren info@sunware.de. SunWare. Retrieved from https://en.sunware.solar/basiswissen/abschattung

Types of Housing and Residential Blocks (2016 April 16). Retrieved from https://www.slideshare.net/slideshow/types-of-housing-and-residintial-blocks/60983980

Vargas-Soplín AD, Prochnow A, Herrmann C, Tscheuschner B, Kreidenweis U (2022) The potential for biogas production from autumn tree leaves to supply energy and reduce greenhouse gas emissions – a case study from the city of Berlin. Resour Conserv Recycl 187:106598. https://doi.org/10.1016/j.resconrec.2022.106598

WBDG (n.d.) Green building standards and certification systems | WBDG – whole building design guide. Retrieved from https://www.wbdg.org/resources/green-building-standards-and-certification-systems

Yvonne (2023, July 23) Designing your house to respond to the sun: an intro to solar orientation. Retrieved from https://yr-architecture.com/an-intro-to-solar-orientation/

Chapter 8
Sustainable Electricity Generation Through Solar Energy Technologies

Roushan Kumar, Varun Pratap Singh, and Mukul Kumar Gupta

Abstract The increasing electricity demand coupled with concerns over environmental degradation has propelled the quest for sustainable energy sources. Solar energy stands out as a favorable solution in terms of abundant availability, scalability, and minimal environmental effect. It explores the advancements in solar energy technologies and their role in achieving sustainable electricity generation. The abstract begins by elucidating the principles of solar energy conversion through solar photovoltaic cells and concentrated solar power (CSP) systems. It discusses the efficiency improvements and cost reductions achieved through technological innovations, such as multi-junction PV cells, thin-film technologies, and next-generation CSP designs. It highlights the importance of energy storage solutions with lithium-ion batteries and molten salt thermal storage, in mitigating intermittency issues and enabling the continuous incorporation of solar power into the grid. Moreover, it examines the socioeconomic consequences of widespread solar adoption which further facilitate different employment creation, energy access in remote areas, and economic growth. It underscores the significance of policy support, incentives, and regulatory frameworks in fostering the deployment of solar energy technologies on a large scale. In conclusion, this abstract emphasizes the pivotal role of solar energy technologies in realizing sustainable electricity generation. Through continued research, innovation, and strategic deployment, solar energy holds the promise of transforming the international energy model toward a cleaner, resilient, and sustainable future.

Keywords Environmental degradation · Sustainable energy · Solar energy · Photovoltaic cells · Concentrated solar.

R. Kumar (✉) · V. P. Singh
Department of Mechanical Cluster, School of Advanced Engineering, UPES, Dehradun, Uttarakhand, India

M. K. Gupta
Department of Electronics and Instrumentation Engineering, MJP Rohilkhand University, Bareilly, Uttar Pradesh, India

S. Kumar, V. K. Singh (eds.), *Renewable Energy Development: Technology, Material and Sustainability*, Clean Energy Production Technologies,
https://doi.org/10.1007/978-981-97-9626-7_8

8.1 Introduction

The contemporary world is witnessing an unprecedented surge in energy consumption driven by population expansion, automation, and technological developments (Ibrahim et al. 2024). This escalating demand for electricity has underscored the urgent need for sustainable electricity generation methods (Das and Kumar 2024). This section explores the multifaceted reasons behind the imperative for sustainable electricity generation (Das et al. 2024). First and foremost, the reliance on fossil fuels for electricity generation poses substantial environmental challenges that cover air and water pollution, deforestation, and most critically, climate change (Firdouse et al. 2023). The combustion of fossil fuels releases greenhouse gases such as carbon dioxide (CO_2), methane (CH_4), and nitrous oxide (N_2O), which can cause global warming and erratic climatic patterns by trapping heat in the earth's atmosphere (Dekkiche et al. 2023). Sustainable electricity generation offers a pathway to mitigate these environmental impacts by cutting dependence on fossil fuels and transitioning to renewable sources with lower carbon footprints (Rawat and Kumar 2023). Moreover, the finite nature of fossil fuel reserves underscores the need for spreading energy sources to ensure long-term energy security and resilience (Liu et al. 2023). As conventional energy sources become increasingly depleted and geopolitically volatile, investing in sustainable electricity generation technologies becomes imperative to mitigate energy supply risks and ensure energy independence (Das et al. 2023). Figure 8.1 illustrates the block diagram that illustrates the flow of electricity from the solar panels through the junction box and inverter, where it can either be consumed immediately by noncritical loads or stored in the battery bank for later use.

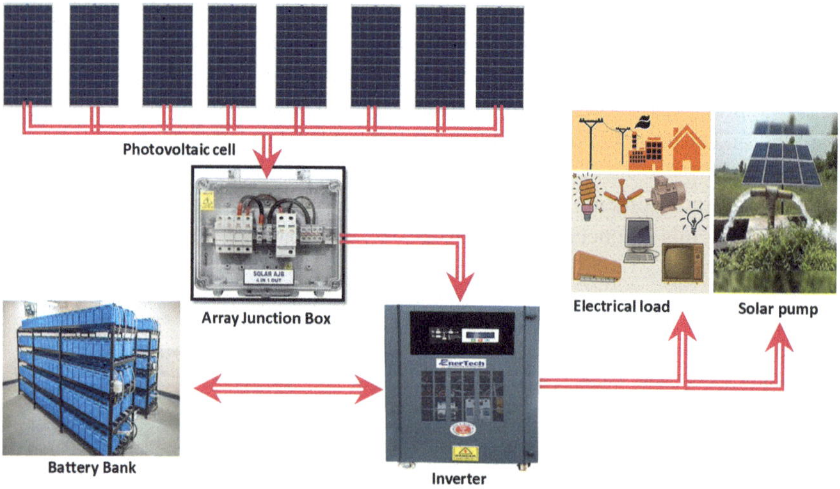

Fig. 8.1 Block diagram of a photovoltaic system with array junction box, battery bank, inverter, and noncritical load

The global pursuit of sustainable electricity generation has into increasingly urgent in the face of climate change, dropping fossil fuel resources, and escalating energy demand. In this context, solar energy has emerged as a beacon of hope, offering a renewable and abundant source of power with minimal environmental impact (Rawat et al. 2023). This introduction sets the stage for exploring the fundamental role of solar energy technologies in achieving sustainable electricity generation. Firstly, the introduction outlines the growing energy demand worldwide, driven by population growth, industrialization, and urbanization. It highlights the necessity for transitioning toward cleaner and more sustainable energy sources to mitigate the adverse effects of greenhouse gas emissions and environmental degradation. Secondly, it introduces the concept of solar energy and its fundamental principles of conversion, highlighting photovoltaic (PV) cells and concentrated solar power systems as the primary technologies for harnessing solar radiation (Das and Kumar 2023a). The scalability and versatility of solar energy make it an attractive solution for both centralized and decentralized electricity generation.

Thirdly, the introduction provides an overview of the rapid advancements in solar energy technologies, including improvements in efficiency, cost reduction, and scalability (Das and Kumar 2023b). It discusses innovations such as multi-junction PV cells, thin-film technologies, and next-generation CSP designs, which have significantly enhanced the competitiveness of solar power compared to conventional fossil fuels (Kumar et al. 2022a). Furthermore, the introduction addresses the challenges associated with the incorporation of solar energy into the existing electricity grid, including intermittency, grid stability, and energy storage. It emphasizes the importance of developing effective storage solutions, such as batteries and molten salt thermal storage, to overcome these challenges and enable the seamless integration of solar power into the grid. Lastly, the introduction underscores the socioeconomic advantages of extensive solar energy implementation, including job creation, energy access in remote areas, and economic growth. It emphasizes the role of supportive policies, incentives, and regulatory frameworks in accelerating the deployment of solar energy technologies and unlocking their full potential (Kumar et al. 2022b). In conclusion, this introduction sets the foundation for exploring the transformative potential of solar energy technologies in realizing sustainable electricity generation. We can set the path for future generations to have a cleaner, more robust, and sustainable energy future by making use of solar energy.

8.2 Solar Energy Technologies for Electricity Generation

Solar energy technologies, including PV systems and CSP plants, offer sustainable electricity generation by directly converting sunlight into electricity or heat. PV systems utilize solar panels to generate electricity, while CSP plants strengthen solar rays to produce heat, driving turbines for energy generation (Kumar et al. 2022c). Advancements in these technologies, coupled with emerging innovations like hybrid systems and floating solar arrays, hold promise for a clean and renewable energy future. This section covers photovoltaic systems, how PV cells work, types of PV panels, applications of PV systems (residential, commercial, industrial), concentrated solar power systems, concentrating solar energy, types of CSP plants, and applications of CSP systems (large-scale electricity generation).

8.2.1 *Photovoltaic Systems*

Photovoltaic systems use the photovoltaic effect to convert solar energy into electrical power. Solar panels directly convert sunlight into electricity. They are usually made of semiconductor materials like silicon. Photons from sunlight excite electrons in the solar cells, causing an electrical current to flow (Kumar et al. 2022d). PV systems come in various forms, including rooftop installations, ground-mounted arrays, and solar farms. Advancements in PV technology have led to increased efficiency, reduced costs, and improved reliability, making solar energy a competitive and widely adopted solution for clean electricity generation. PV cell's work depends on the absorption of sunlight, generation of electron-hole pairs, movement of charge carriers, and collection of charge carriers. Absorption of sunlight hits a PV cell, it is absorbed by semiconductor materials, typically made of silicon. Silicon atoms have four valence electrons that form covalent bonds with neighboring atoms, creating a crystalline lattice structure (Yadav et al. 2022).

The movement of charge carriers in the electric field within the PV cell, created by the internal structure of the material or by an external voltage, causes the free electrons and holes to move in opposite paths. Electrons move toward the negatively charged cell, while the positively charged holes move toward the positively charged side. The electron flow is collected and directed through an external circuit by the charge carriers gathered in the metal contacts on the top and bottom of the photovoltaic cell. This electron flow is known as an electric current, which can be used to power electrical equipment or stored for later use in batteries. The generation of electrical power as electrons flow through the external circuit, creates an electric current that can power electrical devices, such as lights, appliances, or electronic gadgets (Jain et al. 2022). The strength of the sun's rays, the PV cell's efficiency, and the size of the solar panel array are some of the variables that affect how much power is produced. Overall, PV cells harness the energy of sunlight to generate clean and renewable electricity, making them a key technology in the transition toward sustainable energy systems.

8.2.2 *Types of PV Panels*

Several types of photovoltaic panels exist and each one utilizes different materials and technologies for electricity generation using solar irradiance. Common cell-type configuration covers monocrystalline-type silicon PV panels, and it is developed through single-crystal silicon, offering high efficiency and long-term reliability (Yadav et al. 2021). They typically have a consistent black color and their high-power output and space efficiency. Monocrystalline panels are ideal for installations where space is limited, and maximum efficiency is desired. Polycrystalline silicon PV panels are developed using silicon crystals melted together, resulting in a less uniform appearance compared to monocrystalline panels. While they are generally less efficient than monocrystalline panels, polycrystalline panels are more cost-effective to produce, making them a popular choice for residential and commercial installations. Thin-film Photovoltaic material is deposited in thin layers onto a substrate, such as glass, metal, or plastic, to create PV panels. Amorphous silicon, cadmium telluride, and copper indium gallium selenide are examples of thin-film technologies. It is lightweight, flexible, and less expensive to manufacture than crystalline silicon panels, but they typically have lower efficiency and degrade faster over time (Kumar et al. 2021b).

With solar cells on both the front and back surfaces, bifacial photovoltaic panels can collect sunlight from both directions. This design increases energy production by reflecting light onto the rear surface of the panel, especially in environments with high albedo surfaces like snow or white rooftops. Bifacial panels can offer higher efficiency and energy yield compared to traditional single-sided panels. Perovskite solar cells are a relatively innovative type of PV technology that has shown great promise in recent years (Kumar et al. 2021a). These cells are made from a class of materials known as perovskites, which have exceptional light-absorbing properties and can be manufactured using low-cost, solution-based processes. Perovskite solar cells have the potential to achieve high-efficiency levels comparable to traditional silicon cells while offering lower production costs and greater flexibility in design. Different classification of PV panels has their advantages and disadvantages, and the choice of the panel depends on installation location, budget, space constraints, and efficiency requirements. As technology continues to advance, new types of PV panels may emerge, offering even greater efficiency, durability, and cost-effectiveness for solar energy applications (Kumar et al. 2020a).

8.2.3 Concentrated Solar Power Systems

CPS harnesses sunlight to generate electricity through a process that involves concentrating solar radiation onto a receiver. The CSP systems work with solar concentration and CSP systems use mirrors or lenses to strengthen sunlight onto a small area, increasing the intensity of solar radiation (Yilbas et al. 2020). There are several types of CSP technologies, including parabolic troughs, solar power towers, and dish/engine systems, each with its method of concentrating sunlight. Heat generation depends on the concentrated sunlight is directed onto a receiver, typically located at the focal point of the solar concentrator. The receiver absorbs the solar energy and converts it into heat. Depending on the CSP technology used, the receiver may contain a heat transfer fluid (such as oil or molten salt) or directly heat a working fluid (such as water or air) (Kumar et al. 2020b). Heat transfer in CSP systems with heat transfer fluids, the heated fluid circulates through a closed-loop system to transfer thermal energy to a heat exchanger and generate steam by heating water, which then drives a turbine connected to an electricity generator. In systems without heat transfer fluids, the working fluid is directly heated in the receiver and used to produce steam for electricity generation (Huang et al. 2020).

Electricity generation through steam produced by the heat transfer process is employed to direct a turbine attached to an electricity generator, converting the thermal energy into mechanical energy. After the generator transforms the mechanical energy into electrical energy, the grid receives it and distributes it to the end users (Vasconcelos Sampaio et al. 2020). Energy storage of some CSP systems incorporates energy storage to enable electricity generation during periods of low sunlight or high electricity demand. Molten salt thermal storage systems, for example, store excess heat generated during the day for use at night or during cloudy weather, providing dispatchable power and enhancing grid stability. Grid integration with CSP plants is typically connected to the electricity grid, allowing them to supply power to meet local demand (Sharma et al. 2020). They can be integrated with other energy sources, such as fossil fuels or renewable energy, to provide a reliable and dispatchable electricity supply. CSP systems offer several advantages, including the ability to generate electricity even in cloudy conditions or at night (with thermal energy storage), high efficiency, and potential for large-scale deployment in sunny regions. However, they also have challenges, such as high upfront costs, land and water requirements, and environmental impacts. Notwithstanding these obstacles, research and development on CSP technology is still proceeding strong, to raise total system performance and sustainability while cutting costs and increasing efficiency (Senthil and Yuvaraj 2019).

8.2.4 Concentrating Solar Energy

Concentrating solar energy involves using mirrors to converge sunlight to a small area, increasing the intensity of solar radiation. This concentrated sunlight can then be used for various applications, including electricity generation, heating, and thermal energy storage. Solar energy works with solar concentration and concentrating solar energy systems utilize mirrors or lenses to focus sunlight onto a receiver (Hayat et al. 2019). These systems can vary in design, including parabolic troughs, solar power towers, dish/engine systems, and linear Fresnel reflectors. Each design concentrates sunlight onto a specific type of receiver, depending on the application. Receiver operation of the concentrated sunlight is directed onto a receiver, which absorbs the solar energy and converts it into heat. The receiver may contain a heat transfer fluid, such as oil or molten salt, or directly heat a working fluid, such as water or air. In some cases, the receiver may also contain photovoltaic cells to generate electricity directly from concentrated sunlight (Kumar et al. 2018a). Heat transfer in the systems with heat transfer fluids, the heated fluid circulates through a closed-loop system to transfer thermal energy to a heat exchanger. The heat exchanger may generate steam by heating water, which then drives a turbine connected to an electricity generator. Alternatively, the heated fluid can be used for industrial processes or district heating.

In electricity generation in concentrating solar power systems, the steam produced by the heat transfer process is used to drive a turbine connected to an electricity generator (Kumar et al. 2018b). This converts the thermal energy into mechanical energy, which is then converted into electrical energy. The electricity generated can be fed into the electricity grid for distribution to consumers. Thermal energy storage of some concentrating solar energy systems incorporates thermal energy storage to enable electricity generation or heating during periods of low sunlight or high demand. Molten salt thermal storage systems, for example, store excess heat generated during the day for use at night or during cloudy weather, providing dispatchable power and enhancing grid stability. Other applications cover the addition to electricity generation, concentrating solar energy can be used for various other applications, including water desalination, process heat for industrial applications, space heating and cooling, and solar cooking (Kumar et al. 2018c). Overall, concentrating solar energy offers a versatile and sustainable solution for harnessing the power of sunlight for electricity generation, heating, and other applications. Research on concentrating solar energy technologies continues to advance, improving efficiency, reducing costs, and expanding their range of applications.

8.2.5 *Types of CSP Plants*

Concentrated solar power plants utilize various technologies to concentrate sunlight and generate electricity. One of the most established and extensively used CSP technologies, parabolic trough systems are a significant type of CSP plant. These devices focus sunlight into a receiver tube that runs along the trough's focal line using long, curved mirrors, or troughs. Heat transfer fluid (often oil) passing through the receiver tubes is heated by concentrated sunlight (Verma et al. 2017). The fluid transfers the heat to create steam and power a turbine that is connected to an energy generator. Parabolic trough systems can include thermal energy storage to provide electricity during periods of low sunlight. Solar power towers, sometimes referred to as central receiver systems, focus sunlight onto a receiver atop a tower using a collection of mirrors called heliostats. The collected sunlight heats a heat transfer fluid, such as molten salt or water, in the receiver for steam generation and drives a turbine for electricity generation. Solar power towers can achieve higher temperatures and efficiencies compared to parabolic trough systems, making them appropriate for large-scale energy generation (Wu et al. 2017). Figure 8.2 illustrates how these components are integrated into various agricultural applications, showcasing the versatility and benefits of photovoltaic solar systems in the agriculture sector.

Dish or engine systems consist of a parabolic dish-shaped reflector that converges sunlight into different receivers sited at the focal point. To power a Stirling

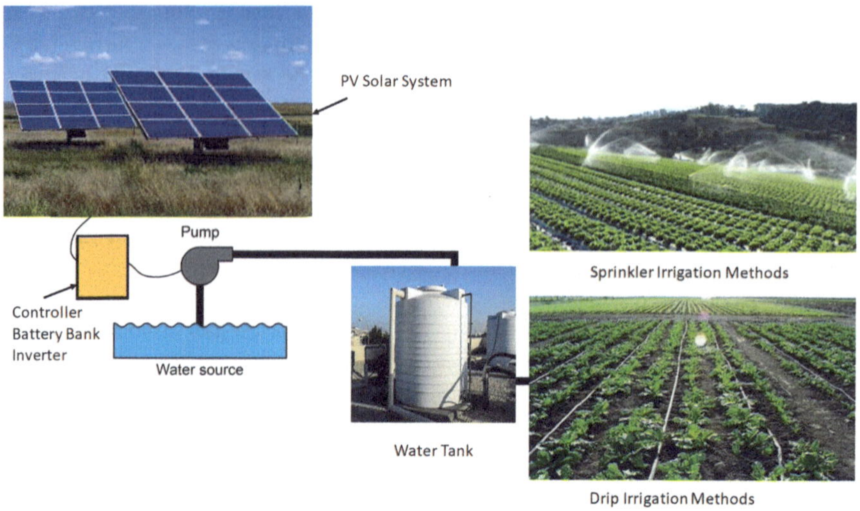

Fig. 8.2 Application of photovoltaic solar system in the agriculture sector

engine or another heat engine, the receiver receives the focused sunlight and transmits the heat to a working fluid (such as hydrogen or helium). Thermal energy is transformed into mechanical energy by the engine, which is subsequently utilized to produce electricity (Das et al. 2017). Dish/engine systems are often used for smaller-scale applications, such as remote power generation or distributed energy systems. Linear Fresnel reflectors offer advantages such as lower cost and simpler construction compared to parabolic trough systems, making them suitable for smaller-scale applications or projects with limited budgets. Various CSP plant types possess unique benefits and drawbacks, contingent on variables including technological advancement, efficacy, expenses, and expandability. The requirements of the project, the features of the site, and financial factors all influence the choice of CSP technology (Kumar et al. 2016).

8.3 Advantages of Solar Power Generation

The advantages of solar power generation cover environmental benefits (reduced emissions, combats climate change), and economic benefits (reduced reliance on fossil fuels, long-term cost savings).

8.3.1 Social Benefits

Solar power generation offers several environmental benefits, contributing to the reduction of greenhouse gas emissions and combatting climate change. One of the main benefits is that solar power generation generates electricity without releasing greenhouse gases like carbon dioxide, methane, or nitrous oxide into the atmosphere (Pal et al. 2015). Solar power uses sunshine to create electricity instead of burning coal, oil, or natural gas like fossil fuel-based power plants do. This results in much reduced emissions of greenhouse gases and air pollutants. Mitigation of climate change by displacing electricity generated from fossil fuels, and solar power helps reduce the carbon footprint of electricity generation. The main greenhouse gas released when fossil fuels are burned, carbon dioxide, plays a significant role in climate change and global warming. We can lessen the effects of climate change by switching to renewable energy sources like solar power, which will also lessen our reliance on fossil fuels and the amount of CO2 and other pollutants we release into the atmosphere. When solar power is generated, there is a reduction in the production of harmful air pollutants like sulfur dioxide (SO2), nitrogen oxides (NOx), particulate matter (PM), and volatile organic compounds (VOC) (Kumar et al. 2015). These pollutants are linked to respiratory illnesses, the formation of smog, and the degradation of the environment.

By replacing polluting fossil fuel-based power plants with clean solar energy, we can improve air quality and public health in communities near power plants and reduce the burden of air pollution-related illnesses (Kumar et al. 2015). Water conservation is important because solar power generation uses no water while it is operating, in contrast to traditional power plants that need a lot of water for cooling and steam production. This is especially helpful in areas that are experiencing drought or water shortages, as conventional power plants increase competition for scarce water supplies. Solar power helps conserve water resources by eliminating the need for water-intensive cooling processes, thereby reducing stress on freshwater ecosystems and ensuring sustainable water management (Meera et al. 2015). Preservation of natural habitats in solar power generation typically involves the installation of solar panels on rooftops, parking lots, or previously disturbed lands, minimizing the impact on natural habitats and ecosystems. Unlike large-scale fossil fuel extraction and power generation projects, which often require extensive land clearing and habitat destruction, solar power facilities can coexist with natural landscapes and wildlife habitats, promoting biodiversity conservation and ecological sustainability. Overall, solar power generation offers substantial environmental advantages by reducing greenhouse gas emissions, mitigating climate change, improving air quality, conserving water resources, and minimizing habitat destruction (Kumar and Kumar 2013).

8.3.2 Economic Benefits

To delve into the economic benefits of solar power generation, focusing on reduced reliance on fossil fuels and long-term cost savings. Dependence on scarce and growingly expensive fossil fuels, such as coal, oil, and natural gas, is decreased when solar power is generated. By harnessing the abundant and renewable energy from the sun, solar power decreases reliance on imported fuels, enhancing energy independence and security (Yau and Rismanchi 2012). This reduced reliance on fossil fuels mitigates risks associated with price volatility, supply disruptions, and geopolitical tensions in energy markets. Long-term financial savings Even though solar power infrastructure may need a large initial investment, solar energy systems can result in significant cost reductions over time. Solar panels have minimal operating costs once installed, as sunlight is a free and abundant resource. Unlike fossil fuel-based power plants that require ongoing fuel procurement and transportation, solar energy systems have low maintenance costs. As solar equipment endures to advance and manufacturing scales up, the cost of solar panels and associated components continues to decline. This trend, coupled with the potential for net metering and government incentives, makes solar power increasingly cost-effective compared to traditional energy sources over the system's lifespan (Kurt and Ozkaymak 2006).

Stable and predictable energy costs in solar power provide stable and predictable energy costs over the long term, as sunlight is consistently available and free of charge. By investing in solar energy systems, businesses, and homeowners can lock in electricity prices and shield themselves from future variations in fossil fuel values and utility rates. Stability in energy costs enhances financial planning, budgeting, and investment decisions for businesses, homeowners, and utilities, contributing to economic stability and resilience. The solar energy business generates jobs in several industries, including manufacturing, installation, maintenance, and research and development, as well as through economic growth. As demand for solar installations grows, so does the need for skilled workers, engineers, technicians, and project managers. This job creation stimulates economic growth, drives innovation, and enhances workforce development in regions investing in solar energy infrastructure. Environmental benefits with economic value are primarily an environmental benefit, the reduction in greenhouse gas emissions associated with solar power generation also has economic value. By mitigating climate change and reducing air pollution, solar power helps avoid the costs associated with healthcare expenses, environmental damage, and climate-related disasters, thereby contributing to overall economic welfare.

8.3.3 Social Benefits

The social benefits of solar power generation, focus on energy independence and job creation. Energy independence is dependent on solar power generation, which encourages energy independence by lowering dependency on imported fossil fuels and diversifying the energy mix. By harnessing the abundant and renewable energy from the sun, communities, regions, and countries can produce their electricity locally, reducing vulnerability to global fuel price fluctuations and supply disruptions (Kumar et al. 2018b). This energy independence enhances national security, economic resilience, and sovereignty, empowering communities to control their energy destiny and make informed decisions about their energy future. The solar energy sector generates employment in several industries, including production, installation, maintenance, and research and development. As demand for solar installations grows, so does the need for skilled workers, engineers, technicians, and project managers. This job creation stimulates economic growth, drives innovation, and enhances workforce development in regions investing in solar energy infrastructure.

Solar power installations also create opportunities for local businesses, contractors, and suppliers, generating economic activity and supporting small and medium-sized enterprises in communities where solar projects are deployed. Skills development and training deals with the growth of the solar energy industry and provides opportunities for skills development and training in renewable energy technologies. Training programs, vocational schools, and educational institutions offer courses and certifications in solar panel installation, system design, maintenance, and solar energy policy and regulation. By investing in workforce development and education, communities can build a skilled workforce capable of meeting the demand for solar energy installations and supporting the transition to clean energy. Community empowerment of solar power generation empowers people to control their energy supply and reduce their environmental footprint. Community-owned solar projects, such as rooftop solar installations, solar cooperatives, and community solar gardens, enable residents to collectively invest in and benefit from solar energy generation. These initiatives foster community engagement, cooperation, and ownership, strengthening social cohesion and promoting sustainable development.

Energy access and equity in solar power can improve energy access and equity by providing electricity to underserved communities, remote areas, and off-grid regions. Off-grid solar systems, microgrids, and solar-powered water pumps propose affordable and clean energy solutions for rural electrification, healthcare facilities, schools, and agricultural activities. By expanding access to clean energy, solar power promotes social inclusion, economic development, and poverty alleviation, enhancing the quality of life and well-being for all members of society (Hayat et al. 2019). Overall, the social benefits of solar power generation, including energy independence, job creation, skills development, community empowerment, and energy access, contribute to sustainable development, social equity, and resilience, fostering a brighter and more inclusive future for communities worldwide.

8.4 Challenges and Prospects of Solar Power Generation

The challenges facing solar power generation, including intermittency, storage limitations, and initial costs, are balanced by promising prospects. Ongoing technological advancements and declining costs make solar energy increasingly competitive, while energy storage solutions offer a path to address intermittency and enhance reliability. Supportive policies and growing global demand further bolster the prospects of solar power. Solar energy has the potential to significantly contribute to reducing global warming, improving energy security, and promoting global economic growth with sustained innovation and investment.

8.4.1 Challenges

Although solar power generation has many benefits, there are several obstacles that could prevent it from being widely used and adopted. Key challenges include intermittency and variability which covers the solar power generation is dependent on sunlight, which varies throughout the day and is affected by factors such as weather, cloud cover, and seasonal changes. Because of its erratic nature, grid integration and reliability may face difficulties. To guarantee a steady supply of electricity during periods of low sunlight, backup power sources or energy storage solutions may be necessary. Because solar power generation is intermittent, efficient energy storage systems are required to store extra energy produced during sunny spells so that it can be used during periods of low sunshine or high demand. A major obstacle to optimizing the dependability and efficiency of solar power systems is energy storage, despite improvements in battery storage technology such as lithium-ion batteries. Land use and environmental impact cover the large-scale solar power installations that require a significant land area for solar panels and associated infrastructure, leading to land use conflicts with agriculture, biodiversity conservation, and other land uses. Furthermore, there may be environmental effects from the extraction of resources, emissions during manufacture, and disposal issues at the end of the product's life.

The cost and financial viability of solar panels have declined significantly in recent years, solar power generation still requires substantial upfront investment for infrastructure and installation. The levelized cost of solar electricity remains competitive with conventional energy sources in many regions, but financial barriers such as access to capital, financing mechanisms, and regulatory uncertainty can hinder the widespread adoption of solar power, especially in developing countries. Infrastructure and grid integration large-scale solar power generating integration into current electrical networks might provide technological difficulties regarding power quality, voltage management, and grid stability. Grid infrastructure upgrades may be necessary to accommodate variable renewable energy sources like solar power and guarantee continuous amalgamation with the existing grid infrastructure.

Policy and regulatory uncertainty frameworks play a crucial role in shaping the deployment and growth of solar power generation. Inconsistent policies, changes in government incentives, and regulatory uncertainty can create barriers to investment and impede the development of solar energy projects. Clear and stable policies, supportive incentives, and long-term planning are essential for fostering a conducive environment for solar power deployment. Resource constraints and supply chain risks in solar power generation rely on raw materials and components such as silicon, metals, and rare earth elements, which may face supply constraints, price volatility, and geopolitical risks. Ensuring a secure and sustainable supply chain for solar energy technologies requires diversification of suppliers, investment in recycling and reuse initiatives, and development of alternative materials and technologies. Addressing these challenges requires concerted efforts from policymakers, industry stakeholders, researchers, and communities to advance technological innovation, improve regulatory frameworks, enhance grid infrastructure, and promote sustainable practices across the solar energy value chain. Despite these challenges, the continued growth and development of solar power generation are critical for transitioning to a more sustainable, resilient, and low-carbon energy future. Figure 8.3 illustrates the challenges and prospects of sustainable electricity generation through solar energy technologies are addressed. It outlines various challenges such as intermittency, energy storage limitations, high initial costs, land use conflicts, and environmental impacts associated with solar energy. However, it also explores prospects for overcoming these challenges, including advancements in

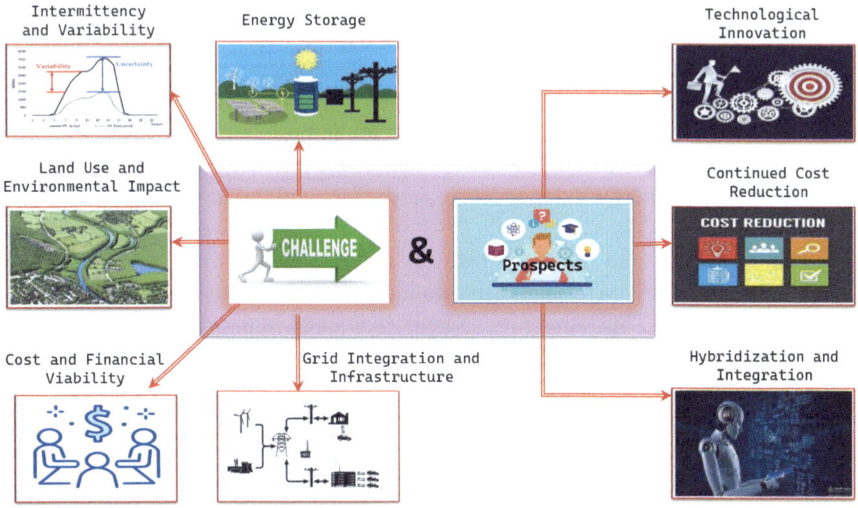

Fig. 8.3 Challenges and prospects of sustainable electricity generation through solar energy technologies

solar panel efficiency, energy storage technologies, grid integration strategies, and policy initiatives.

8.4.2 Prospects

The outlook of sustainable energy appears promising, with ongoing advancements in technology, declining costs, and increasing global adoption driving its growth and development. Here are some key trends and opportunities shaping the future of solar energy and continued cost reductions. The price of PV systems has been declining steadily over time, driven by economies of scale, technological innovations, and increased manufacturing efficiency. As solar panel prices continue to fall and installation costs decrease, solar energy becomes increasingly competitive with conventional fossil fuel-based electricity generation. Technological innovation is ongoing research and development efforts leading to technological advancements in solar photovoltaic and concentrating solar power technologies. The efficiency, scalability, and performance of solar energy systems are being enhanced by innovations like bifacial modules, thin-film solar panels, higher-efficiency solar cells, and sophisticated tracking systems.

Energy storage integration, such as batteries and molten salt thermal storage, with solar power generation, enables grid flexibility, enhances reliability, and enables dispatchable solar electricity. Advances in battery technology, cost reductions, and supportive policies are driving the deployment of energy storage systems alongside solar installations. Hybridization and integration of solar energy systems that combine PV with other renewable energy sources, such as wind, hydroelectric, or geothermal power, offer synergies and complementarity in energy generation. Integrated energy systems that combine solar with storage, smart grid technologies, and demand-side management optimize energy use, improve grid stability, and enhance overall efficiency. Decentralization and distributed generation trend toward decentralized energy generation and distributed solar installations is transforming the energy landscape, empowering consumers, communities, and businesses to produce their electricity locally. Rooftop solar, community solar projects, and microgrids enable energy independence, resilience, and equitable access to clean energy resources.

Global expansion and market growth of solar energy deployment is expanding rapidly worldwide, driven by supportive policies, declining costs, and growing environmental awareness. Sustainability and environmental impact concerns about climate change and environmental sustainability grow, solar energy's low carbon footprint, minimal environmental impact, and renewable nature make it an attractive option for reducing greenhouse gas emissions and mitigating climate change. Sustainable practices in solar panel manufacturing, recycling, and end-of-life management further enhance solar energy's environmental credentials.

Policy support and market incentives deal with supportive policies, financial incentives, and regulatory frameworks that play a crucial role in driving solar energy deployment and market growth. Governments worldwide are implementing renewable energy targets, feed-in tariffs, tax incentives, and renewable energy auctions to stimulate investment in solar power generation and accelerate the transition to clean energy. Overall, the future of solar energy is bright, with the potential to play a central role in the global energy transition toward a more sustainable, resilient, and equitable energy system. By capitalizing on technological advancements, market opportunities, and policy support, solar energy can contribute significantly to meeting the world's growing energy demand while reducing carbon emissions, improving energy access, and fostering economic development.

8.5 Conclusions

In conclusion, solar energy technologies offer a sustainable solution for electricity generation, with numerous environmental, economic, and social benefits. By harnessing the abundant and renewable energy from the sun, solar power reduces greenhouse gas emissions, mitigates climate change, and promotes energy independence. The future of solar energy looks promising, with ongoing advancements in technology, declining costs, and increasing global adoption driving its growth and development. However, challenges such as intermittency, energy storage, and grid integration need to be addressed to maximize the potential of solar power generation. Policymakers, industry stakeholders, and communities must collaborate to overcome these challenges and create an enabling environment for solar energy deployment. By investing in research, innovation, infrastructure, and supportive policies, we can unlock the full potential of solar energy and transition toward a more sustainable, resilient, and low-carbon energy future. In summary, sustainable electricity generation through solar energy technologies offers a pathway to achieve our energy goals while preserving the planet for future generations. With dedication, collaboration, and commitment, we can harness the power of the sun to power a brighter and more sustainable future for all.

References

Das K, Kumar R (2023a) Assessment of electric two-wheeler ecosystem using novel Pareto optimality and TOPSIS methods for an ideal design solution. World Electr Veh J 14(8):215

Das K, Kumar R (2023b) Electric vehicle battery capacity degradation and health estimation using machine-learning techniques: a review. Clean Energy 7(6):1268–1281

Das K, Kumar R (2024) A comprehensive review of categorization and perspectives on state-of-charge estimation using deep learning methods for electric transportation. Wirel Pers Commun:1–20

Das AS, Dwivedi PK, Mondal AK, Kumar R, Reddy RM, Kumar A (2017) Storage optimization of automated storage and retrieval systems using breadth-first search algorithm. In: Proceedings of the international conference on nano-electronics, circuits & communication systems. Springer Singapore. https://doi.org/10.1007/978-981-10-2999-8_18

Das K, Kumar R, Krishna A (2023) Supervised learning and data intensive methods for the prediction of capacity fade of lithium-ion batteries under diverse operating and environmental conditions. Water Energy Int 66(1):53–59

Das K, Kumar R, Krishna A (2024) Analyzing electric vehicle battery health performance using supervised machine learning. Renew Sust Energ Rev 189:113967

Dekkiche M, Tahri T, Denai M (2023) Techno-economic comparative study of grid-connected PV/reformer/FC hybrid systems with distinct solar tracking systems. Energy Convers Manag X 18. https://doi.org/10.1016/j.ecmx.2023.100360

Firdouse Z, Ashraf SM, Rahman K, Rehman H (2023) Hybrid solar PV system for electric vehicles battery charging. In: 2023 international conference on power, instrumentation, energy and control, PIECON 2023. Institute of Electrical and Electronics Engineers Inc. https://doi.org/10.1109/PIECON56912.2023.10085883

Hayat MB, Ali D, Monyake KC, Alagha L, Ahmed N (2019) Solar energy – a look into power generation, challenges, and a solar-powered future. Int J Energy Res 43(3):1049–1067. https://doi.org/10.1002/er.4252

Huang Q, Huang C, Liu S, Wang F, Guo C (2020) Solar vapor generation optimization of a carbon-black/wood-flour system with strength enhanced by polystyrene. Int J Energy Res 44(5):3687–3696. https://doi.org/10.1002/er.5154

Ibrahim NA, Wan Alwi SR, Abd Manan Z, Mustaffa AA, Kidam K (2024) Climate change impact on solar system in Malaysia: techno-economic analysis. Renew Sust Energ Rev 189, no. PA:113901. https://doi.org/10.1016/j.rser.2023.113901

Jain A et al (2022) Smart communication using 2D and 3D mesh network-on-Chip. Intell Autom Soft Comput 34(3):2007–2021. https://doi.org/10.32604/iasc.2022.024770

Kumar R, Kumar A (2013) Design and hardware development of power window control mechanism using microcontroller. In: 2013 international conference on signal processing and communication (ICSC). IEEE, pp 361–365. https://doi.org/10.1109/ICSPCom.2013.6719813

Kumar R, Dwivedi PK, Praveen Reddy D, Das AS (2015) Design and implementation of hydraulic motor based elevator system. In: 2014 IEEE 6th India international conference on power electronics (IICPE). IEEE, pp 1–6. https://doi.org/10.1109/IICPE.2014.7115821

Kumar R, Ahuja NJ, Saxena M, Kumar A (2016) Modelling and simulation of object detection in automotive power window. Indian J Sci Technol 9(43). https://doi.org/10.17485/ijst/2016/v9i43/104393

Kumar A, Kumar R, Jain A, Pandey A, Thapliyal S, Sharma A (2018a) Automatic street lighting system, vol 624. Springer Singapore. https://doi.org/10.1007/978-981-10-5903-2_92

Kumar A, Sharma P, Gupta MK, Kumar R (2018b) Machine learning based resource utilization and pre-estimation for network on Chip (NoC) communication. Wirel Pers Commun 102(3):2211–2231. https://doi.org/10.1007/s11277-018-5376-3

Kumar R, Ahuja NJ, Saxena M (2018c) Improvement and approval of impediment recognition and activity for power window. In: Intelligent communication, control and devices. Springer, pp 855–864

Kumar R, Ahuja NJ, Saxena M, Kumar A (2020a) Automotive power window communication with DTC algorithm and hardware-in-the loop testing. Wirel Pers Commun 114(4):3351–3366

Kumar A, Bansal K, Kumar D, Devrari A, Kumar R, Mani P (2020b) FPGA application for wireless monitoring in power plant. Nucl Eng Technol 53(4):1167–1175. https://doi.org/10.1016/j.net.2020.09.003

Kumar R, Divyanshu, Kumar A (2021a) Nature based self-learning mechanism and simulation of automatic control smart hybrid antilock braking system. Wirel Pers Commun 116(4):3291–3308. https://doi.org/10.1007/s11277-020-07853-7

Kumar R, Bansal K, Kumar A, Yadav J, Gupta MK, Singh VK (2021b) Renewable energy adoption: design, development, and assessment of solar tree for the mountainous region. Int J Energy Res 46:1–17. https://doi.org/10.1002/er.7197

Kumar R, Gupta N, Bharadwaj D, Dutt D, Joshi A (2022a) Design and development of electronic clutch control unit for manual transmission. Mater Today Proc 69:240–245. https://doi.org/10.1016/j.matpr.2022.08.470

Kumar R, Kumar Gupta M, Kumar A, Sharma P, Deorari R (2022b) Analysis of electronic clutch control unit for manual transmission vehicle oriented toward safety. Mater Today Proc 69:255. https://doi.org/10.1016/j.matpr.2022.08.473

Kumar R, Pachauri RK, Badoni P, Bharadwaj D, Mittal U, Bisht A (2022c) Investigation on parallel hybrid electric bicycle along with issuer management system for mountainous region. J Clean Prod 362(June):132430. https://doi.org/10.1016/j.jclepro.2022.132430

Kumar R, Kumar A, Gupta MK, Yadav J, Jain A (2022d) Solar tree-based water pumping for assured irrigation in sustainable Indian agriculture environment. Sustain Prod Consum 33:15–27. https://doi.org/10.1016/j.spc.2022.06.013

Kurt H, Ozkaymak M (2006) Performance evaluation of a small-scale sodium carbonate salt gradient solar pond. Int J Energy Res 30(11):905–914. https://doi.org/10.1002/er.1193

Liu X, Liu XC, Xie C, Ma X (2023) Impacts of photovoltaic and energy storage system adoption on public transport: a simulation-based optimization approach. Renew Sust Energ Rev 181. https://doi.org/10.1016/j.rser.2023.113319

Meera CS, Sunny S, Singh R, Sairam PS, Kumar R, Emannuel J (2015) Automated precise liquid transferring system. In: 2014 IEEE 6th India international conference on power electronics (IICPE). IEEE, pp 1–6. https://doi.org/10.1109/IICPE.2014.7115831

Pal A, Kumar R, Sanal Kumar VR (2015) Conceptual design of an automatic fluid level controller for aerospace applications. In: 2015 international conference on soft-computing and networks security (ICSNS). IEEE, pp 1–8. https://doi.org/10.1109/ICSNS.2015.7292433

Rawat KS, Kumar R (2023) Insight into the novel low cost green air pollution monitoring and control systems: a technological solution from concept to market. J Clean Prod 427:139218. https://doi.org/10.1016/j.jclepro.2023.139218

Rawat K et al (2023) Structural optimisation of go kart chassis with basic electronic driver assistance systems. In: 2023 2nd international conference on paradigm shifts in communications embedded systems, machine learning and signal processing (PCEMS 2023), p 1–6. https://doi.org/10.1109/PCEMS58491.2023.10136045

Senthil R, Yuvaraj S (2019) A comprehensive review on bioinspired solar photovoltaic cells. Int J Energy Res 43(3):1068–1081. https://doi.org/10.1002/er.4255

Sharma H, Kumar R, Kumar A, Gupta MK, Kaundal V (2020) Implementation of HMI using VISUAL BASIC .NET for induction hardening process. In: Advances in intelligent systems and computing. Springer. https://doi.org/10.1007/978-981-13-8618-3_96

Vasconcelos Sampaio PG et al (2020) Prospecting technologies for photovoltaic solar energy: overview of its technical-commercial viability. Int J Energy Res 44(2):651–668. https://doi.org/10.1002/er.4957

Verma V, Kumar R, Kaundal V (2017) Implementation of ladder logic for control of pipeline inspection robot using PLC. In: Proceeding of international conference on intelligent communication, control and devices: ICICCD 2016, pp 965–971

Wu L, Wen C, Ren H (2017) Reliability evaluation of the solar power system based on the Markov chain method. Int J Energy Res 41(15):2509–2516. https://doi.org/10.1002/er.3816

Yadav J, Kurre SK, Kumar A, Kumar R (2021) Nonlinear dynamics of controlled release mechanism under boundary friction. Results Eng 11:100265. https://doi.org/10.1016/j.rineng.2021.100265

Yadav J, Maity D, Kurre SK, Kumar A, Kumar R, Chouhan S (2022) Identification of damping prohibiting stick-slip form friction influenced dynamic system. Proc Inst Mech Eng Part C J Mech Eng Sci:1–8. https://doi.org/10.1177/09544062221135517

Yau YH, Rismanchi B (2012) A review on cool thermal storage technologies and operating strategies. Renew Sust Energ Rev 16(1):787–797. https://doi.org/10.1016/j.rser.2011.09.004

Yilbas BS, Al-Sharafi A, Sahin AZ (2020) Solar energy harvesting and a water droplet cleaning of micropost arrays surfaces. Int J Energy Res 44(3):2072–2083. https://doi.org/10.1002/er.5063

Chapter 9
Biomass Energy Conversion Technologies to Produce Electrical Power, Green Hydrogen, Bio-Oil and Biochar Through Biomass Gasification and Pyrolysis

P. Raman

Abstract The demand for energy grows continuously along with population and economic growth. Growing energy demand results in the exploitation of fossil fuels, like coal, diesel, petrol, natural gas, etc.; the fossil fuel used in different applications have a large carbon footprint (CFP) and emits a large quantity of greenhouse gases (GHG), which results in global warming and climate change. Biomass wastes like agricultural residues and MSW are a potential resource for sustainable energy management. Unlike solar photovoltaic (PV) or windmill-based power generation, biomass power generation plant can work round the clock as per demand of the power requirement. Green hydrogen production through advanced biomass-steam gasification technology is the most efficient technology of hydrogen when compared to other technologies like hydrogen from power produced by PV or windmills. Hydrogen production through biomass-steam gasification is the lowest production cost when compared to other sources of green hydrogen production. Biomass gasifier system can be used to meet different types of thermal energy requirement of the industries.

Keywords Fossil fuels · Greenhouse gases · Global warming · Biomass wastes · Solar photovoltaic · Hydrogen production

9.1 Introduction

The demand for energy grows continuously along with population and economic growth. Growing energy demand results in the exploitation of fossil fuels like coal, diesel, petrol, natural gas, etc. The fossil fuel used in different applications has a

P. Raman (✉)
Energy Efficiency and Environment P. Ltd, New Delhi, New Delhi, India
e-mail: praman@cleanenergyee.com

S. Kumar, V. K. Singh (eds.), *Renewable Energy Development: Technology, Material and Sustainability*, Clean Energy Production Technologies,
https://doi.org/10.1007/978-981-97-9626-7_9

large carbon footprint (CFP) and emits a large quantity of greenhouse gases (GHG), which results in global warming and climate change. Extreme summer, unpresidential climate conditions delayed monsoons and cloud bust. For environment-friendly and sustainable growth, biomass as a renewable energy source is a suitable alternative fuel. For example, biomass from agricultural residues (coconut shell, groundnut shell, cotton stalk, mustered stalk, etc.) and biomass from municipal solid waste can be used to produce hydrogen. Syngas produced from biomass can be used for electric power generation and industrial process heat (thermal applications). Green hydrogen produced from biomass is carbon neutral and environment friendly and does not contribute to greenhouse gas emissions. The conventional biomass-steam gasification systems produce the syngas with impurities like tar. Presence of tar is a great barrier to separate the hydrogen using pressure swing adsorbers (PSA). Recent development in gasification technology an advanced three-stage (ATS) biomass-steam gasification system produces high-quality syngas without tar, which makes possibility of production of green hydrogen from biomass. Green hydrogen produced from biomass can be used for transport applications (hydrogen-fueled vehicles) and in fertilizer industries for ammonia production (green ammonia). Green hydrogen produced from biomass is a potential resource to use in oil refineries to substitute the hydrogen produced from natural gas.

Biomass wastes (agricultural and municipal solid waste) can be used for power generation using advanced three-stage (ATS) biomass gasification technology. The conventional gasifiers have several technology barriers (such as impurities like tar) due to gas quality. Impurities in the gas were not allowing the technology adoptions and upscaling. The recent development in gasification technology enables to produce electricity through biomass gasification without any technical barrier. Biomass is a potential resource to meet the industrial process heat requirement. Similar to power generation, biomass gasifiers can be used for thermal applications/heat generation as per the industrial process requirement. Biomass also can be converted into user-friendly fuels like bio-oil and biochar through a process called pyrolizer. Bio-oil can be used to replace furnace oil or diesel.

Hence, biomass is a potential resource to meet several types of energy needs in industrial and domestic sectors; there are different processes of biomass energy conversion to use biomass efficiently for different applications. Different processes adopted for biomass energy conversion and its applications are shown in Fig. 9.1.

9.2 Biomass Gasifier Systems for Power Generation

Globally 675 million people are without access to electricity. The growing need for energy and economic development demand access to modern energy like electricity. In developing countries, there are millions of telecom towers, which use diesel generator for its power generation. Producing power using locally available sources, like biomass, will be an ideal option. In its present form, biomass gasification

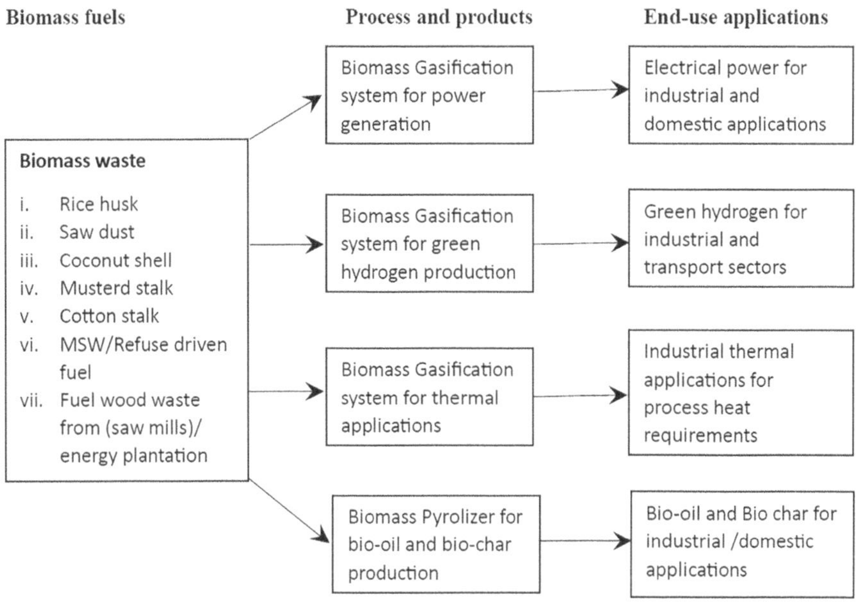

Fig. 9.1 Biomass conversion process and its applications in different sectors

technology needs to be upgraded to make it technically, economically, and practically viable.

9.2.1 *Advanced Three-Stage Biomass Gasifier System: Concept and Components*

Biomass is generally stored in a roofed shed to protect it from rain. Biomass is fed into the gasifier using skip hoist/bucket elevator. The ATS gasifier consists of three reactors to produce high-quality syngas without tar. Hot syngas is removed through a blower from the bottom of the gasifier (below the ash pit). The gas comes out of the gasifier at 350–400° C. Hot gas is cooled to 125° C using a heat exchanger. The sensible heat of the hot gas is used to preheat the air supplied to the gasifier for gasification. Ambient air is heated to 250° C and injected to the gasification reactors. Ash is removed using a rotating grate ash removal system connected through a programmable timer switch. The ash removal rate can be set as per the ash content of the fuel. The gas exiting the heat exchanger I is further cooled and cleaned before feeding to the engine. The system uses dry gas cooling and cleaning equipment. Hence, no wastewater is produced from the system and eliminates the conventional wet scrubbers. The process flow diagram of the ATS biomass gasifier system for power generation with details components is shown in Fig. 9.2.

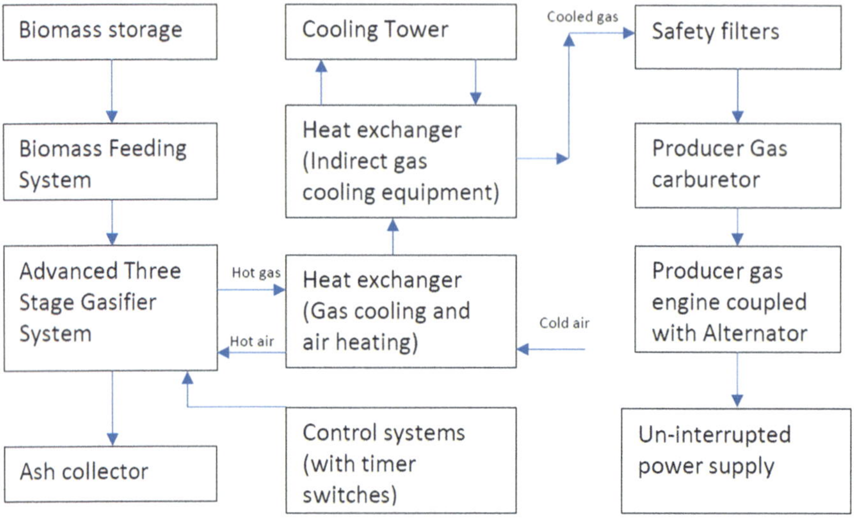

Fig. 9.2 Process flow diagram with detail components of ATS biomass gasifier system with gas cleaning, cooling equipment, and other accessories

9.2.2 *The Advanced Three-Stage (ATS) Biomass Gasifier System*

The advanced three-stage (ATS) biomass gasifier system developed by EEEPL produces high-quality producer gas without tar. The ATS gasifier system is a patented technology. Basically, a tar cracking reactor is added to the gasifier. The ATS gasifier system can be operated with multiple types of fuel like coconut shell, agricultural residues, fuelwood, briquettes, etc. ATS gasifier system produces gas without tar, and it does not use any water scrubbers as the case of conventional system. Due to the gas quality without presence of tar, the engine operation is smooth and trouble free. The specific fuel consumption (SFC) for power generation is 1 kg/kWh. The cold gas efficiency of the gasifier system is 88%. The details about the gas quality and the system are given below.

The ATS biomass gasifier system is suitable to supply the required high-quality gas for running internal combustion (IC) engines for power generation, using 100% producer gas. Details of the components and their features are given in Table 9.1.

9.2.3 *Producer Gas Engine*

For biomass to power, 100% producer gas engine controlled by an electronic governor coupled with a 120 kW_e alternator, along with a control panel, is used in biomass gasifier-based power generation systems. The details of the IC engines used for biomass-based power generation system is presented in Tables 9.2 and 9.3.

Table 9.1 Producer gas components and heating value of the gas

Components	Unit	Value
Hydrogen	%	22
Methane	%	1
Carbon monoxide	%	20
Carbon dioxide	%	11
Nitrogen	%	46
Calorific value of gas	Kcal/Nm^3	

Table 9.2 The key components of the ATS Biomass gasifier system and their details

S. no	Component	Details of the component
1.	Main gasifier unit (ATSG), with three-stage reactors. The gasifier unit is designed to gasify any woody biomass, which has a density of above 250 kg/m^3 and an ash content of less than 5%	The main gasifier unit will have a biomass hoper made of mild steel and three numbers of reactor components made of high temperature stainless steel with multiple layers of insulation
2.	Ash removal system	A well-insulated and timer-operated ash removal system made of high-temperature stainless steel will be used for continuous operation of the system. The ash removal system is provided with an ash storage unit to enable long duration and uninterrupted operation of the system
3.	Heat exchanger I	The heat exchanger one will be used to cool the gas and to heat the ambient air. The hot air will be used for gasification. The heat exchanger is a shell and tube type made of stainless steel (SS) with double layers of insulation
4.	Cyclone filter	A cyclone filter made of SS will be used to remove the core dust of the gas
5.	Heat exchangers II and III	There will be two exchangers connected in series to cool the gas near to ambient temperature. The gas will be cooled using a cooling tower. Both of the heat exchangers are made of SS
6.	Mist separator	A mist separator made of SS will be used to remove the mist formed during the process of gas cooling
7.	Safety filter	A safety filter made of SS shell and cartridge filter is used for removing the fine dust, present in the gas
8.	Other accessories	Gas and air handling blowers, cooling tower pump, flare burner (for initial starting of the gasifier), connecting pipes valves, and control panel

Table 9.3 Details of the producer gas engine and power generator are given in the table below

S. no	Component	Details of the component
1.	A 100% producer gas engine	A 100% producer gas engine will be used for the 120 kW_e power generation system
2.	Ignition system	The engine will be having electrically controlled spark ignition (SI) system, to enable the engine to work on 100% producer gas
3.	Governor system	The engine will be having an electronic governor system will sense the RPM of the engine to maintain the frequency and power quality
4.	Engine cooling	Engine will be cooled through a radiator cooling system
5.	Alternator (power generator) 120 kW_e	Capacity 120 KW (150 KVA), 210 AMP, three-phase, 415 volts, 50 Hz, 1500 RPM, 0.8 P.F. Brushless with AVR, double bearing type coupled with engine
6.	Control panel	Power control panel: Digital volt meter, AMP meter, frequency meter (combined), hour meter, MCB, etc. Engine panel: Containing battery indicator, temperature meter with sensor, oil pressure meter with sensor, battery, AMP meter, starting key, etc.

Table 9.4 The details of the automated fuel feeding system and biomass dryer are given in the table

S. no	Component	Details of the component
1.	Automated fuel feeding system	A large fuel storage tank (made of mild steel) will be provided with an automated fuel feeder. The automatic fuel feeder will be connected to the fuel hopper of the gasifier system. The fuel level sensors will be used to control the fuel feeding system. Fuel feeding will be done through a low-rpm screw conveyer
2.	Biomass dryer	A biomass dryer (made of mild steel) will be included in the system to dry the biomass using the engine exhaust heat. The dryer will be insulated for an efficient drying of biomass fuel. The biomass for the 120 kWe system should not exceed 75 mm from form any dimension of length, breadth, and height (l, b, h). The moisture content of the biomass fed to the gasifier should be less than 10%

9.2.4 Automated Fuel Feeding System with Biomass Dryer

The automated biomass feeding system and biomass drier are additional equipment included in the system as per the requirement. The details of the biomass feeding and dryer units are given in Table 9.4.

9.2.5 Biomass Gasifier Systems in Field

Pictures of the biomass power generation system installed are shown in this chapter. The gasifier systems are installed for power generation. The systems are designed to operate on fuelwood and bamboo waste in Thailand (Figs. 9.3 and 9.4).

9.2.6 *Innovative Aspects of Advanced Three-Stage (ATS) Biomass Gasification System*

Biomass gasification technology is a century-old technology. Due to several reasons, the technology is unable to reach a technically and commercially viable product, which will be useful to the society. The proposed system is aimed to overcome the barriers of the existing system. The innovativeness of the proposed advanced biomass-based power generation system is:

Fig. 9.3 A view of a 125 kW_e gasifier system-based power generation system using bamboo waste as fuel

Fig. 9.4 A view of a 125 kW_e gasifier-based power generation system used as a decentralized power generation system

(i) Advanced gasification reactor to produce the gas with low impurities.
(ii) Eliminate the hassles related to the gas cleaning equipment.
(iii) Gasifier-based power generation system, which will be environmentally friendly by elimination of water pollution and soil pollution.
(iv) Ease of operation and maintenance.
(v) Increased reliability and economic viability.
(vi) A viable solution for distributed decentralized power generation.

9.2.7 Key Features and Advantages of the Advanced Three-Stage Gasification (ATSG) System for Power Generation Using Biomass as Fuel

The ATS biomass gasifier system is designed to overcome several problems associated with conventional gasifiers, to have a long durable and trouble-free operation. The entire gasifier ATSG system is made of long durable and high temperature resistant stainless steel, which ensures a trouble-free long duration operation. The ATS gasifier system is designed to have flexibility in the use of biomass fuels, unlike other gasifier systems, which are designed to operate with a selected type of fuel (fuelwood/briquette/loos biomass, etc.). The ATS biomass gasifier system is compact and environment friendly, due to use of dry gas cleaning equipment. The dry gas cleaning equipment eliminates the conventional use of water scrubbers and necessity of water treatment plant.

For a better understanding, some of the key features of ATS biomass gasifier system are compared with the conventional gasifier systems in Table 9.5.

Table 9.5 The key features and advantages of the advanced three-stage gasifier system, over the conventional gasifier systems

S. no	Conventional gasifiers system	Advanced three-stage gasifier (ATSG) system
1.	The maximum heating value of the gas is about 1200 kcal/M^3	The maximum heating value is above 1300 kcal/M^3
2.	Air supply for the gasification is at an ambient temperature of around 30 °C	Hot air is used for gasification at 200 °C (air is preheated using waste heat of the producer gas exiting from the reactor)
3.	Tar content in producer gas is at 200–1000 mg per cubic meter	No (detectable) tar in the gas due to the tar cracker integrated gasification system
4.	Specific fuel consumption (SFC) is 1.2 kg/kWh	Specific fuel consumption is 1.0 kg/kWh (at and above 80% of the design power output)
5.	The gasifiers are designed to take a particular fuel (they are fuel-specific)	Have a large fuel flexibility. It can be operated with furl wood or agro-residues like ground nutshell, cotton stalk, etc. (bulk density above 250 kg/m3)
6.	Employs a sequence of gas cleaning system/scrubber/baghouse filter to clean the gas	There is no scrubber or bag-house filter is used. Only a safety filter is used as the gas produced have no tar
7.	Large quantity of water consumption in scrubbers and requirement of water treatment system to clean the contaminated water from the scrubbers	Completely dry gas cleaning system with indirect cooling systems through heat exchangers
8.	Frequent and complex maintenance of gas cleaning system due to presence of tar in the gas	Reduced and simple maintenance of gas cleaning system due to cleanliness the gas, without tar
9.	Complex and elaborate gas cleaning equipment and takes a large area of footprint for assembling the equipment	Compact and simple gas cleaning equipment and takes a little (about 40% only) area of footprint for assembling the equipment
10.	Consists of individual components assembled on the floor and occupies a large area of footprint	Modular system consists of main gasifier unit, gas cooling and cleaning system, and the power generator. Compact and occupies a lesser area of footprint

9.2.8 Biomass Gasification-Based Power Generation Using Rice Husk as Fuel

There are several developing countries lack of grid power. Biomass gasifier is a potential solution for distributed power generation (DDG) to meet the power requirement of the local. In the case of DDG with surplus biomass potential, the excess power generated can be injected into the power grid. A rice husk-based biomass gasifier system was installed in a remote village near Siem Reap, Cambodia. The capacity of the system is 125 kW electricity output. The power plant is used to provide power to 30 villages to meet their domestic and industrial power requirements.

A dual fired advanced biomass gasifier system suitable for electric power generation is installed in Cambodia for power generation using rice husk. The system is equipped with dry gas cleaning system. The new gasifier design uses a minimum quantity of water for cooling the gas.

Rice husk has several disadvantages as a fuel for gasification when compared to fuelwood. Rice husk is a low calorific value fuel, due to high ash content. When the ash content of the fuelwood is less than 2%, the ash content of rice husk is close to 20%. It is ten times higher than the ash content of fuelwood. The bulk density of rice husk is 150 kg/M3, whereas the fuelwood has 350 kg/M3. The bulk density of the rice husk is less than half of the bulk density of the fuelwood. Bulk density of the fuel plays a significant role in fuel movement, within the gasifier system. The rice husk gasifier system is designed with a suitable fuel feeding system and ash removal system for an uninterrupted operation of the system. The ash removal system and fuel feeding systems are operated and controlled by a programable timer switch. A view of the rice husk-based biomass gasifier system for power generation is shown in Fig. 9.5.

The design of the biomass gasifier system is done considering the influencing factors of gasification referred to in Table 9.6.

Fig. 9.5 A view of the rice husk-based biomass gasifier system for power generation

Table 9.6 Properties of fuelwood rice husk a comparison key component

Component details	Unit	Fuelwood	Rice husk
Bulk density	Kg/m^3	350	150
Volatile mater	%	76.8	63.9
Fixed carbon	%	21.4	13.2
Ash content	%	1.8	22.9
Calorific value	$Kcal/Nm^3$	4300	3250

9.3 Hydrogen Production from Biomass

9.3.1 Present Status of Hydrogen Production

At present, more than 98% of hydrogen is produced using conventional fuel as source. Globally, 76% of hydrogen is produced from natural gas by SMR, with 22% produced through coal gasification and 2% from electrolysis. Hydrogen produced from uncontrolled fossil fuels is referred to as *gray hydrogen*. The use of gray hydrogen contributes to GHG emission and climate change. Globally, there is an effort to use of green hydrogen to reduce the carbon foot print. As biomass is a carbon-neutral RE source, production of hydrogen from biomass is a better alternative in the point of cost and GHG reduction, when compared to green hydrogen production through electrolysis. The proposed proposal is for demonstration of hydrogen production through biomass-steam gasification at a cost of less than 2$/kg a competitive price to the conventional hydrogen produced fusing "Natural Gas (by SMR)." It is much less than the hydrogen production cost of 6–8 $/kg by electrolysis process.

9.3.2 Challenges in Conventional Gasification Technologies

There are several challenges associated with the conventional gasification system to produce green hydrogen from biomass. Indian Institute of Science, Bangalore, worked on a fixed bed downdraft biomass gasification system to produce hydrogen-rich syngas. Analysis of tar obtained from hydrogen-rich syngas generated from a fixed bed downdraft biomass gasification system is reported (Rakesh and Dasappa 2018). It is reported that the tar content in raw syngas is 268 mg/Nm^3. The presence of tar components in syngas after cleaning is about 37 mg/Nm^3. Cleaning of tar is complicated process and disposal of materials used to clean have environmental concerns. Tar content of 37 mg/Nm^3 may look as a small and negligible quantity, but in cumulative terms, the accumulation of tar will be very much high. A 10 kg/h hydrogen production system with a tar content of 37 mg/Nm^3 can contribute to a cumulative accumulation of tar to the order of 222 g/d and 6.7 kg/month. The PSA equipment used for hydrogen separation cannot work with the tar loading rate of

6.7 kg/month. Challenges related to tar content in the syngas are reported by Narnaware et al., from "Maharana Pratap University of Agriculture and Technology, Udaipur" (Narnaware and Panwar 2022). Hence, irrespective of hydrogen content in syngas, the hydrogen separation or hydrogen production from syngas becomes impossible. In the conventional biomass-steam gasification system, the presence of tar in syngas is a greater challenge for green hydrogen production. The proposed ATS biomass-steam gasification system produces hydrogen-rich syngas without tar making it possible to produce green hydrogen from biomass.

9.3.3 National Green Hydrogen Mission (NGHM)

In India, to make the hydrogen as future energy, a national green hydrogen mission is in place. The "National Green Hydrogen Mission" aims to make India the Global Hub for the production, usage, and export of green hydrogen and its derivatives. This will contribute to India's aim to become Atmanirbhar through clean energy transition. The mission will lead to significant decarbonization of the economy and reduced dependence on fossil fuel imports and enable India to assume technology and market leadership in green hydrogen.

NGHM Consists of Five Nectar Elements, "Panchamrit," to Deal with Climate Challenge

(i) India will take its non-fossil energy capacity to 500 GW by 2030.
(ii) India will meet 50% of its energy requirements from renewable energy by 2030.
(iii) India will reduce the total projected carbon emissions by 1 billion tonnes from now till 2030.
(iv) By 2030, India will reduce the carbon intensity of its economy by more than 45%.
(v) By the year 2070, India will achieve the target of net zero.

These "Panchamrits" will be an unprecedented contribution of India to climate action. Depending on electrolysis alone may not achieve the goals of "Panchamrit." Hydrogen production through electrolysis is three to four times costlier than hydrogen production from biomass. Hence, this proposal aims at the establishment of techno-economic viability and demonstration of hydrogen production from biomass and the use of hydrogen-based transport.

The advanced three-stage (ATS) biomass-steam gasification system is developed by EEEPL to produce hydrogen-rich syngas (H2 > 50%) without tar, which is at TRL of 6. The VPSA to separate the hydrogen from syngas will be outsourced and it is at TRL of 7 or 8. The technology of hydrogen-fueled IC engines is at the TRL of 7. Integration of ATS gasifier system with the PSA equipment and hydrogen-fueled internal combustion engine can be an appropriate solution for transport applications and to avoid the GHG emissions of the transport sector.

9.3.4 Hydrogen-Rich Syngas Production from Biomass

Hydrogen-rich syngas is produced using an advanced three-stage (ATS) biomass-steam gasification system developed by EEEPL. The ATS biomass-steam gasifier system has several advantages over conventional systems. The syngas cleaning equipment and reactors are made of high-quality heat-resistant stainless-steel material, which ensures a long life of gasifier system. The ATS biomass-steam gasifier produces high-quality syngas without tar. Syngas without tar makes it possible to separate the hydrogen using VPSA equipment. Production of clean syngas enables a simplified syngas cleaning system and eliminates the baghouse filters/ wet scrubbers, etc. Due to the tar elimination, it reduces the maintaining cycle of the gas cleaning equipment significantly. Tar-free hydrogen-rich syngas from ATS gasifier system enables production of green hydrogen from biomass. The clean hydrogen-rich syngas is passed through a "vacuum pressure swing adsorption" (VPSA) equipment, which separates the hydrogen from syngas.

9.3.5 Process Flow and Concept of Green Hydrogen Production from Biomass

A process flow diagram of biomass to hydrogen production plant is given below with details of components. Biomass is fed to the ATS biomass-steam gasifier system using a semi-automatic fuel feeding system. The reactor of the ATS gasification system works in the range of 950–1100 °C. The gasifier reactor is provided with preheated oxygen-enriched air that is supplied to the gasification reactors. Steam is injected into ATS biomass-steam gasification reactor to produce hydrogen-rich syngas. The syngas through biomass-steam gasification produces syngas with a hydrogen content of about 50%. The syngas produced from ATS reactor is cleaned and cooled (to remove the fine dusts and moisture). A pressure swing adsorption (VPSA)-based hydrogen separation system will be used to separate the hydrogen from syngas. The residual gas is used to generate power using a 100% gas-driven IC engine. The power generator will be used to meet the power requirement of the ATS gasifier system and VPSA. A process flow diagram of the ATS biomass-steam gasifier system is presented in Fig. 9.6. The components of the ATS biomass-steam gasifier system are presented in Table 9.7.

Green hydrogen production through advanced biomass-steam gasification technology is the most efficient technology of hydrogen when compared to other technologies like hydrogen from power produced by PV. An electrolysis-based hydrogen production system will need about 55–60 kWh of electrical power. Biomass-steam gasification-based hydrogen production system needs about 10 kg of biomass to produce 1 kg of hydrogen. A biomass-steam gasifier system with a capacity of 500 kg of biomass consumption per hour can produce 1oo kg of green hydrogen per

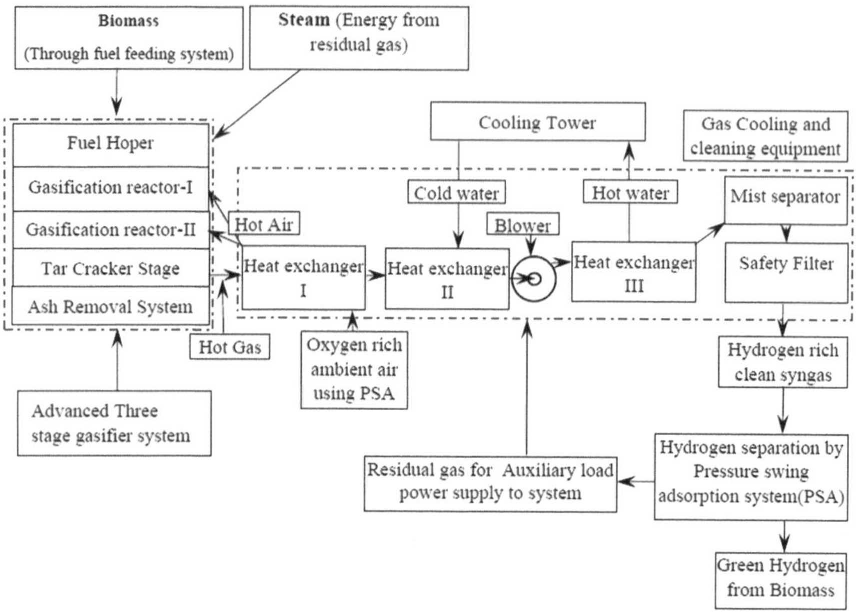

Fig. 9.6 A process flow diagram of green hydrogen production through biomass-steam gasification, along with components

Table 9.7 The key components of the biomass-based green hydrogen production plant

S. no	Name of equipment	Remarks/function
1.	Advanced three-stage biomass-steam gasification system with steam generator	A biomass-steam gasification system to produce hydrogen-rich clean syngas (without tar)
2.	A pressure swing adsorption (PSA) equipment for the separation of nitrogen and production of oxygen-rich air	The system will use oxygen-rich air for gasification to increase the hydrogen content in syngas. The auxiliary power required for this PSA will be generated from the residual gas
3.	An advanced ash removal system	An efficient ash removal system will be incorporated with the gasification reactor for high-quality syngas production
4.	Hot air generator cum gas cooling heat exchanger with cooling tower	Hot gas is cooled and the thermal energy of the hot gas is used to heat the air supplied for gasification
5.	Heat exchangers to cool the gas close to ambient temperature	Gas is cooled close to ambient temperature by indirect dry gas cooling and cleaning system
6.	Cooling towers	
7.	Gas cleaning equipment	Gas cleaning and mist separator and safety filters are part of the gas cleaning equipment to clean the gas suitable for hydrogen separation using PSA
8.	Fuel feeding system	A skip hoist to feed the biomass into the gasifier
9.	Pressure swing adsorption (PSA) equipment for hydrogen separation from syngas	PSA is used to separate hydrogen from syngas

(continued)

Table 9.7 (continued)

S. no	Name of equipment	Remarks/function
10.	A syngas-based electric power generator (25 kWe)	A producer gas engine will be used for power generation using the residual syngas (after separation of hydrogen for auxiliary power)
11.	High RPM blowers	High RPM blowers are used for air handling and gas handling in the gasification process
12.	Control panel	To control and operate several equipment in the system

hour. About 2.4 tonne per day. The cost of hydrogen production using biomass is about 2 $/kg, whereas hydrogen production using electricity will be costing about 5 $/kg.

9.3.6 Hydrogen Separation from Syngas

An equipment that works on vacuum pressure swing adsorption is used (VPSA) to separate hydrogen from syngas, which produces green hydrogen from syngas with a purity level of 99.9%. A VPSA equipment is used to separate hydrogen from syngas. The cost of VPSA equipment is the major expense of the project. The VPSA cost is almost 50% of the total equipment cost. VPSA is integrated with ATS biomass-steam gasification system to produce hydrogen from biomass. A process flow diagram of the hydrogen separation from the syngas produced from ATS biomass-steam gasifier system is shown in Fig. 9.7. The syngas with 50% of hydrogen content is passed through the VPSA. VPSA works at about 80% of efficiency. The residual syngas after separation of (80% of) hydrogen is suitable to generate power using an IC engine. Hence, the system will generate the power required to operate several components in the gasifier system and VPSA unit.

9.3.7 Vacuum Pressure Swing Adsorption (VPSA) for Hydrogen Separation from Syngas

Indian Institute of Sciencc, Bangalore, worked on a fixed bed downdraft biomass gasification system to produce hydrogen-rich syngas. Analysis of tar obtained from hydrogen-rich syngas generated from a fixed bed downdraft biomass gasification system is reported (Rakesh and Dasappa 2018). It is reported that the tar content in raw syngas is 268 mg/Nm^3. The presence of tar components in syngas after cleaning is about 37 mg/Nm^3. Cleaning of tar is a complicated process, and disposal of materials used to clean has environmental concerns. Tar content of 37 mg/Nm^3 may look like a small and negligible quantity, but in cumulative terms, the accumulation of tar will be very much high. A 10 kg/h hydrogen production system with a tar content of

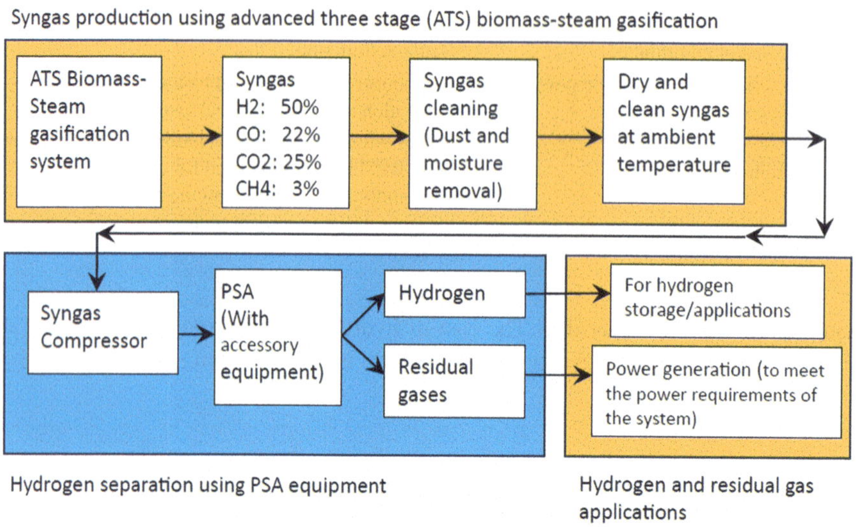

Fig. 9.7 A process flow diagram showing the syngas production, hydrogen separation from syngas, and supply of hydrogen to end-use applications

37 mg/Nm^3 can contribute to a cumulative accumulation of tar to the order of 222 g/d and 6.7 kg/month. The PSA equipment used for hydrogen separation cannot work with the tar loading rate of 6.7 kg/month. Challenges related to tar content in syngas is reported by Narnaware et al., from "Maharana Pratap University of Agriculture and Technology, Udaipur" (Narnaware and Panwar 2022). Hence, irrespective of the hydrogen content in syngas, hydrogen separation or hydrogen production from syngas becomes impossible. In the conventional biomass-steam gasification system, the presence of tar in syngas is a greater challenge for green hydrogen production. The proposed ATS biomass-steam gasification system produces hydrogen-rich syngas without tar making it possible to produce green hydrogen from biomass.

Hydrogen separation from syngas using vacuum pressure swing adsorption (VPSA) is reported by Prof. Dasappa's team, IISc, Bangalore (Singh et al. 2023). It is possible to produce hydrogen from syngas to a hydrogen purity level of 99.9%. Pressure swing adsorption-based hydrogen separation units are commercially available to separate the hydrogen from syngas.

From Japan, Umeki et al. (2010), in their study of wood-steam gasification, reported 55% of hydrogen content in syngas. However, the reported efficiency was 60%. Though the hydrogen content is 55, the tar content of the gasifier is 5200 mg/Nm^3. High tar content is a challenge to separate the hydrogen using PSA. It may be noted that the proposed ATS biomass-steam gasifier works at an efficiency of 85%, without tar with hydrogen content of >50% in the syngas.

From China, Zang et al. (2015) reported on biomass steam gasification. Syngas production with 49% hydrogen content is reported using CaO and MgO as catalyst to reduce tar in syngas. The proposed system is tar-free syngas; hence, it is more practical for the production of green hydrogen from biomass.

A study (from the Institute of Plasma Physics ASCR, Prague, Czech Republic, and Ghent University, Ghent, Belgium) of plasma gasification of refuse derived fuel (RDF) in a single-stage system using different gasifying agents is reported by Agon et al. (2016). An average of 350 mg of Nm3 of syngas is reported. Presence tar needs an elaborate tar cleaning system. Even the residual tar in the syngas will affect the PSA and presence of it is a great challenge to separate hydrogen from syngas. The plasma gasification system works with a cold gas efficiency of 70%, whereas the ATS gasifier works with a cold gas efficiency of 85%. The parasitic load of plasma gasification systems is very high. According to Materazzi et al., a plasma gasification system with a biomass gasifying capacity of 50 kg/h will be having an electrode load of 100 kW to create the plasma arc (Materazzi et al. 2015).

It is important to select a suitable pathway for the production of hydrogen. The JRC technical report of the European Commission, titled "Well-to-wheels analysis of future automotive fuels and powertrains in the European context," (Edwards et al. 2004) quotes "Turning usable electricity into hydrogen is unlikely to make sense from a global energy or GHG point of view." This is basically about creating a favorable GHG reduction in circumstances it would make sense to use them. Hence, apart from the production cost carbon footprint is also an important factor to be considered for the production of green hydrogen. As biomass is a carbon-neutral RE source, the production of hydrogen from biomass is a better alternative in the point of cost and GHG reduction.

Toyota has done research into developing hydrogen fuel cars using internal combustion engines. Several features and problems were presented by Toyota (YouTube n.d.).

The ATS biomass-steam gasification system makes hydrogen production from biomass a technologically and economically viable option.

9.4 Biomass Gasifier System for Thermal Applications

Biomass gasifier for thermal applications Energy Efficiency and Environment Pvt. ltd. New Delhi-110,077. Biomass gasification is an efficient process for converting the solid biomass fuel into gaseous fuel known as producer gas. The energy conversion efficiency of gasification process is 80–85%. Biomass is a carbon neutral fuel and does not contribute to GHG emission. Thermal gasifiers are designed to produce high-quality producer gas (without tar), the best technology among the existing gasifiers across the world. The design of the burners is very much important to achieve a maximum thermal efficiency of the system. Gasifiers for thermal applications can be used with capacity ranging from 2 kW_{th} to 1 MW_{th}. The burners work at 55–60% efficiency according to the applications and capacity. The calorific value of the producer gas is in the range of 1200–1300 kcal/m^3, depending on the CV and

Fig. 9.8 A view of the inner surface of the furnace with the view of the glowing insulation layer, when the furnace is above 1200 °C

moisture content of the biomass. The maximum furnace temperature can be reached up to 1200 °C. The furnace temperature can be lowered as per the need of the applications and processing temperature. A view of the producer gas-fired furnace when it was at more than 1200 °C is shown in Fig. 9.8. Thermal gasifiers can be used for power generation using steam turbines, industrial process heating (steel rerolling, ceramic industries), and hot air generation to use for drying applications (e.g., tea industries). MSW-based power generation is one of the potential applications to add value to waste (through steam turbine).

9.5 Value Addition to Biomass Waste Through Pyrolysis Process

A large amount of biomass waste generated from agricultural forming and municipal solid waste (MSW) goes unutilized. Agricultural residues and MSW are potential resource to produce value-added fuel using pyrolysis technology. A biomass pyrolizer operating at 500–550 °C is used to produce bio-oil and biochar from biomass. Bio-oil can be used in industrial furnace to replace furnace oil or diesel. The pyrolizer will be using processed biomass, which means the biomass will be dried and shredded to a smaller size (about 4–5 mm particles). The biomass to be dried to have the moisture content >10%. In the pyrolysis process, biomass is converted into gaseous fuel (20%), liquid fuel (bio-oil: 60%), and solid fuel (bio-char: 20%) (Akthar and Amin 2012; Tripathi et al. 2016). The components of a biomass pyrolizer system are presented in Table 9.8.

Table 9.8 Components of a biomass pyrolizer system

S. no	Name of the component
1.	Main pyrolizer unit 10 TPD. Consisting of furnace, pyrolizer with screw conveyor
2.	Biomass dryers/pretreatment components
3.	Pyrolizer gas cooling components/ condensers
4.	Bio-oil collecting tanks
5.	Biochar collecting tanks
6.	A thermal gasifier for pyrolizer furnace heating
7.	Cooling tower
8.	Reduction gears
9.	Timer switches and control panels
10.	Pipeline network

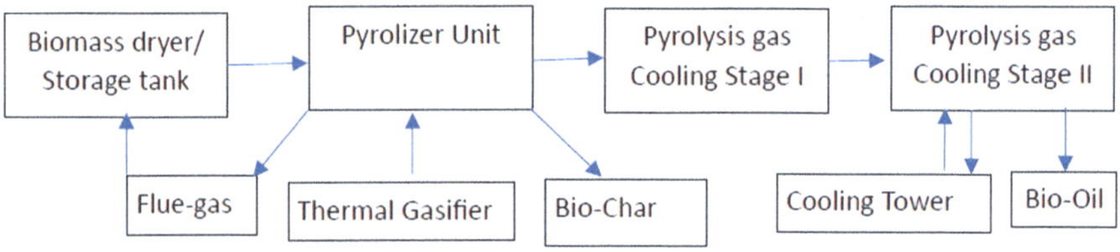

Fig. 9.9 A process flow showing the methodology and component A process flow showing the methodology and component

A process flow showing the methodology and component is shown in the following Fig. 9.9. The total area requirement for a 10 TPD pyrolizer plant would be about 8 × 6 m (48 sq. m). Processed and preheated biomass is fed into the pyrolizer. The pyrolizer is maintained at the required temperature of 550 C. A screw auger is used to convey the biomass across the pyrolizer. The screw conveyor is programmed to the required biomass retention time into the pyrolizer reactor. Biochar, combustible gas, and condensable oil vapors are formed during the pyrolysis process. The oil vapor is cooled by using condensers and a cooling tower. The combustible gas produced is used to heat the pyrolizer reactor. Biochar is cooled. Biochar obtained from pyrolizer can be converted as a value-added product by converting biochar into activated carbon.

9.6 Conclusions

Biomass wastes, like agricultural residues and MSW, are a potential resource for sustainable energy management. Unlike solar photovoltaic (PV) or windmill-based power generation, biomass power generation plant can work round the clock as per demand of the power requirement. Green hydrogen production through advanced biomass-steam gasification technology is the most efficient technology of hydrogen when compared to other technologies like hydrogen from power produced by PV or windmills. Hydrogen production through biomass- steam gasification is the lowest production cost when compared to other sources of green hydrogen production. Biomass gasifier system can be used to meet different types of thermal energy requirement of the industries. Biomass pyrolizer is the best technology to add value to biomass waste like MSW through the conversion of the MSW into bio-oil and biochar. In total, biomass could be a potential renewable energy source to reduce the use of fossil fuels and reduction in greenhouse gas mitigation.

Competing Interest Declaration The authors have no conflicts of interest to declare that are relevant to the content of this book.

References

Agon N, Hrabovský M, Chumak O, Hlína M, Kopecký V, Masláni A, Bosmans A, Helsen L, Skoblja S, Van Oost G, Vierendeels J (2016) Plasma gasification of refuse derived fuel in a single-stage system using different gasifying agents. Waste Manag 47:246–255

Akthar J, Amin NS (2012) A review on operating parameters for optimum liquid oil yield in biomass pyrolysis. Renew Sust Energ Rev 16:5101–5109

Edwards R, Mahieu V, Griesemann JC, Larivé JF, Rickeard DJ (2004) Well-to-wheels analysis of future automotive fuels and powertrains in the European context. SAE Trans:1072–1084

Materazzi M, LettieriP ML, Taylor R, Chapman C (2015) Reforming of tars and organic sulphur compounds in a plasma-assisted process for waste gasification. Fuel Process Technol 137:259–268

Narnaware SL, Panwar NL (2022) Biomass gasification for climate change mitigation and policy framework in India: a review. Bioresour Technol Rep 17:100892

Rakesh N, Dasappa S (2018) Analysis of tar obtained from hydrogen-rich syngas generated from a fixed bed downdraft biomass gasification system. Energy Convers Manag 167:134–146

Singh A, Shivapuji AM, Dasappa S (2023) VPSA process characterization for ISO quality green hydrogen generation using two practical multi-component biomass gasification feeds. Sep Purif Technol 315:123667

Tripathi M, Sahu JN, Ganesan P (2016) Effect of process parameters on production of biochar from biomass waste through pyrolysis: a review. Renew Sust Energ Rev 55:467–481

Umeki K, Yamamoto K, Namioka T, Yoshikawa K (2010) High temperature steam-only gasification of woody biomass. Appl Energy 87:791–798

YouTube (n.d.) Toyota's developing a hydrogen combustion engine! Available at: https://www.youtube.com/watch?v=3IPR50-soNA. Visited on 18 June 2024

Zang B, Zhang L, Yang Z, Yan Y, Pu G, Guo M (2015) Hydrogen-rich gas production from wet biomass steam gasification with CaO/MgO. Int J Hydrog Energy 40(29):8816–8823

Chapter 10
Hydrogen Energy: Innovation in Production, Storage, and Diverse Applications

Nazrul Hsan, Santosh Kumar, and Joonseok Koh

Abstract Hydrogen is emerging as a promising energy carrier in the global quest for sustainable and clean energy sources. This chapter provides a comprehensive overview of hydrogen energy sources, discussing their production methods, storage technologies, and various applications. Additionally, it highlights the current developments and future prospects of hydrogen as a key player in the global energy transition. The aim is to present a thorough analysis of hydrogen energy, addressing its potential benefits, challenges, and opportunities for achieving a sustainable and low-carbon future.

Keywords Hydrogen · Emerging energy · Storage technologies · Low carbon

10.1 Introduction

10.1.1 Background

Concerns about climate change and the depletion of fossil fuel reserves have driven the demand for clean and sustainable energy sources in recent years (Ji and Wang 2021). Hydrogen is a potential alternative energy source that can be utilized to address these issues. It is a versatile and abundant element that can be produced from various renewable sources such as water and biomass and is used in various applications, including electricity generation, transportation, and industrial

N. Hsan · J. Koh (✉)
Department of Materials Science and Engineering, Konkuk University, Seoul, Republic of Korea
e-mail: ccdjko@konkuk.ac.kr

S. Kumar
Department of Chemistry, School of Basic and Applied Sciences, Harcourt Butler Technical University, Kanpur, Uttar Pradesh, India

S. Kumar, V. K. Singh (eds.), *Renewable Energy Development: Technology, Material and Sustainability*, Clean Energy Production Technologies,
https://doi.org/10.1007/978-981-97-9626-7_10

processes (Hassan et al. 2023). Hydrogen is considered one of the most promising clean energy sources for the future, and its development has been a topic of interest for several decades (Hosseini and Wahid 2020). The British scientist Henry Cavendish was the first to identify hydrogen as a separate substance in 1766. Hydrogen is the lightest and most abundant element in the Universe. Because of its flammability, Cavendish dubbed it "inflammable air." Throughout the nineteenth and twentieth centuries, hydrogen was used in many industrial processes, such as the manufacture of ammonia for fertilizers, the refining of petroleum, and the use of hydrogen as a lifting gas in airships (Panda et al. 2023).

From midway in the twentieth century, interest in hydrogen as a fuel source increased, particularly when it was employed as a rocket fuel during the space race (Dawood et al. 2020). The use of hydrogen as a clean energy source has attracted considerable research interest. Several methods, such as electrolysis, steam methane reforming, and thermochemical processes can be used to produce hydrogen (Batgi and Dincer 2024). Over time, these techniques have improved in terms of efficiency and environmental impact. Because hydrogen does not exist in large quantities in nature, it is frequently used as an energy carrier rather than an energy source. It is a flexible energy vector owing to its effective energy storage and transportation capabilities. Hydrogen produced using carbon capture and storage (blue hydrogen) or renewable energy (green hydrogen) is regarded as from clean energy sources (Batgi and Dincer 2024). It emits only water vapor when used as a fuel, making it a desirable choice for lowering greenhouse gas emissions (Dufour et al. 2011).

The high cost of production, energy loss during conversion, and infrastructure requirements for storage and delivery are some of the challenges associated with hydrogen energy. Another major obstacle is guaranteeing an inexpensive and sustainable supply of hydrogen. Hydrogen fuel cells have attracted attention in the transportation industry, particularly for fuel cell vehicles (FCVs) (Aminudin et al. 2023). These vehicles use hydrogen to generate electricity and power electric motors with zero tailpipe emissions. Combined heat and power (CHP) systems, off-grid or distant power sources, and backup power for vital infrastructure are just a few of the stationary power production applications of hydrogen fuel cells (Ma et al. 2021). Numerous nations and regions have established initiatives to support hydrogen development. Examples include Japan's commitment to a "hydrogen society" and the European Union's Hydrogen Strategy.

Research is ongoing to identify new uses for hydrogen, such as in the chemical and steel sectors, and increase the efficiency of hydrogen generation, storage, and transportation. With increasing investments in infrastructure, research, and the formation of products and technologies related to hydrogen, the hydrogen market has expanded significantly. Hydrogen is one of the most important components in the shift toward a low-carbon and more sustainable energy system. Its development is ongoing, with a focus on green hydrogen generated from renewable energy sources. Overcoming adoption barriers is also becoming increasingly important.

10.1.2 Objectives

The goal of this chapter is to present a thorough overview of hydrogen energy sources and their evolution, an analysis of their applications across various industries, an evaluation of the opportunities and challenges that come with their widespread adoption, and an assessment of their potential to help create a future in which energy is more sustainable and low-carbon.

10.2 Hydrogen as an Energy Carrier

10.2.1 Properties of Hydrogen

Hydrogen is the most prevalent and lightest element in the Universe. It is an extremely flammable, colorless, and odorless gas (Yang et al. 2023). The essential characteristics of hydrogen are as follows:

Energy content: Hydrogen is a desirable fuel source because of its high energy content per unit mass. It releases a large amount of energy when burned or utilized in fuel cells (Stambouli 2011).

Versatility: Hydrogen has a wide range of applications, including in power production, industrial processes, and transportation. It can be used in fuel cells to produce energy or directly burned in internal combustion engines (Hwang et al. 2023).

Clean combustion: Hydrogen burns efficiently, leaving behind only water vapor. When utilized as fuel, it does not produce carbon dioxide, particulate matter, or other hazardous pollutants, making it a potential low-emission energy source (Roy and Pramanik 2023).

High diffusivity: Compared to other gases, hydrogen diffuses more quickly because of its low density. Certain transportation and containment issues are associated with this asset (Durbin and Malardier-Jugroot 2013).

Storage considerations: Compared to fossil fuels, hydrogen has a lower energy density per unit volume; hence, effective transport and storage methods must be developed (Singh et al. 2015).

10.2.2 Environmental Benefits of Hydrogen

When used as an energy source, hydrogen has the following environmental advantages:

Reduced emissions of greenhouse gases: Hydrogen can be a carbon-neutral or even carbon-negative fuel when produced from renewable sources through electrolysis using electricity from renewable energy sources, such as solar or wind power. By preventing the release of greenhouse gases, the procedure—known as "green hydrogen production"—helps to slow down global warming (Loreti et al. 2023).

Reduction of air pollution: Pollutants such as sulfur oxides (SOx) and nitrogen oxides (NOx), which are linked to air pollution and health problems are not produced during the combustion of hydrogen. Local air quality can be enhanced using hydrogen as fuel, particularly in places with a high population density.

Potential to decarbonize numerous industries: Hydrogen can significantly reduce carbon emissions from the transportation, industrial, and heating sectors. Carbon emissions have been significantly reduced in several areas by replacing fossil fuels with hydrogen as energy source (Oliveira et al. 2021).

10.2.3 The Role of Hydrogen in Energy Systems

Energy storage: Hydrogen has the potential to serve as a medium for storing energy, particularly from sporadic renewable energy sources such as solar and wind energy (Apostolou and Enevoldsen 2019). Rough electrolysis process and excess power produced during periods of low demand or high renewable energy output can be converted into hydrogen. When the supply of renewable energy is low, hydrogen can be stored for later use to produce power for various applications.

Fuel for transportation: Fuel cell electric vehicles can operate on hydrogen. Fuel cells provide a car with power by converting hydrogen and oxygen into electricity, with the only byproduct being water vapor (Manoharan et al. 2019). Cell electric vehicles (CEVs) are an alternative to traditional internal combustion engine vehicles owing to their extended driving range and quick refueling periods.

Industrial processes: Steel production, ammonia production, and petroleum refining are the few industrial processes in which hydrogen can be employed. Emissions can be significantly decreased by substituting fossil fuels with hydrogen.

Power generation: Hydrogen fuel cells can produce energy for stationary power applications. This can involve generating power on the grid scale or even supplying electricity to buildings or backup power systems (Hossain et al. 2023).

10.2.4 Hydrogen Economy Concept

According to the concept of the hydrogen economy, conventional fossil fuel-based energy systems will be supplemented or replaced by hydrogen as a major energy carrier in the future. The idea is to produce hydrogen using different energy sources, such as renewable energy, and use it in a variety of industries. Hydrogen is produced sustainably in a hydrogen economy using low-carbon techniques, such as electrolysis, driven by renewable electricity sources. Subsequently, hydrogen is delivered to end users via a system of pipes or specialized containers. The idea is to incorporate hydrogen into the power generation, transportation, and industrial sectors of the current energy infrastructure. This calls for widespread use of hydrogen fuel cell automobiles, the adoption of hydrogen to cut emissions in industrial operations, and the implementation of hydrogen-based power production technologies. The goals of the hydrogen economy concept are to reduce greenhouse gas emissions, improve energy security through source diversification, and move toward a more robust and sustainable energy system. However, there are several drawbacks, including the need to create infrastructure for distribution and storage, develop effective ways to produce hydrogen, and guarantee cost competitiveness compared to traditional energy sources.

10.3 Hydrogen Production Methods

10.3.1 Steam Methane Reforming

Hydrogen production via steam methane reforming (SMR) is the most popular technique. Carbon monoxide and hydrogen are produced when methane reacts with steam in the presence of a catalyst (McConnachie et al. 2023). The following equation can be used to illustrate this reaction:

$$CH_4 + H_2O \rightarrow CO + 3H_2$$

where CH_4, H_2O, CO, and H_2 stand for methane, steam, carbon monoxide, and hydrogen, respectively.

Fossil fuels, primarily natural gas, are the feedstocks for SMR. Although this is a well-established and economically feasible technology, it contributes to greenhouse gas emissions by producing carbon dioxide as a byproduct. Technologies for carbon capture and storage (CCS) can also be used to reduce emissions. Some advantages and challenges of SMR are summarized in Table 10.1.

10.3.2 Electrolysis

Water is split into hydrogen and oxygen through electrochemical reactions in a process known as electrolysis, which is powered by electricity (Wang et al. 2014). Alkaline electrolysis and proton exchange membrane (PEM) electrolysis are the two main forms of electrolysis. An alkaline solution, typically potassium hydroxide, is employed as the electrolyte in alkaline electrolysis. By contrast, PEM electrolysis uses a solid polymer membrane as the electrolyte. The electrolysis process, powered

Table 10.1 Summary of hydrogen production methods

Method	Description	Advantages	Challenges	References
Steam methane reforming (SMR)	Methane reacts with steam at high temperatures to produce H_2, CO, and CO_2	Established technology, high hydrogen yield, economically viable	CO_2 emissions, dependency on fossil fuels	McConnachie et al. (2023)
Electrolysis	Uses electricity to split water into hydrogen and oxygen. Types: PEM, alkaline, solid oxide	Can use renewable energy, no CO_2 emissions at point of use	High energy consumption, cost	Ursúa et al. (2012), Schmidt et al. (2017)
Biomass gasification	Converts organic materials into H_2, CO, and CO_2 through high-temperature processes	Utilizes renewable resources, can reduce waste	Technologically complex, requires efficient feedstock management	Tijmensen et al. (2002)
Photobiological and photoelectrochemical	Uses biological organisms or semiconductor materials to convert sunlight into hydrogen	Direct use of solar energy, potential for low-cost production	Low efficiency, early-stage technology	Kudo and Miseki (2009)
Thermochemical water splitting	Uses high temperatures and chemical reactions to split water into hydrogen and oxygen	High efficiency can be integrated with solar or nuclear heat sources	High temperatures required, complex technology	Kodama (2003)
High-temperature electrolysis	Electrolysis conducted at high temperatures, often using solid oxide electrolytes	Higher efficiency than low-temperature electrolysis, can use waste heat	High operational temperatures, material stability issues	He et al. (2023)
Solar-thermal water splitting	Uses concentrated solar power to achieve high temperatures needed for water splitting	Utilizes abundant solar energy, high potential efficiency	Requires large solar collectors, intermittent energy supply	Diver et al. (2008)
Nuclear-powered hydrogen production	Uses heat from nuclear reactors for thermochemical processes or high-temperature electrolysis	High efficiency, continuous power supply	Nuclear safety concerns, high capital costs	Forsberg (2003)

by energy derived from renewable sources such as solar or wind power, requires an electrical supply. Because electrolysis produces no greenhouse gas emissions during operation, it is a clean and effective technique for producing hydrogen (McHugh et al. 2020). Although research and technological developments are constantly enhancing its viability, electrolysis faces difficulties in terms of cost and scalability. The advantages and challenges of electrolysis method are summarized in Table 10.1.

10.3.3 Biomass Gasification

Biomass feedstocks, such as wood chips or agricultural waste, are transformed into syngas or synthetic gas during biomass gasification (Maitlo et al. 2022). Hydrogen, carbon monoxide, carbon dioxide, and trace amounts of other gases constitute the majority of syngas. Hydrogen can then be extracted from the syngas via additional processing. The gasification process usually involves heating biomass in a regulated setting with a restricted amount of air or oxygen (Sansaniwal et al. 2017). This procedure yields fuel gas with a high hydrogen content that can be refined and cleaned to remove contaminants. Because biomass gasification uses organic waste materials, it is considered a renewable technique for hydrogen production. Gasification reactors and gas clean-up systems must be carefully managed during this process, which may be complicated. The advantages and challenges of biomass gasification are summarized in Table 10.1.

10.3.4 Photobiological and Photoelectrochemical Processes

Sunlight is used in photobiological and photoelectrochemical processes to directly transform water into hydrogen (Prince and Kheshgi 2005; Saraswat et al. 2018). Certain microorganisms such as bacteria and algae are employed in photobiological processes to capture sunlight and perform photosynthesis. Consequently, the bacteria produce hydrogen. Subsequently, the hydrogen can be extracted and collected. Specialized materials, such as semiconductors, are employed in photoelectrochemical processes to collect solar radiation and accelerate the process of splitting water into hydrogen and oxygen. These substances function as photoelectrodes and are frequently coupled with catalysts to increase the reaction efficiency. One advantage of both the photobiological and photoelectrochemical processes is that their principal input is renewable solar energy. These processes are still in the research and development phase; therefore, there are issues with cost-effectiveness, scalability, and efficiency.

10.3.5 Other Hydrogen Production Methods

The following other approaches of producing hydrogen are being investigated and developed:

(i) *Thermochemical water splitting*: This process separates water into hydrogen and oxygen using high temperature and chemical processes.
(ii) *High-temperature electrolysis*: This method is more efficient than ordinary electrolysis because it operates at higher temperatures.
(iii) *Solar-thermal water splitting*: In this technique, high temperatures are created for the thermochemical water-splitting process by utilizing concentrated sun energy.
(iv) *Nuclear-powered hydrogen production*: Nuclear reactors can generate high-temperature steam, which can be employed in high-temperature electrolysis or SMR to produce hydrogen. The summary of thermochemical water splitting, high-temperature electrolysis, solar-thermal water splitting, and nuclear-powered hydrogen production is given in Table 10.1.

10.3.6 Comparison and Analysis of Production Methods

Several factors, including cost, availability of feedstock, scalability, environmental impact, and efficiency, should be considered when evaluating different hydrogen production techniques. Because natural gas feedstock is readily available and SMR is highly efficient, it is currently the most extensively used technology. However, this depends on fossil fuel and leads to carbon emissions. Although electrolysis can use renewable energy sources and is a clean technology, its current cost is higher than that of SMR. It is anticipated that the growing technological developments and falling costs of renewable energy sources will increase their competitiveness. Although biomass gasification offers a renewable method, it requires a consistent supply of biomass feedstock. Although this can reduce carbon emissions, careful handling is necessary. Also, although photobiological and photoelectrochemical processes are still in their infancy and have difficulties in terms of efficiency and scalability, they have the benefit of the use of solar energy. The efficiency of different hydrogen production methods is given in Fig. 10.1.

Additional research is required to maximize the efficiency, economy, and scalability of other approaches, such as thermochemical water splitting, high-temperature electrolysis, solar-thermal water splitting, and nuclear-powered hydrogen production. Currently, these approaches are in the developmental stages. In the end, in order to meet the increasing demand for hydrogen while guaranteeing cost competitiveness and minimizing environmental impacts, a combination of various hydrogen generation processes may be required. Several variables, such as

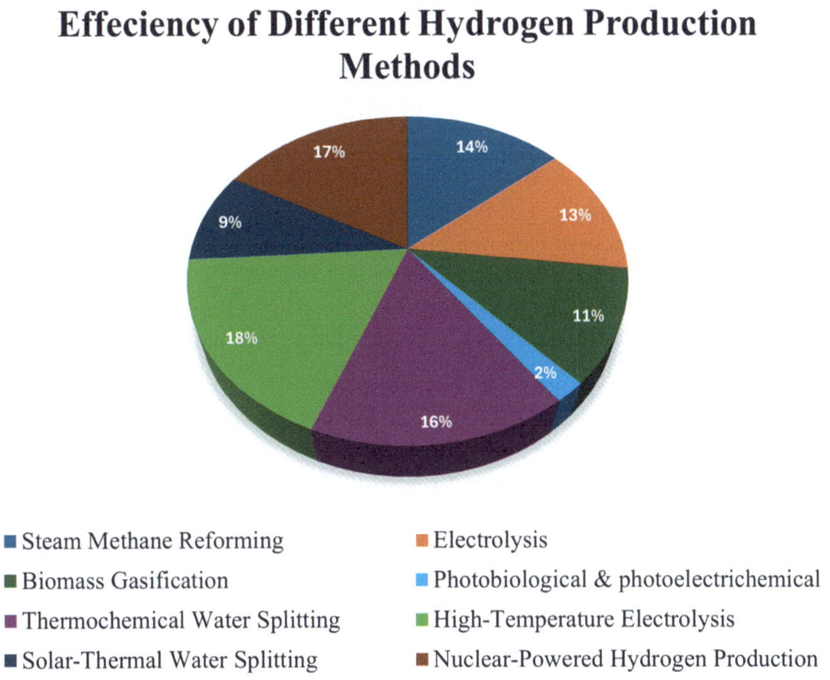

Fig. 10.1 The efficiency of different hydrogen production methods

policy concerns, the availability of energy, and local resources, influence the method of choice.

10.4 Hydrogen Storage Technologies

10.4.1 Gaseous Hydrogen Storage

Gaseous hydrogen storage refers to the storage of hydrogen in a gaseous state. Hydrogen gas must be stored in high-pressure containers at pressures between 5000 and 10,000 psi (pounds per square inch) (Cristello et al. 2023). To survive high pressures, these containers are usually composed of sturdy materials, such as metal alloys or carbon fiber composites. Gaseous hydrogen storage has the benefit of being comparatively inexpensive and having a quick fill and discharge time. However, its poor energy density, which requires a large volume to contain a sufficient amount of hydrogen, is a drawback. Furthermore, the high pressure required for storage may pose safety risks, necessitating the implementation of additional safety precautions.

10.4.2 Liquid Hydrogen Storage

To turn hydrogen gas into a liquid state, it must be cooled to an exceptionally low temperature (−253 °C, or −423.67 °F). Compared with gaseous hydrogen, liquid hydrogen has a higher energy density, enabling more compact storage (Rivard et al. 2019). Usually, it is stored at low temperatures in double-walled, vacuum-insulated tanks to reduce heat transfer. The higher energy density of liquid hydrogen allows for the storage of more hydrogen in a given container, which is one of its benefits. However, sophisticated cryogenic equipment is required to sustain low temperatures, which can be costly and energy-intensive. Furthermore, there is a chance of hydrogen boil-off, which occurs when hydrogen evaporates, and thus pressure relief devices become necessary.

10.4.3 Solid-State Hydrogen Storage

Solid-state hydrogen storage involves chemically fixing hydrogen to a solid substance, usually by adsorption or absorption. Metal hydrides, complex metal hydrides, carbon-based materials, and other porous materials are examples of solid-state materials used for hydrogen storage (Faye et al. 2022). Compared with gaseous or liquid storage, solid-state storage has the potential to provide higher hydrogen densities and increased safety. The creation of materials with appropriate thermodynamics and kinetics, effective reversible hydrogen uptake and release processes, and the capacity to store and release hydrogen under realistic circumstances are obstacles. The summary of gaseous, liquid, and solid-state hydrogen storage is given in Table 10.2.

10.4.4 Hydrogen Storage Challenges and Innovations

Several obstacles in hydrogen storage must be addressed, and engineers and researchers are working hard to find solutions. These difficulties include the following:

Energy density: Owing to its low energy density per unit volume, hydrogen must be stored in large quantities or under high pressures. To increase the efficiency of energy storage, innovations have been focused on creating materials and systems that can store hydrogen at higher densities.

Safety: Because hydrogen is extremely combustible, safe handling can be difficult, particularly at high pressures or freezing temperatures. Scholars are currently engaged in enhancing safety protocols and creating materials and containment systems that are resilient to spills and mishaps.

Table 10.2 Summary of hydrogen storage technologies

Aspect	Gaseous hydrogen storage	Liquid hydrogen storage	Solid-state hydrogen storage
Description	Hydrogen gas compressed at high pressures (typically 350–700 bar)	Hydrogen cooled to cryogenic temperatures (−253 °C) to become a liquid	Hydrogen stored in solid materials such as metal hydrides, chemical hydrides, or adsorbed onto porous materials
Storage method	Compressed gas cylinders or underground storage	Insulated cryogenic tanks	Metal hydrides, chemical hydrides, metal-organic frameworks (MOFs), carbon nanotubes (CNTs)
Energy density	~5.6 MJ/L (at 700 bar)	~8.5 MJ/L	Varies by material, generally higher than gaseous but lower than liquid
Pressure/ temperature	High pressure (350–700 bar) at ambient temperature	Cryogenic temperatures (−253 °C) at near ambient pressure	Ambient or moderate temperatures and pressures, depending on material
Volume efficiency	Lower volume efficiency, large tanks required	Higher volume efficiency but requires cryogenic conditions	High volumetric density, compact storage
Weight efficiency	Lightweight storage tanks (carbon fiber) but lower overall efficiency due to gas density	Heavier due to insulation and cooling system	Heavier storage systems, depending on material
Cost	Moderate cost, dependent on pressure vessel materials and compression energy	High cost due to cryogenic technology and energy for liquefaction	Varies, can be high due to material costs and processing
Energy requirements	High energy for compression	Very high energy for liquefaction and maintaining cryogenic temperatures	Moderate energy for absorption/desorption processes
Safety considerations	Risk of high-pressure leaks and explosions	Risk of rapid boil-off, cryogenic burns, and explosions	Generally safer at ambient conditions, but material-specific risks (e.g., reactivity, toxicity)
Infrastructure requirements	Requires high-pressure compressors, reinforced storage vessels, and distribution networks	Requires cryogenic storage tanks, liquefaction plants, and specialized transport	Requires specific storage materials and potentially integrated heat management systems
Applications	Fuel cell vehicles, industrial hydrogen supply, portable hydrogen storage	Long-distance hydrogen transport, space applications, large-scale storage	Fuel cell vehicles, portable electronics, backup power systems
Advantages	Established technology, straightforward storage and dispensing	High energy density, efficient for transport and large-scale storage	High volumetric density, potentially safer, and more efficient for certain applications

Cost: Technologies for hydrogen storage, including cryogenic systems and high-pressure tanks, are prohibitively expensive. Thus, universally accepted and affordable storage options are required.

Hydrogen release: For real-world applications, efficient and controlled release of hydrogen from storage materials is essential. The goal of this research is to create materials and systems that can store hydrogen and release it when needed.

Technological advancements in hydrogen storage are intended to address these issues. Advanced materials such as carbon nanotubes, metal-organic frameworks (MOFs) (Suh et al. 2012), and nanomaterials are a few examples (Chronopoulos et al. 2022; Gangu et al. 2019). New concepts for hydrogen storage include chemical hydrogen storage and transport. Furthermore, efforts are being made to enhance the overall viability, safety, and efficiency of hydrogen storage systems, with progress made in material engineering, catalysis, and system integration.

10.5 Applications of Hydrogen

10.5.1 Transportation

With many applications in the transportation industry, hydrogen is a viable way to reduce fossil fuel consumption and greenhouse gas emissions. Hydrogen is transported in several ways:

Hydrogen fuel cell vehicles: One significant way in which hydrogen is used in transportation is through hydrogen fuel cell vehicles (FCVs). Fuelled by hydrogen gas, FCVs produce energy by reacting hydrogen with oxygen to produce water. As water vapor is the only exhaust output from these vehicles, they provide zero-emission mobility. Compared with battery electric vehicles, FCVs offer longer driving ranges and require less time to refill, making them a viable option for long-distance travel.

Internal combustion engines with hydrogen: With some changes, hydrogen can be used as fuel in internal combustion engines (ICEs). ICEs fueled by hydrogen produce only water vapor and can potentially replace traditional petrol and diesel engines. Nevertheless, FCVs are more widespread than this technology.

Hydrogen as a fuel additive: To improve combustion efficiency and reduce emissions, hydrogen can be blended with traditional fossil fuels. It can be added to petrol or diesel engines to increase fuel efficiency, lower carbon emissions, and reduce pollutants including particulate matter and NOx.

Aviation and hydrogen: Hydrogen is being investigated as a potential aircraft fuel to reduce pollution. The advantage of hydrogen-powered aircraft is that they do not emit carbon dioxide; however, logistical and technological issues must be resolved before they are widely used.

Hydrogen in marine transportation: Boats and ships can operate using hydrogen. Electric propulsion systems can be powered by hydrogen fuel cells, offering a sustainable and effective alternative to traditional maritime propulsion systems.

Hydrogen for refueling stations: Hydrogen refueling stations are required to support vehicles running on hydrogen. Similar to regular petrol stations, these stations also deliver hydrogen to cars. To encourage the use of hydrogen-powered vehicles, a crucial first step is the construction of hydrogen refueling infrastructure. It is important to remember that obstacles to the broad use of hydrogen in transportation include the high cost of producing hydrogen and the scarcity of infrastructure for refueling, and there is the requirement for technological breakthroughs to boost productivity and cut expenses. Ongoing research and development initiatives have concentrated on resolving these issues and expanding the application of hydrogen in the transportation sector.

10.5.1.1 Hydrogen FCVs

Hydrogen FCVs are one way in which hydrogen is transported. These vehicles operate on hydrogen, which fuels fuel cell reactions that generate power. Vehicles that use hydrogen fuel cells offer several benefits including the following:

Zero emissions: Because water vapor is the only consequence of chemical reactions, hydrogen FCVs have zero tailpipe emissions. Therefore, they are desirable for reducing greenhouse gas emissions and air pollution.

Greater range: When compared with battery-electric vehicles, hydrogen FCVs often have a greater driving range. Thus, there is a greater capacity for energy storage because hydrogen may be maintained in high-pressure tanks.

Quick refueling: Filling a hydrogen fuel cell car requires approximately the same amount of time as filling a car with petrol or diesel. Filling a hydrogen tank takes a few minutes, but charging an electric car can take much longer.

Versatility: Vehicles powered by hydrogen fuel cells can be used in various settings, such as passenger cars, lorries, buses, and even trains. This provides adaptable options for various transportation requirements.

10.5.1.2 Hydrogen Infrastructure

The establishment of hydrogen infrastructure is essential for the mass use of hydrogen fuel cell automobiles. This entails establishing a network of facilities for producing, storing, and refueling hydrogen. The key uses of the hydrogen infrastructure include the following:

Production of hydrogen: Steam methane reforming, electrolysis, and biomass gasification are processes that can be used to produce hydrogen. To guarantee a steady supply of hydrogen fuel, both centralized and decentralized hydrogen production facilities must be established.

Storage of hydrogen: Typically, liquid or high-pressure tanks are used to store hydrogen. Refueling stations with access to hydrogen must have sufficient storage facilities.

Hydrogen refueling stations: Hydrogen refueling stations are required to provide a practical and accessible way to refuel hydrogen FCVs. To provide adequate coverage for car users, these stations should be strategically positioned along transportation routes.

Safety measures: Creating guidelines and regulations for the safe handling, storage, and transportation of hydrogen is essential. This entails resolving issues related to the flammability of hydrogen and ensuring that workers engaged in the hydrogen infrastructure receive appropriate training. It is feasible to overcome the restrictions imposed on hydrogen as a fuel and establish a sustainable ecology for hydrogen-powered vehicles by building strong hydrogen infrastructure.

10.5.2 Power Generation

Hydrogen has a multitude of applications in power generation, providing a viable means of producing sustainable and clean energy. The following are some uses of hydrogen in electricity production:

Hydrogen fuel cells: Fuel cells use an electrochemical mechanism to transform the chemical energy of hydrogen and oxygen into electrical energy. Hydrogen fuel cells can produce electricity for various applications, such as stationary power generation. Instead of conventional combustion-based power generation, they offer a clean and effective method of powering houses, buildings, and other establishments.

Combined heat and power (CHP) systems: Hydrogen can be employed in combined heat and power (CHP) systems, which are often referred to as cogeneration. Hydrogen can be used as a fuel in CHP systems to power combustion engines or fuel cells that produce energy. The overall energy efficiency can be increased by capturing waste heat generated during the production of electricity and using it for industrial processes, such as heating.

Hydrogen turbines: Hydrogen can be consumed directly in gas turbines to generate power. Similar to natural gas turbines, hydrogen gas turbines operate primarily on hydrogen. This application is particularly beneficial when hydrogen is produced in a carbon-neutral manner using renewable energy sources.

Energy storage: The integration of renewable energy sources into the grid is made possible by the use of hydrogen for energy storage. Through electrolysis, the surplus electricity produced by renewable energy sources, such as solar or wind power, can be converted into hydrogen. When the supply of renewable energy is limited, stored hydrogen can be used to generate electricity, acting as a counterbalance to intermittent energy sources.

Backup power systems: Hydrogen fuel cells can be used as backup power systems to supply dependable, emission-free electricity in isolated locations or during grid outages. Hospitals, data centers, and telecom buildings are examples of essential infrastructures that can benefit from fuel cell backup power systems, which provide a steady supply of electricity.

The production of hydrogen power generates clean energy with low greenhouse gas emissions because water vapor is the only consequence of hydrogen combustion in fuel cells. However, there are still issues such as the price of hydrogen production, storage, and distribution. Hydrogen is anticipated to become a more attractive alternative for power generation as long as hydrogen technologies continue to progress and renewable energy sources are deployed more widely.

10.5.3 Industrial Applications

The unique qualities and adaptability of hydrogen make it useful in a wide range of industrial areas. The following are a few examples of how hydrogen is used in the industry:

A. *Manufacturing of chemicals*: Hydrogen is an essential feedstock in the chemical industry. It is employed in several procedures, including hydrocracking, methanol synthesis, ammonia generation, and petroleum refining. Hydrogen is a necessary ingredient in the production of fertilizers, polymers, solvents, and other chemical products because it functions as a reducing agent and is important in a variety of chemical reactions.
B. *Hydrogenation*: The chemical process known as "hydrogenation" uses hydrogen to change unsaturated molecules into saturated ones. It is commonly used in the food sector to produce different edible fats, oils, and hydrogenated oils, such as margarine. The chemical and pharmaceutical sectors also employ hydrogenation to produce certain chemicals and alter the characteristics of existing materials.
C. *Metal production*: Several metallurgical processes utilize hydrogen. It is employed in the reduction process to create iron and steel from metal ores such as iron ore. To avoid oxidation and achieve particular material qualities, hydrogen can also be used as a protective environment during annealing and heat treatment procedures.
D. *Welding and metal cutting*: Hydrogen is used in industrial welding operations, such as atomic hydrogen welding for metal cutting and welding. This process combines metals using heat produced by atomic hydrogen recombination. Hydrogen is also used in plasma-cutting procedures, where it is used as a fuel gas to produce high-temperature plasma that is effective for cutting metal.
E. *Power generation*: Hydrogen is used in power generation, including industrial power generation, as previously indicated. Combustion engines and hydrogen

fuel cells are two options for industrial establishments seeking a clean and dependable power source for their operations.

F. *Hydrogen as a reducing agent*: Hydrogen is used as a reducing agent in several industrial processes. It can be used to remediate specific waste types such as chlorinated chemicals and to extract oxygen from metal oxides. Because of its reducing qualities, hydrogen can be used to clean and purify processes in the manufacturing of chemicals, electronics, and semiconductors.

It is crucial to remember that the majority of the hydrogen used in industrial settings comes from fossil fuels, which may increase carbon emissions. Ongoing endeavors are being made to devise sustainable and renewable techniques for hydrogen production, such as the use of renewable electricity sources for electrolysis. This supports the role of hydrogen in attaining sustainable industrial applications and reducing the carbon footprint associated with manufacturing. With its many advantages in terms of energy supply, heating, and storage, hydrogen also has potential applications in commercial and residential contexts. The following are examples of the applications of hydrogen in both private and business settings:

(i) *Heating and cooking*: In stoves and heating systems for homes and businesses, hydrogen can be used directly instead of natural gas. Appliances that operate on hydrogen, such as cooktops and boilers, offer dependable and effective heating and cooking choices. This makes it possible to decarbonize heating and cooking operations because the sole consequence of hydrogen combustion is water vapor.

(ii) *Combined heat and power (CHP) systems*: Hydrogen can be used in CHP systems in both residential and commercial buildings to produce heat and electricity simultaneously. The waste heat produced during the operation is collected and used for space heating, hot water, or other heating requirements. These systems generate electricity by using fuel cells or hydrogen-powered motors. CHP systems can reduce dependency on conventional power networks and increase energy efficiency.

(iii) *Microgrids*: Microgrids are small-scale electrical systems that function either independently or in conjunction with larger power grids. Hydrogen has the potential to contribute to this development. Microgrids can generate, store, and use hydrogen to balance intermittent renewable energy sources and support energy supply.

It is important to remember that widespread hydrogen adoption in residential and commercial settings is still in its infancy and faces obstacles such as the construction of necessary infrastructure, concerns about cost-effectiveness, and safety issues. Ongoing research and development initiatives have concentrated on resolving these issues and establishing hydrogen as a practical and sustainable energy source for both residential and commercial use.

10.6 Environmental and Economic Considerations

10.6.1 Carbon Neutrality and Climate Change Mitigation

Hydrogen generated from low-carbon or renewable sources can help achieve carbon neutrality and slow global warming. Green hydrogen is produced by electrolysis using renewable energy; therefore, it does not directly emit greenhouse gases. In many industries, green hydrogen can replace fossil fuels, lowering carbon dioxide emissions and reliance on fossil fuels. However, the carbon neutrality of hydrogen varies, depending on its production method. Currently, SMR is used to manufacture large amounts of hydrogen from natural gas, which emits CO_2. Hydrogen generation must switch to renewable energy sources to become carbon-neutral.

10.6.2 Economic Viability of Hydrogen Energy

The cost of hydrogen production, development of infrastructure, and technological improvements are some of the elements that determine whether hydrogen energy is economically viable. Currently, producing hydrogen from renewable resources is more expensive than traditional techniques. However, it is anticipated that the cost of producing green hydrogen will drop, increasing its economic competitiveness, as the cost of renewable energy sources declines and electrolyzer technology advances. Furthermore, research and development expenditures, economies of scale, and government policies promoting hydrogen energy are factors affecting its cost-effectiveness. Economies of scale and technological development are expected to reduce costs, making hydrogen a more economically feasible energy option as the sector matures and deployment grows.

10.6.3 Policy and Regulatory Frameworks

Efficient legislative and regulatory frameworks have significantly influenced hydrogen development and use. Globally, governments are implementing plans and regulations to encourage the use of hydrogen, including tax breaks, R&D support for research and development, and carbon pricing schemes. Expansion of infrastructures, consumer demand, and hydrogen production are the goals of these initiatives. To guarantee the safe handling, storage, and transportation of hydrogen, regulations are also important. Standards and regulations pertaining to hydrogen production, distribution, and usage have been established to guarantee the safe integration of hydrogen into current energy systems.

10.6.4 Social Acceptance and Public Perception

The adoption of hydrogen energy is influenced by various factors, including social acceptance and public perception. Public understanding of the advantages and possible drawbacks of hydrogen energy influences its acceptance. Campaigns for knowledge and education can assist allay worries and increase public confidence in the dependability and safety of hydrogen technology.

Involving stakeholders in the decision-making process ensures that their viewpoints and concerns are considered. These stakeholders include communities, businesses, and environmental organizations. Clear and open communication regarding the advantages of hydrogen energy for the environment, business prospects, and safety precautions can help increase societal acceptance and cultivate favorable public opinion. In general, the effective integration of hydrogen energy into energy systems and the achievement of a sustainable and low-carbon future depend on addressing environmental issues, guaranteeing economic feasibility, implementing supportive laws, and fostering social acceptability.

10.7 Current Developments and Case Studies

10.7.1 Hydrogen Infrastructure Development

The development of hydrogen infrastructure is a key component of the hydrogen economy. Infrastructures for production, storage, transportation, and distribution must be developed to facilitate the broad use of hydrogen technology. Here are a few noteworthy advancements:

Hydrogen refueling stations: To facilitate the use of FCVs, nations such as the United States, Japan, and Germany are developing national networks of hydrogen refueling stations. Firms are investing in infrastructure for hydrogen refueling, including Shell, Total, and Air Liquide.

Hydrogen pipelines: Several initiatives are being developed to create long-distance hydrogen transportation pipelines. Through a vast network of hydrogen refueling stations and pipelines, the European H2Haul project seeks to prove that using hydrogen for long-haul vehicles is feasible.

Port infrastructure: The use of hydrogen as a fuel for marine transportation is being explored. Ports are investigating the construction of hydrogen infrastructures to facilitate the use of hydrogen as a marine fuel. One such port is the Port of Rotterdam in the Netherlands.

10.7.2 Challenges and Barriers

Despite the advances, a number of obstacles and challenges still exist:

Cost: Compared to fossil fuels, the cost of producing hydrogen is still very high, especially when it comes from renewable sources. Additional research and development are required to reduce the price and increase the economic viability of hydrogen.

Infrastructure: One major obstacle is the construction of hydrogen infrastructure, which includes pipelines, storage facilities, and refueling stations. This necessitates extensive funding and cooperation among the stakeholders.

Scalability: Increasing the production and use of hydrogen to satisfy the needs of many industries is challenging. To facilitate broad adoption, the production capacity and supply chain logistics must be increased.

Safety: It is crucial to ensure that hydrogen is safely handled, stored, and transported. Regulatory frameworks and public safety issues must be addressed to increase the public confidence in hydrogen technology.

Policy and regulations: To encourage investment in hydrogen technology, policies, and regulatory frameworks must be dependable and supportive. Policies must be uniform and clear in order to promote innovation and expand markets.

Governments, businesses, and research organizations must work together to address these issues. Public acceptance, encouraging legislation, and ongoing technological developments are crucial for removing obstacles and accelerating the shift to a hydrogen-based economy.

10.8 Future Prospects and Outlook

10.8.1 Technological Advancements and Research Directions

It is anticipated that continuous research initiatives combined with technological developments will propel the future prospects of hydrogen. The crucial areas of attention are as follows:

A. *Electrolysis efficiency*: One of the main areas of research is the efficient separation of water into hydrogen and oxygen at a lower cost. This process is referred to as electrolysis. PEM electrolysis and solid oxide electrolysis are two examples of electrolyzer technologies that have advanced to improve durability and efficiency of electrolysis.

B. *Hydrogen storage*: New approaches to secure and effectively store hydrogen are the focus of current research. This covers the development of materials for storing hydrogen, such as metal hydrides and carbon-based compounds, and substitutes for hydrogen carriers, such as ammonia.

C. *Fuel cell technologies*: Fuel cell technologies are being improved, including the creation of more robust and efficient catalysts and fuel cell membranes. This can lower the cost and improve the performance of fuel cells, thereby increasing their appeal for a range of applications.
D. *Carbon capture and utilization*: Studies are being conducted to investigate the possibility of using hydrogen in processes related to carbon capture and utilization. Using hydrogen in these processes, greenhouse gas emissions can be decreased by converting the captured carbon dioxide into useful chemicals or synthetic fuels.

10.8.2 Scaling Up Hydrogen Production and Distribution

The infrastructure for production and distribution must be scaled up for hydrogen to reach its full potential. Prospects for future research include the following:

(i) *Renewable hydrogen*: Increasing the amount of hydrogen produced by expanding the use of renewable energy sources, such as wind and solar energy, will entail building large electrolysis plants and combining hydrogen production with renewable energy sources.
(ii) *Hydrogen pipelines and transport*: The development of specific hydrogen pipelines, shipping, and trucking are some of the ways in which the infrastructure for delivering and distributing hydrogen is expanding. This will encourage the broad use of hydrogen and ease its trading on regional and international scales.
(iii) *Localized production*: Developments in small-scale and decentralized hydrogen production technologies can enable localized hydrogen production by eliminating the need for a large transportation network and ensuring energy self-sufficiency in particular areas or industries.

10.8.3 Integration of Hydrogen into Renewable Energy Systems

Hydrogen incorporation into renewable energy systems has the potential to satisfy energy demands and balance intermittent renewable energy supplies. Prospects for future research include the following:

Power-to-gas: Power-to-gas is the process of electrolyzing excess renewable electricity to produce hydrogen for later use in various applications, such as transportation, heating, and power generation.

Grid management of renewable energy: Grid balancing services and the storage of excess renewable energy can be accomplished using hydrogen. To provide a steady and dependable energy system, hydrogen can be used to generate power at times of high demand or when the availability of renewable energy is limited.

Sector coupling: Hydrogen can facilitate sector coupling, which is the process of producing hydrogen from surplus renewable energy for use in industries such as transportation, heating, and manufacturing. This integration makes it possible to use renewable energy in various sectors and contributes to the development of a more sustainable energy system.

10.8.4 Potential Role of Hydrogen in Decentralized Energy Systems

Hydrogen can play a significant role in decentralized energy systems by offering the following benefits:

Energy independence: Communities, businesses, and isolated locations can become energy-independent by producing hydrogen through on-site electrolysis or local renewable energy sources, which will lower their dependency on centralized power grids.

Local energy storage: Hydrogen has the potential to function as a local energy storage medium, facilitating the assimilation of intermittent renewable energy sources into decentralized energy systems. Within localized grids, it can aid in balancing the supply and demand of energy.

Grid resilience: Fuel cells and backup power systems are examples of hydrogen infrastructures that can offer decentralized energy system resilience, guaranteeing a steady supply of electricity in the event of an emergency or grid outage.

10.8.5 International Collaboration and Partnerships

Partnerships and international cooperation are essential for the future prospects of hydrogen production. Cooperation between nations, academic institutions, and business stakeholders can hasten the exchange of knowledge and technological progress. International trade and application of hydrogen technology can be facilitated through cooperation in creating uniform standards, laws, and safety procedures. International cooperation can draw on investments, support the growth of a strong supply chain, and help establish a global market for hydrogen technologies. By exchanging best practices, lessons learned, and policy frameworks, nations can encourage the uptake of hydrogen technologies and facilitate the creation of favorable regulatory frameworks. The International Partnership for Hydrogen and Fuel Cells in the Economy and the Hydrogen Council are two international projects seeking to promote cooperation and hasten the global adoption of hydrogen technologies.

The future of hydrogen appears bright because of continued scientific improvements, expanding infrastructure for production and delivery, integrating hydrogen with renewable energy sources, and international cooperation. These initiatives are essential to achieve the full potential of hydrogen as a flexible and sustainable energy source.

10.9 Conclusion

10.9.1 Summary of Key Findings

Numerous important conclusions on the energy transition have been drawn from this research. The conclusions are as follows:

Growing need for renewable energy: Concerns about climate change, energy security, and sustainable development are driving the growing demand for renewable energy sources.

Technological Developments: Owing to their increased cost-effectiveness and efficiency, renewable energy sources, such as wind, solar, and energy storage, are becoming increasingly competitive compared with fossil fuels.

Governmental and regulatory support: Encouraging the deployment of renewable energy sources and propelling energy transition require strong and constant governmental support, including incentives and restrictions.

Decentralization and energy democracy: Energy transition is defined as a move toward decentralized energy systems that enable people to produce their own renewable energy and participate in the energy market.

Difficulties with integration: Intermittency, variability, the need for grid upgrading, and the infrastructure required for storage make the integration of renewable energy into current networks difficult.

Economic opportunities: Several economic opportunities are associated with energy transition, including the creation of jobs, investments in clean energy technologies, and the growth of new markets and industries.

10.9.2 Implications for Energy Transition

The following are some implications of the findings on energy transition:

Accelerated transition: In terms of combating climate change and maintaining energy security, the findings emphasize the necessity of prioritizing decarbonization of the energy industry and accelerating the transition to renewable energy sources.

Policy and regulatory support: To encourage the expansion of the markets for renewable energy governments and legislators should maintain steady and strong support in the form of laws, rules, and financial incentives to encourage the expansion of renewable energy markets.

Investment in research and development: Funding is required to continue advancing grid integration solutions, energy storage systems, and renewable energy technologies.

Infrastructure development: To enable the integration of renewable energy sources, investments must be made to modernize and expand the energy infrastructure. This includes the creation of smart grids and energy storage facilities.

International cooperation: To exchange best practices, encourage technology transfer, and establish an atmosphere favorable for global energy transition, international collaboration is essential.

10.9.3 Recommendations for Further Research and Action

The following are suggestions for additional study and action based on the findings:

Long-term energy planning: There is a need to create thorough plans for energy that consider the viability and long-term potential of renewable energy sources and how to integrate them into the current energy infrastructures.

System flexibility: Investigation and the creation of novel approaches, such as demand-side management and cutting-edge grid technologies, are required to improve the resilience and flexibility of energy systems and support larger proportions of renewable energy.

Community involvement: Community involvement in energy transition by supporting cooperative models, inclusive decision-making procedures, and community-led renewable energy projects should be encouraged.

Just transition strategies: There is a need to create and execute transition plans that consider the economic and social ramifications of the energy shift, guaranteeing a just and equitable allocation of advantages and aiding impacted areas.

Policy evaluation and learning: To develop future policy frameworks, it is important to continuously assess the efficacy of renewable energy policies and draw lessons from their successful implementation.

Cross-sectoral integration: To optimize the advantages of renewable energy deployment and create a more sustainable and effective energy system, there is a need to investigate possibilities for cross-sectoral integration between the energy, transportation, and other sectors.

Societies may take advantage of these opportunities and overcome the obstacles posed by energy transition by putting these suggestions into practice and acting proactively, ultimately leading to a more sustainable and cleaner energy future.

References

Aminudin MA, Kamarudin SK, Lim BH, Majilan EH, Masdar MS, Shaari N (2023) An overview: current progress on hydrogen fuel cell vehicles. Int J Hydrog Energy 48:4371–4388. https://doi.org/10.1016/j.ijhydene.2022.10.156

Apostolou D, Enevoldsen P (2019) The past, present and potential of hydrogen as a multifunctional storage application for wind power. Renew Sust Energ Rev 112:917–929. https://doi.org/10.1016/j.rser.2019.06.049

Batgi SU, Dincer I (2024) A study on comparative environmental impact assessment of thermochemical cycles and steam methane reforming processes for hydrogen production processes. Comput Chem Eng 180:108514. https://doi.org/10.1016/j.compchemeng.2023.108514

Chronopoulos DD, Saini H, Tantis I, Zbořil R, Jayaramulu K, Otyepka M (2022) Carbon nanotube based metal–organic framework hybrids from fundamentals toward applications. Small 18. https://doi.org/10.1002/smll.202104628

Cristello JB, Yang JM, Hugo R, Lee Y, Park SS (2023) Feasibility analysis of blending hydrogen into natural gas networks. Int J Hydrog Energy 48:17605–17629. https://doi.org/10.1016/j.ijhydene.2023.01.156

Dawood F, Anda M, Shafiullah GM (2020) Hydrogen production for energy: an overview. Int J Hydrog Energy 45:3847–3869. https://doi.org/10.1016/j.ijhydene.2019.12.059

Diver RB, Miller JE, Allendorf MD, Siegel NP, Hogan RE (2008) Solar thermochemical water-splitting ferrite-cycle heat engines. J Sol Energy Eng Trans ASME 130:0410011–0410018. https://doi.org/10.1115/1.2969781

Dufour J, Serrano DP, Gálvez JL, Moreno J, González A (2011) Hydrogen production from fossil fuels: life cycle assessment of technologies with low greenhouse gas emissions. Energy Fuels 25:2194–2202. https://doi.org/10.1021/ef200124d

Durbin DJ, Malardier-Jugroot C (2013) Review of hydrogen storage techniques for on board vehicle applications. Int J Hydrog Energy 38:14595–14617. https://doi.org/10.1016/j.ijhydene.2013.07.058

Faye O, Szpunar J, Eduok U (2022) A critical review on the current technologies for the generation, storage, and transportation of hydrogen. Int J Hydrog Energy 47:13771–13802. https://doi.org/10.1016/j.ijhydene.2022.02.112

Forsberg CW (2003) Hydrogen, nuclear energy, and the advanced high-temperature reactor. Int J Hydrog Energy 28:1073–1081. https://doi.org/10.1016/S0360-3199(02)00232-X

Gangu KK, Maddila S, Mukkamala SB, Jonnalagadda SB (2019) Characteristics of MOF, MWCNT and graphene containing materials for hydrogen storage: a review. J Energy Chem 30:132–144. https://doi.org/10.1016/j.jechem.2018.04.012

Hassan Q, Sameen AZ, Salman HM, Jaszczur M, Al-Jiboory AK (2023) Hydrogen energy future: advancements in storage technologies and implications for sustainability. J Energy Storage 72:108404. https://doi.org/10.1016/j.est.2023.108404

He S, Zou Y, Chen K, Jiang SP (2023) A critical review of key materials and issues in solid oxide cells. Interdiscip Mater 2:111–136. https://doi.org/10.1002/idm2.12068

Hossain MB, Islam MR, Muttaqi KM, Sutanto D, Agalgaonkar AP (2023) Advancement of fuel cells and electrolyzers technologies and their applications to renewable-rich power grids. J Energy Storage 62:106842. https://doi.org/10.1016/j.est.2023.106842

Hosseini SE, Wahid MA (2020) Hydrogen from solar energy, a clean energy carrier from a sustainable source of energy. Int J Energy Res 44:4110–4131. https://doi.org/10.1002/er.4930

Hwang J, Maharjan K, Cho HJ (2023) A review of hydrogen utilization in power generation and transportation sectors: achievements and future challenges. Int J Hydrog Energy 48:28629–28648. https://doi.org/10.1016/j.ijhydene.2023.04.024

Ji M, Wang J (2021) Review and comparison of various hydrogen production methods based on costs and life cycle impact assessment indicators. Int J Hydrog Energy 46:38612–38635. https://doi.org/10.1016/j.ijhydene.2021.09.142

Kodama T (2003) High-temperature solar chemistry for converting solar heat to chemical fuels. Prog Energy Combust Sci 29(6):567–597. https://doi.org/10.1016/S0360-1285(03)00059-5

Kudo A, Miseki Y (2009) Heterogeneous photocatalyst materials for water splitting. Chem Soc Rev 38:253–278. https://doi.org/10.1039/b800489g

Loreti G, Facci AL, Ubertini S (2023) Combining renewable sources towards negative carbon emission hydrogen. Int J Hydrog Energy 48:20875–20888. https://doi.org/10.1016/j.ijhydene.2023.03.177

Ma S, Lin M, Lin TE, Lan T, Liao X, Maréchal F et al (2021) Fuel cell-battery hybrid systems for mobility and off-grid applications: a review, renew. Sustain Energy Rev 135. https://doi.org/10.1016/j.rser.2020.110119

Maitlo G, Ali I, Mangi KH, Ali S, Maitlo HA, Unar IN, Pirzada AM (2022) Thermochemical conversion of biomass for syngas production: current status and future trends. Sustainability 14. https://doi.org/10.3390/su14052596

Manoharan Y, Hosseini SE, Butler B, Alzhahrani H, Senior BTF, Ashuri T, Krohn J (2019) Hydrogen fuel cell vehicles; current status and future prospect. Appl Sci 9(11):2296

McConnachie M, Konarova M, Smart S (2023) Literature review of the catalytic pyrolysis of methane for hydrogen and carbon production. Int J Hydrog Energy 48:25660–25682. https://doi.org/10.1016/j.ijhydene.2023.03.123

McHugh PJ, Stergiou AD, Symes MD (2020) Decoupled electrochemical water splitting: from fundamentals to applications. Adv Energy Mater 10:1–21. https://doi.org/10.1002/aenm.202002453

Oliveira AM, Beswick RR, Yan Y (2021) A green hydrogen economy for a renewable energy society. Curr Opin Chem Eng 33:100701. https://doi.org/10.1016/j.coche.2021.100701

Panda PK, Sahoo B, Ramakrishna S (2023) Hydrogen production, purification, storage, transportation, and their applications: a review. Energ Technol 11:1–11. https://doi.org/10.1002/ente.202201434

Prince RC, Kheshgi HS (2005) The photobiological production of hydrogen: potential efficiency and effectiveness as a renewable fuel. Crit Rev Microbiol 31:19–31. https://doi.org/10.1080/10408410590912961

Rivard E, Trudeau M, Zaghib K (2019) Hydrogen storage for mobility: a review. Materials (Basel) 12. https://doi.org/10.3390/ma12121973

Roy A, Pramanik S (2023) A review of the hydrogen fuel path to emission reduction in the surface transport industry. Int J Hydrog Energy 49:792–821. https://doi.org/10.1016/j.ijhydene.2023.07.010

Sansaniwal SK, Rosen MA, Tyagi SK (2017) Global challenges in the sustainable development of biomass gasification: an overview. Renew Sust Energ Rev 80:23–43. https://doi.org/10.1016/j.rser.2017.05.215

Saraswat SK, Rodene DD, Gupta RB (2018) Recent advancements in semiconductor materials for photoelectrochemical water splitting for hydrogen production using visible light. Renew Sust Energ Rev 89:228–248. https://doi.org/10.1016/j.rser.2018.03.063

Schmidt O, Gambhir A, Staffell I, Hawkes A, Nelson J, Few S (2017) Future cost and performance of water electrolysis: an expert elicitation study. Int J Hydrog Energy 42:30470–30492. https://doi.org/10.1016/j.ijhydene.2017.10.045

Singh S, Jain S, Ps V, Tiwari AK, Nouni MR, Pandey JK, Goel S (2015) Hydrogen: a sustainable fuel for future of the transport sector. Renew Sust Energ Rev 51:623–633. https://doi.org/10.1016/j.rser.2015.06.040

Stambouli AB (2011) Fuel cells: the expectations for an environmental-friendly and sustainable source of energy. Renew Sust Energ Rev 15:4507–4520. https://doi.org/10.1016/j.rser.2011.07.100

Suh MP, Park HJ, Prasad TK, Lim DW (2012) Hydrogen storage in metal–organic frameworks. Chem Rev 112(2):782–835

Tijmensen MJA, Faaij APC, Hamelinck CN, Van Hardeveld MRM (2002) Exploration of the possibilities for production of Fischer Tropsch liquids and power via biomass gasification. Biomass Bioenergy 23:129–152. https://doi.org/10.1016/S0961-9534(02)00037-5

Ursúa A, Gandía LM, Sanchis P (2012) Hydrogen production from water electrolysis: current status and future trends. Proc IEEE 100:410–426. https://doi.org/10.1109/JPROC.2011.2156750

Wang M, Wang Z, Gong X, Guo Z (2014) The intensification technologies to water electrolysis for hydrogen production – a review. Renew Sust Energ Rev 29:573–588. https://doi.org/10.1016/j.rser.2013.08.090

Yang N, Deng J, Wang C, Bai Z, Qu J (2023) High pressure hydrogen leakage diffusion: research progress. Int J Hydrog Energy 50:1029–1046. https://doi.org/10.1016/j.ijhydene.2023.08.221

Chapter 11
Hydrogen Horizons: Advancements, Applications, and Global Initiatives

Kruthika Eswaran

Abstract This book chapter offers an overview of the importance of hydrogen energy, especially green hydrogen, as a solution to the pressing need for transitioning to sustainable and renewable energy sources in the context of climate change. The chapter also highlights the properties and advantages of hydrogen as a clean energy carrier, explores its historical context and technological advancements, and delves into its diverse applications in different sectors. Additionally, the chapter discusses the global initiatives, strategies, and challenges associated with adopting hydrogen energy, emphasizing the importance of overcoming these challenges for successfully integrating hydrogen into the global renewable energy landscape.

Keywords Hydrogen energy · Renewable energy · Climate change · Global initiatives · Challenges

11.1 Introduction

Climate change is not just a "burning" (pun intended) issue; it's a crisis that demands our immediate attention. The World Economic Forum's "Global Risks Perception Survey 2023–2024" underscores this urgency, with five of the top ten global risks over a 10-year period related to the environment, including the consequences of climate change and the urgent need for climate mitigation and adaptation (World Economic Forum n.d.). This crisis necessitates a swift and effective transition to sustainable and renewable energy sources.

It is crucial to understand that the risks we face are not isolated but interconnected across multiple categories. Addressing the risks caused by climate change will not only affect the environmental risks category but also have a ripple effect on

K. Eswaran (✉)
Lead Sustainability and Climate Consultant, Net Zero Think Pvt. Ltd., Bengaluru, Karnataka, India

S. Kumar, V. K. Singh (eds.), *Renewable Energy Development: Technology, Material and Sustainability*, Clean Energy Production Technologies,
https://doi.org/10.1007/978-981-97-9626-7_11

other risk categories. This interconnectedness is further evident by the interconnections between various risks, as shown in Fig. 11.1, highlighting the complexity of the climate crisis.

With the world rapidly approaching the critical temperature rise of 1.5 °C, the urgency to transition to cleaner, more sustainable energy sources has never been more pressing. Hydrogen is gaining prominence as a versatile and environmentally friendly energy carrier in this quest. The increasing interest in renewable energy sources is sparking a new wave of energy revolution, driving the demand for green hydrogen. Hydrogen has long been commonly used in industries like refining and chemicals due to its abundance in the universe. However, introducing green hydrogen, produced by renewable-powered electrolysis, is a game-changer. Green hydrogen, a sustainable alternative to conventionally produced hydrogen, is a key player in decarbonizing various sectors, from transportation to heavy industry.

This chapter will delve into the global perspective of hydrogen and green hydrogen energy, exploring their significance, applications, and the burgeoning green hydrogen revolution. Our investigation will cover the strategies and initiatives adopted by leading nations, the technological advancements driving this transition, and the multifaceted applications of hydrogen in different sectors. However, it is crucial to address the global challenges that must be overcome, such as policies, infrastructure development, and societal acceptance. By the end of this chapter, we

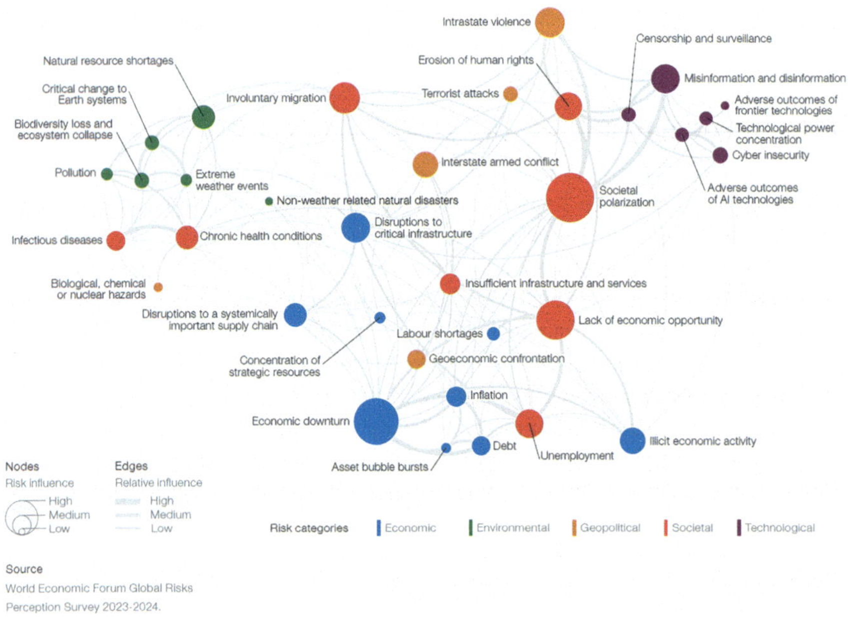

Fig. 11.1 Global risks landscape: an interconnections map. (Reproduced with permission from World Economic Forum n.d. Source: World Economic Forum Global Risks Perception Survey 2023–2024)

aim to provide a comprehensive understanding of how hydrogen energy, mainly green hydrogen, is shaping the renewable energy landscape worldwide and the importance of overcoming these challenges in its adoption.

11.2 Understanding Hydrogen

Hydrogen is an adaptable and environmentally friendly energy carrier, making it a key focus in the global change toward a more sustainable future. Hydrogen (H_2) is a simple and unique molecule with two hydrogen atoms covalently bonded, resulting in no net charge. This composition makes it an ideal energy carrier. When burned or reacted in fuel cells with oxygen, hydrogen produces energy with pure water (H_2O) as the only byproduct. This chemical reaction is free of harmful emissions. Hydrogen has a higher energy density per unit mass compared to traditional fossil fuels like gasoline. It takes 2.8 kilograms of petrol or 2.1 kilograms of natural gas to contain the same amount of energy as 1 kilogram of hydrogen. This means that hydrogen provides about three times more energy per unit of weight than traditional fuels, making it potentially useful for energy storage and transportation, especially in cases where weight and volume limitations are critical. Although hydrogen is crucial for decarbonizing various sectors, it is also extremely flammable, with a broad range of flammability at normal atmospheric pressure and low ignition energy. In addition to these disadvantages, problems arise due to the ease of leaking as a gaseous fuel and the high density as a liquid fuel, along with extremely low temperatures to maintain the liquid form, making it essential to consider safety guidelines in producing, storing, and transporting hydrogen.

Hydrogen has a rich and diverse history as an energy source, spanning centuries. Today, it is recognized as a renewable energy solution. Henry Cavendish first discovered hydrogen in the late eighteenth century, and it was initially used in ballooning and lighting. Its low density and high buoyancy made it an ideal choice for lifting gas in airships. However, the high flammability nature of hydrogen, low ignition energy, and leakage problems made it very dangerous, leading to historical events like the Hindenburg disaster. Hydrogen also played a crucial role in the mid-twentieth-century Space Race. It was used as rocket fuel, mainly as liquid hydrogen and liquid oxygen, to power the engines of rockets, enabling human spaceflight and space exploration. This era advanced hydrogen technology and showed its ability as a high-energy, clean-burning fuel. Figure 11.2 shows the use of liquid hydrogen as a rocket fuel during the Space Race.

Understanding the sources and eco-friendly hydrogen production methods is essential as the global energy landscape evolves. This involves examining the emerging hydrogen industry, identifying sources and production methods, and focusing on the shift toward greener hydrogen production technologies. By doing so, we can better understand the opportunities and challenges this emerging industry presents and its role in the shift toward a cleaner energy future.

Fig. 11.2 Using liquid hydrogen as a rocket fuel during the Space Race

1. Grey Hydrogen: For several years, hydrogen has been produced using conventional techniques such as steam methane reforming (SMR). The hydrogen generated by SMR, using natural gas as the source, is known as "grey hydrogen." The byproduct generated during production contains significant carbon dioxide (CO_2) directly released into the atmosphere. Notably, fossil fuels are used as feedstock in this process, which is both nonrenewable and environmentally unfriendly. Hence, there is a growing need to shift toward more sustainable production methods. Additionally, it is essential to note that during the production of 1 kg of hydrogen, 10 kg of CO_2 is released, which significantly impacts the environment.
2. Green Hydrogen: This variety of hydrogen is generated through renewable-powered electrolysis, which uses electricity from renewable sources like solar, wind, or hydropower to separate hydrogen from water. Green hydrogen is recognized as a carbon-free and sustainable energy solution with the potential to considerably decrease greenhouse gas emissions in the energy sector. Its production method aligns with the goals of achieving a cleaner, more sustainable energy landscape.

Between grey and green hydrogen, there is a "rainbow" of colors. This includes blue hydrogen, generated from SMR technology and carbon capture, and pink hydrogen, generated through nuclear power electrolysis (Table 11.1).

Table 11.1 The colors of hydrogen and hydrogen production (Técnicas Reunidas n.d.)

Colour	Technology	Energy Source	GHG Emissions
Brown or Black	Gasification	Black/Brown Coal (lignite)	High
Grey	Steam Gas Reforming	Natural Gas	Medium
Blue	Steam Gas Reforming & CCUS, Gasification & CCUS	Natural Gas, Coal	Low
Turquoise	Pyrolysis	Natural Gas	Solid Carbon (by-product)
Yellow	Electrolysis	Mixed-origin grid energy	Medium
Pink	Electrolysis	Nuclear	Minimal
Green	Electrolysis	Wind, Solar, Hydro, Geothermal, Biomass (in a few cases)	Minimal

11.3 Hydrogen Production Technologies

Considering the primary feedstocks used, hydrogen production technologies can be divided into three main categories:

1. Fossil-based: Fossil fuels are used as feedstock.
2. Water-based: Water is used as the feedstock.
3. Bio-based: Biomass is used as feedstock.

The production technologies, along with the different feedstocks and subcategories, are shown in Fig. 11.3.

This section will cover three primary methods for hydrogen production: SMR, coal gasification, and electrolysis using water as a feedstock.

11.3.1 SMR

Hydrogen is typically produced using the steam methane reforming (SMR) process. The process has high efficiency and low operational cost. The most frequently used feedstocks are natural gas, methanol, and other hydrocarbons. In this energy-intensive process, methane reacts with steam in the presence of a catalyst under 3–25 bar pressure. This reaction produces carbon monoxide and hydrogen, which then undergo a water-gas shift reaction to convert the carbon monoxide to carbon dioxide and additional hydrogen. After removing impurities through pressure-swing

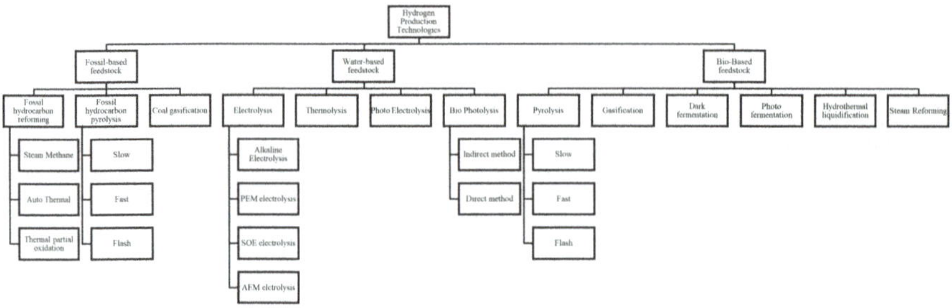

Fig. 11.3 The different hydrogen production technologies. (Reproduced with permission from Megía et al. 2021)

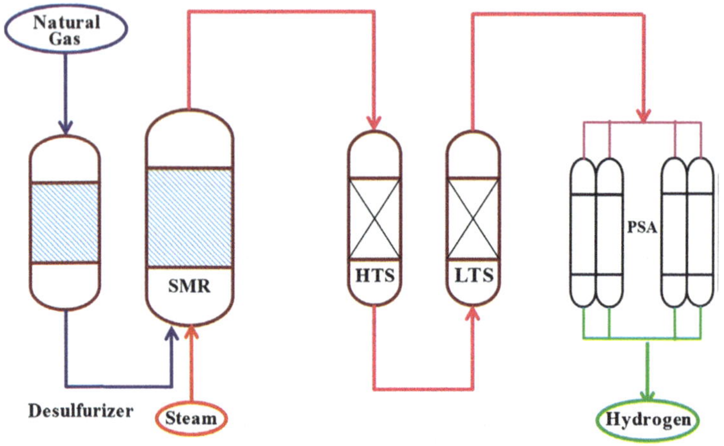

Fig. 11.4 Process diagram of the SMR method of hydrogen production. (Reproduced with permission from Xu et al. 2023)

adsorption, pure hydrogen is obtained. However, a significant amount of carbon dioxide is produced, which contributes to environmental degradation (Hydrogen Production: Natural Gas Reforming n.d.; Xu et al. 2023). Figure 11.4 shows a simplified flow diagram of the SMR production of hydrogen.

There are four basic steps in the SMR process. To begin, natural gas containing sulfur must undergo desulfurization to prevent damage to the equipment and the catalyst used in the next step. After desulfurization, the natural gas, primarily methane (CH_4), is combined with steam (H_2O) and passed through furnaces loaded with catalysts to create a gas mixture consisting of hydrogen, carbon monoxide (CO), carbon dioxide (CO_2), and unconverted methane. The steam reforming process requires external energy input for preheating the feedstock, as the reactions are highly endothermic. These reactions are typically conducted under 1.4–2 MPa pressure, utilizing high-efficiency nickel-based catalysts at temperatures ranging from 800 to 1000 °C. In the third step, in the presence of catalysts, carbon monoxide

(CO) reacts with steam to produce CO_2 and additional H_2. To speed up the hydrogen production, this step involves a high-temperature shift (HTS) followed by a low-temperature shift (LTS) due to the exothermic nature of the water-gas shift reaction. The temperatures used are 350 °C for the high-temperature shift and 190–210 °C for the low-temperature shift. Finally, to obtain high-purity H2 (~99%), the effluent gas, post water-gas shifting reaction, undergoes a physical separation process known as pressure swing adsorption (PSA) to eliminate significant amounts of CO, CO_2, and CH_4 (Xu et al. 2023).

11.3.2 Coal Gasification

Coal is an abundant and low-cost fossil fuel used for producing chemicals, fuels, and other products. In the hydrogen industry, coal gasification is especially crucial for large-scale manufacturing plants, contributing to 18% of global hydrogen production. Figure 11.5 shows the process flow diagram of coal gasification process (Xu et al. 2023).

Coal gasification is a process in which pulverized coal is combined with oxygen and steam in a special gasifier at high temperature and moderate pressure. This results in a gas mixture containing carbon monoxide, hydrogen, and small amounts of carbon dioxide, methane, and other components. To obtain high-purity H2, additional processes such as water-gas shift and purification are used. Carbon capture and storage (CCS) technologies are potential solutions to mitigate heavy carbon emissions from coal gasification plants. Coal gasification can be categorized into three main types based on the gasifier used: fluidized bed, fixed bed, and entrained flow gasifiers (Xu et al. 2023).

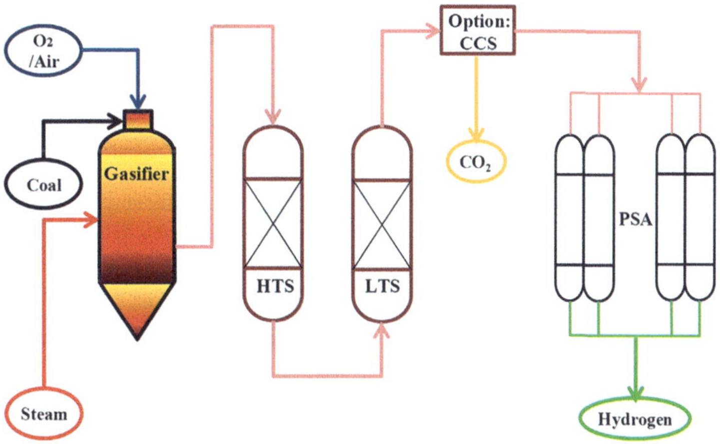

Fig. 11.5 Process flow diagram of coal gasification process. (Reproduced with permission from Xu et al. 2023)

11.3.3 Electrolysis

With the need for a sustainable and energy-efficient future, there has been a flurry of activities and searches for green and clean solutions in the past few years. While hydrogen has always been a clean energy carrier, the type of process in its production makes hydrogen "dirty" (fossil-fuel feedstock) or "green" (water feedstock and renewable-fueled electrolysis).

To produce green hydrogen, an electrolyzer powered by renewable electricity and deionized water is used in the electrolysis process. Depending on the type of electrolyzer technology used, it can take 45–52 kWh of renewable electricity and 9 kgs of deionized water to produce 1 kg of green hydrogen (IRENA 2020).

During electrolysis, water is split into hydrogen and oxygen as shown in Fig. 11.6 (Seddon 2022). After being captured and stored, hydrogen gas is utilized for various applications. The electrolysis process can be classified into four distinct types: alkaline electrolysis, polymer electrolyte membrane (PEM) electrolysis, solid oxide electrolysis (SOE), and anion exchange membrane (AEM) electrolysis. Table 11.2 summarizes the technical aspects of the four processes (IRENA 2020; Seddon 2022; Bhandari et al. 2013; Miller et al. 2020).

Green hydrogen production through electrolysis only leaves behind water vapor, making it a sustainable alternative to fossil fuels. Its potential as a clean energy source has sparked growing interest in using it to lower greenhouse gas emissions in hard-to-abate sectors, like heavy industry.

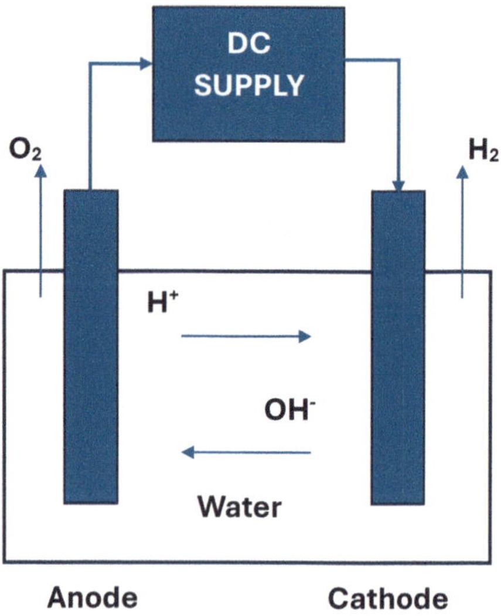

Fig. 11.6 Water electrolysis process for producing hydrogen

Table 11.2 Summary of the technical aspects of the four electrolysis processes

Parameter	Unit	Alkaline electrolysis	PEM electrolysis	SOE electrolysis	AEM electrolysis
Electrolyte	–	Aqueous potassium hydroxide (KOH)	Proton exchange ionomer	Zirconium doped yttrium oxide (Y_2O_3)	Anion exchange ionomer
Cathode	–	Nickel, nickel molybdenum	Platinum, palladium	Nickel-zirconium dioxide	Nickel, nickel alloys
Anode	–	Nickel, nickel cobalt alloys	Ruthenium oxide, iridium oxide	Lanthanum strontium manganate	Nickel, iron, cobalt oxides
Catalyst	–	NaOH, NaCl	Ir, Pt	Pt, Pt/C, BSCF (barium-strontium-cobalt-ferro)	Ni, Ni-Fe alloys, C (graphene)
Current density (current travelling per unit cross-section area)	A/cm^2	0.2–0.4	0.6–2.0	0.3–1.0	0.2–1.0
Cell voltage	V	1.8–2.4	1.8–2.2	0.95–1.3	1.8–2.2
Operating temperature	°C	60–80	50–80	700–800	50–60
Operating pressure	Bar	1–30	30–76	1–30	1–30
Hydrogen production rate for 5 MWe	Nm^3/h	-200	-200	-200	-200
Hydrogen purity	Vol%	>99.5	>99.9999	>99.99	>99.99
Stack lifetime	Hours	60–100 k	20–60 k	<40 k	<3 k
Water requirement	Liter/kg of H_2	9–10.8	18.04	9.1	8.8
Power requirement	kWh/kg of H_2	50.5–74.1	54.6	36.14	53.9–58.4
Availability		Commercially available, 5 MWe and above			

11.4 Global Hydrogen Strategies and Initiatives

The possible benefits of green hydrogen as a clean energy carrier have led 41 governments, representing almost 80% of global energy-related CO_2 emissions, to develop their hydrogen (green and blue) strategies and initiatives. Table 11.3 summarizes some of the significant hydrogen initiatives taking place worldwide (IEA 2023).

Recognizing the global nature of the hydrogen economy, countries have increasingly participated in international collaborations and agreements to expedite the

Table 11.3 Summary of the significant hydrogen initiatives taking place worldwide

Country	Overview
Japan	The "Basic Hydrogen Strategy," announced by Japan in 2017, outlines a roadmap for achieving a hydrogen-based society. Key elements include hydrogen infrastructure development, fuel cell vehicle (FCV) promotion, and international collaboration (Basic Hydrogen Strategy – 経済産業省 2023). The Global Hydrogen IEA report (IEA 2023) mentions that Japan aims to reach a hydrogen demand of 12 million tonnes (Mt) by 2040 and have 15 GW of electrolysis installed worldwide. Moreover, the focus has moved from FCVs to using hydrogen in industries such as steel and petrochemicals
S. Korea	The Korean government, via the hydrogen economy roadmap launched in 2019, has set ambitious goals of producing 5.26 Mt./yr. of hydrogen for KRW 3000/kg by 2040. In February 2021, the Ministry of Trade, Industry, and Energy introduced the Hydrogen Economy Promotion and Hydrogen Safety Management Act to support the advancement of hydrogen-related activities. This act encompasses provisions for certifying, developing, and aiding hydrogen-specialized businesses and ensuring safety in the management of fuel cells, water electrolyzers, and hydrogen facilities. Additionally, the act emphasizes the importance of installing hydrogen stations and fuel cells to encourage the use of hydrogen in the transportation sector (ICDK n.d.). The new target is to achieve 2.1% and 7.1% of total electricity production using hydrogen and ammonia by 2030 and 2036, respectively (IEA 2023)
India	India launched the National Green Hydrogen Mission (NGHM) in January 2023 to pursue a clean energy transition and achieve zero emissions by 2070. The mission aims to establish India as the global hub for producing, utilizing, and exporting green hydrogen and its derivatives. By 2030, India plans to produce 5 Mt. of green hydrogen and add 125 GW of renewable capacity. The goal is to capture about 10% of the global market share by 2030 by exporting approximately 10 Mt. of green hydrogen/green ammonia annually. The Strategic Interventions for Green Hydrogen Transition (SIGHT) program is a significant financial initiative under the National Green Hydrogen Mission, with a budget of ₹17,490 crore. Under the central government's provision, the SIGHT program allocates ₹4440 crores for domestic electrolyzer manufacturing (Component-1) and ₹13,050 crores for green hydrogen production (Component-2) (National Green Hydrogen Mission | Ministry of New and Renewable Energy | India n.d.)
Australia	Australia set ambitious goals in its National Hydrogen Strategy, leveraging its abundant renewable resources (Australia's National Hydrogen Strategy n.d.). The strategy focuses on three key elements: 1. Hydrogen export industry: Australia aims to be a significant hydrogen exporter, especially to Asian countries 2. Renewable hydrogen production: The main emphasis is on producing renewable hydrogen on a large scale using solar and wind energy resources 3. Domestic Utilization: The strategy includes integrating hydrogen into domestic energy systems, transport, and industrial applications. Australia aims to establish a global-leading clean hydrogen industry by 2030. The strategy includes 57 actions to scale-up the industry rapidly, review the legal framework, and ensure safety. In July 2020, Australia adopted eight international standards for safe hydrogen use, transport, and trade. The 2030 hydrogen goals and initiatives for each state can be found in (State of Hydrogen 2022)

(continued)

Table 11.3 (continued)

Country	Overview
Chile	The National Green Hydrogen Strategy was published in November 2020 by the Ministry of Energy to diversify the country's energy sources, boost local industries, and position Chile in the global market. The goal is to produce the most affordable green hydrogen by 2030 as part of achieving carbon neutrality by 2050. The strategy outlines plans to tap into Chile's vast renewable energy potential of over 1800 GW and invest \$5 billion to construct a green hydrogen production capacity of 200,000 tons/year and an electrolysis capacity of 5 GW by 2025. It also aims to promote the export industry for green hydrogen (H2V) and its derivatives, dividing the strategy into three stages From 2020 to 2025, the first stage aims to achieve widespread domestic consumption by targeting various sectors, including oil refineries, green ammonia production, mining trucks, heavy-duty vehicles, and integrating hydrogen into gas grids. The second stage, from 2025 to 2030, aims to expand the use of green hydrogen in the transport sector and begin exporting it. Lastly, the third stage, after 2030, concentrates on decarbonizing heavy-duty transport, specifically maritime and aviation (National Green Hydrogen Strategy Chile n.d.)
The United States	The United States aims to produce 10 Mt. of "clean" hydrogen by 2030, 20 Mt. by 2040, and 50 Mt. by 2050 as per the Clean Hydrogen Strategy and Roadmap (U.S. National Clean Hydrogen Strategy and Roadmap n.d.). The Inflation Reduction Act (IRS 2024) has allocated USD 369 billion in incentives and introduced 10-year tax credits for clean hydrogen projects. The law allows producers to receive up to \$3.00 per kilogram of hydrogen. Additionally, the Bipartisan Infrastructure Law (Energy.gov 2021) has invested USD 9.5 billion in clean hydrogen projects over 5 years, with a maximum of \$7 billion for regional hydrogen hubs (H2Hubs). These hubs, located in carbon-producing centers, will cut approximately 25 million metric tons of CO2 from end-uses annually and produce 3 million metric tons of hydrogen annually. They will also support clean hydrogen production, storage, delivery, and end-use while reducing emissions from hard-to-decarbonize industrial sectors. Seven regional hubs were selected for the BIL funding in October 2023 (Energy.gov n.d.-a)
Germany	Germany, which plays a vital role in the European Union's energy transition, introduced its hydrogen strategy in 2020. The strategy plans to utilize hydrogen as a means of decarbonizing various sectors, with a particular emphasis on three key areas—expansion of electrolysis capacity, industrial integration where hydrogen will be integrated into heavy industries to reduce carbon emissions, and research and innovation (National Hydrogen Strategy Update n.d.). The goals also include achieving a hydrogen production capacity of 5 GW by 2030, and an additional 5 GW by 2035, with the latest target being 2040 (Basic Hydrogen Strategy – 経済産業省 2023). Germany is currently aiming to install 10 GW of electrolyzers by 2030 and to increase demand in industry and transport, aiming to replace larger shares of natural gas with over 1800 km of pipeline infrastructure (IEA 2023)

(continued)

Table 11.3 (continued)

Country	Overview
The European Union (EU)	The EU's hydrogen strategy, adopted in 2020, outlined policy actions in five key areas: investment support, production and demand support, creation of a hydrogen market and infrastructure, research and cooperation, and international cooperation. The strategy aims to install a minimum of 6 GW of renewable hydrogen electrolyzers by 2024 and 40 GW by 2030. In March 2022, the REPowerEU initiative was introduced with the goal of ending Europe's reliance on Russian fossil fuels by 2030. The initiative also aims to build 10 Mt./yr. of renewable hydrogen production capacity in the EU and through imports by the same year. The European Commission has approved two "Important Projects of Common European Interest (IPCEI)" competitions, namely, "Hy2Tech" and "HyUSE." These competitions will receive significant public funding of up to €5.4 billion and €5.2 billion, respectively, to support a wide range of hydrogen-related projects. The Green Deal Industrial Plan, launched in February 2023, includes the creation of the EU Hydrogen Bank. This bank aims to provide subsidies for renewable energy-based hydrogen production over the course of 10 years. The plan, along with the revised Renewable Energy Directive, sets out goals for hydrogen technologies. This includes achieving a 42% share of green hydrogen in industries by 2030, demonstrating the EU's dedication to integrating sustainable hydrogen across various sectors (European Commission 2024)

development, production, and utilization of hydrogen. These collaborations aim to foster knowledge exchange and technology transfer and establish a robust global hydrogen market. Some of the initiatives are shown below Table 11.4.

11.5 Applications Across Sectors

The twenty-first century has witnessed a paradigm shift in how we perceive and harness hydrogen energy. Hydrogen's versatility across various applications has placed it at the forefront of renewable energy development. As the world deals with the urgent need to reduce greenhouse gas emissions and shift toward more sustainable energy systems, hydrogen, particularly green hydrogen, has emerged as a beacon of hope in striving for a cleaner and greener future. This has resulted in a skewed view of hydrogen being the "one" solution for decarbonization across all sectors—*The Swiss Army Knife* (Fig. 11.7).

One must remember the following: like the Swiss Army Knife, hydrogen could be a potential decarbonization solution in some cases; however, there are also better, more efficient, and cost-effective alternatives. In 2023, Liebreich Associates updated their well-known "Clean Hydrogen Ladder," demonstrating different ways hydrogen can be used in a particular order of importance (Liebreich n.d.-b). The graphic as shown in Fig. 11.8 illustrates the important role that clean hydrogen will play in reaching a net-zero future. It indicates the transition from the current use of grey or polluting hydrogen to better available alternatives like direct electrification and batteries. Additionally, it aims to consider various factors that will impact the

Table 11.4 International collaborations on clean hydrogen

International collaboration/agreement	Objective/purpose	Key initiatives	References
Hydrogen Council's global collaboration	Facilitate global cooperation and advocacy for hydrogen integration in the energy transition	Knowledge exchange among 150+ member companies from various industries	About the Council | Hydrogen Council (2023)
European clean hydrogen Alliance	The alliance promotes investments and clean hydrogen production to accelerate industry decarbonization in line with the EU's climate change objectives	Cross-border collaboration among European countries. Public-private partnerships for research and innovation	European Clean Hydrogen Alliance(n.d.)
Mission innovation clean hydrogen Mission	The mission is a global initiative involving 23 countries and the European Commission. It aims to develop the clean hydrogen economy and reduce the cost of clean hydrogen to \$2/kg by 2030	The Mission's activities focus on: 1. Promote research, development, and innovation in clean hydrogen production and applications 2. Demonstrate production, storage, and transportation methods with stakeholders to create concentrated demand centers 3. Identify efforts to create demand and deploy solutions, facilitating knowledge dissemination and engaging stakeholders	Mission Innovation (n.d.)
Australia-Germany hydrogen accord	Strengthen collaboration between Australia and Germany in advancing the development of a global renewable hydrogen industry	1. HyGATE (hydrogen innovation and technology incubator) is a project that aims to support real-world initiatives throughout the hydrogen supply chain 2. Cooperation between industries on demonstration projects in Australian hydrogen hubs 3. Trade of Australian hydrogen and its derivatives produced from renewable energy sources, including through H2Global	DCCEEW (n.d.)

(continued)

Table 11.4 (continued)

International collaboration/ agreement	Objective/purpose	Key initiatives	References
US-EU-Japan hydrogen and fuel cells agreement	Foster joint efforts in advancing hydrogen and fuel cell technologies	1. Applying technologies and aligning regulations, codes, and standards 2. Encouraging information sharing and international collaboration in research and development to ensure the safety of hydrogen and its supply chain infrastructure 3. Researching and assessing hydrogen's potential across various sectors, including its ability to reduce CO2 and other emissions 4. Communicating, educating, and reaching out to increase awareness about the benefits of hydrogen	F & F (2019)

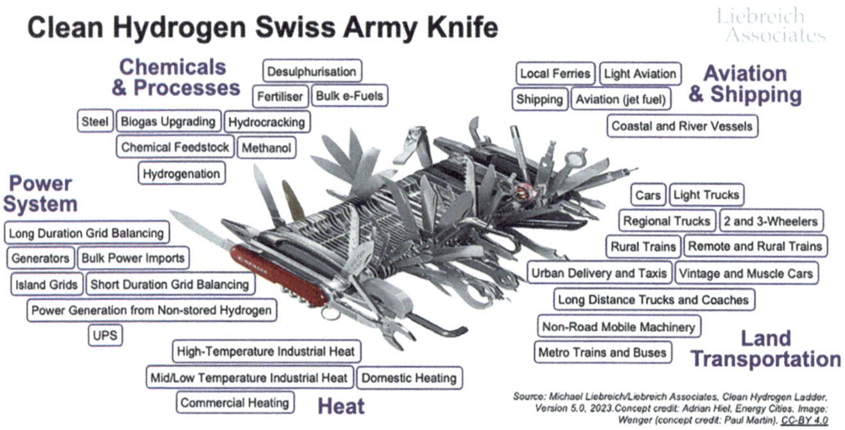

Fig. 11.7 Hydrogen as the *Swiss Army Knife* solution for decarbonization across all sectors. (Reproduced with permission from Liebreich n.d.-a)

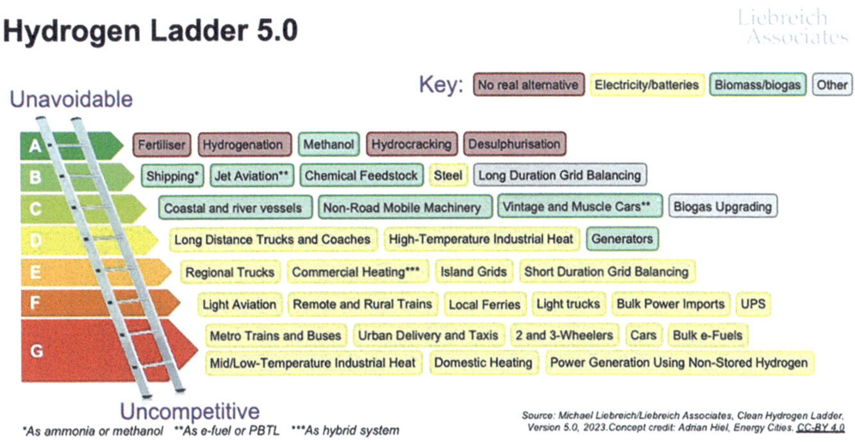

Fig. 11.8 Use Cases of Clean Hydrogen in an order of merit

success or failure of a clean hydrogen solution, such as thermodynamics, economics, safety, human behavior, resilience, and geopolitics.

The above diagram is just one concept that shows the importance of accepting hydrogen as a solution only on a case-by-case basis since usually there is a cheaper, safer, and more efficient alternative present. Regional differences will be based on policy framework, resource availability, technological advancements, cost competitiveness, infrastructure development, and policy support. Similar studies have discussed the potential end-use cases for clean hydrogen based on maturity and necessity (IRENA 2022; CSTEP n.d.; Whitehead et al. 2023; Kukreja and Narain 2024). The following Table 11.5 shows a comprehensive list of applications for hydrogen across different sectors, along with possible alternatives.

Overall, the end-use applications for hydrogen can be classified into three types:

1. Sectors with limited alternatives and high decarbonization potential:

   - Heavy-duty transport: Trucks, ships, and airplanes. Batteries struggle with range and weight limitations in these applications, making hydrogen a prime contender for achieving carbon neutrality. E-fuels and biofuels also are potential solutions.
   - Industrial processes: Processes like steel and cement rely heavily on fossil fuels and have limited electrification options. Hydrogen can enable clean heat and chemical reactions in these industries.
   - Long-duration energy storage: Hydrogen can store excess renewable energy for periods longer than batteries, providing grid stability and flexibility.
   - Remote off-grid applications: Hydrogen fuel cells or stored hydrogen offer a clean and reliable solution for power generation in areas without access to centralized grids or where there are no renewable sources of energy to generate electricity.

Table 11.5 Hydrogen end-use applications across different sectors along with potential alternatives

Sector	Application	Maturity	Priority	Competitiveness	Alternatives	Prioritization?	Advisable?	References
Industrial	Feedstock replacement (ammonia, methanol)	Moderate	High	Moderate (cost, infrastructure)	Fossil fuels (natural gas, coal)	Yes, crucial for decarbonization	Yes	The Future of Hydrogen – Analysis – IEA (n.d.), Artrocker (n.d.)
	Direct heating processes (glass, metals)	Early	High	Moderate (cost, technology)	Furnace oil, natural gas	Yes, for high-temperature applications	Yes, with research and development	Hydrogen Production I Hydrogen Program (n.d.)
	Hydrocracking	Early	Moderate	Low (technology, cost)	Fossil fuels	Yes (reduced reliance on fossils)	With research development	IEA (2023)
	Hydrotreating	Moderate	Moderate	Moderate (cost, infrastructure)	Conventional methods	Yes (cleaner emissions)	With infrastructure development	Hydrogen Production I Hydrogen Program (n.d.)
	Hydrogenation of fats and oils	Early	Moderate	Low (technology, cost)	Fossil-derived chemicals	Yes (renewable chemicals)	With research pilot projects	Hydrogen Europe Home (2024)
	Pulp and paper production	Early	Moderate	Low (technology, cost)	Chlorine bleaching	Yes (cleaner process)	With research policy support	Hydrogen (n.d.)
	Food processing	Early	Moderate	Low (technology, cost)	Conventional methods	Yes (food stabilization, flavor)	With research market adoption	Ammonia Technology Roadmap – Analysis – IEA (n.d.)

Power generation	Long-term energy storage	Early	High	Moderate (cost, infrastructure)	Pumped hydro, batteries	Yes, for grid flexibility and resilience	Yes, with policy support, infrastructure development	IEA (2023)
	Peak power generation	Moderate	Moderate	Moderate (cost, infrastructure)	Natural gas-fired peakers	Yes (grid flexibility)	With policy support, infrastructure development	IEA (2023)
	Microgrids	Early	Moderate	Low (technology, cost)	Diesel generators, renewables	Yes (remote areas, off-grid)	With research pilot projects	Clean Air Task Force (2023)
Buildings	Direct heating (individual homes)	Early	Moderate	Low (cost, infrastructure)	Heat pumps, electric boilers,	Not yet, unless policy incentives exist	No, currently uneconomical	IEA (2023, Hydrogen Europe Home (2024)
	Blending with natural gas (existing infrastructure)	Moderate	Moderate	Moderate (cost, infrastructure)	Existing natural gas infrastructure	Yes, as a near-term decarbonization solution	Yes, with infrastructure optimization policy support	International Council on Clean Transportation's "Global FCEV Market Guide 2023" (2023), REUTERS (n.d.)
	Cooking	Early	Moderate	Low (technology, cost)	Natural gas stoves, electric cooktops	Yes (clean alternative)	With research policy support	Publications Office of the EU (2016), Fuel Cell & Hydrogen Energy Association (n.d.)
	Combined heat and power (CHP)	Early	Moderate	Low (technology, cost)	Conventional boilers generators	Yes (energy efficiency)	With research pilot projects	IEA DHC (n.d.)

(continued)

Table 11.5 (continued)

Sector	Application	Maturity	Priority	Competitiveness	Alternatives	Prioritization?	Advisable?	References
Transportation	Light-duty vehicles (FCEVs)	Moderate	Moderate	Low (cost, range, infrastructure)	Battery electric vehicles (BEVs)	Not yet; BEVs currently dominate	No, unless the cost infrastructure improves significantly	EERE: Alternative Fuels Data Center Home Page (n.d.), Energy.gov (n.d.-b), CHFCC (n.d.), Clean Trucks and Buses (n.d.)
	Heavy-duty vehicles (trucks, ships, airplanes)	Early	High	Low (technology, infrastructure)	Diesel, batteries (limited range)	Yes, for the decarbonization of hard-to-electrify sectors	Yes, with significant research development	Hydrogen Council's, (2023), Home | Clean Aviation (n.d.)
	Forklifts industrial vehicles	Moderate	Moderate	Moderate (cost, infrastructure)	Battery electric vehicles, LPG	Yes (material handling, warehouses)	With infrastructure development policy support	IEA (2023), Fuel Cell & Hydrogen Energy Association (n.d.), Hydrogen Council's, (2023), IEA (n.d.-a)
	Trains	Moderate	Moderate	Moderate (cost, infrastructure)	Diesel locomotives	Yes (alternative to diesel)	With infrastructure investment policy support	IEA (2023), International Council on Clean Transportation's "Global FCEV Market Guide 2023" (2023), Enking (2023), IEA (n.d.-b)
	Motorcycles and scooters	Early	Low	Low (technology, cost, infrastructure)	Battery electric vehicles	Not yet, unless significant advancements in policy support		Hydrogen Europe Home (2024), Hydrogen Newsletter (2023)

Portable power	Backup power and off-grid applications	Moderate	Moderate	Moderate (cost infrastructure)	Batteries, diesel generators	Yes, for specific applications requiring high energy density	Yes, with cost technology improvements	Fuel Cell & Hydrogen Energy Association (n.d.)
District heating	Heating networks (urban areas)	Early	Moderate	Low (cost, infrastructure)	Heat pumps, existing district heating systems	Not yet, unless significant infrastructure investment is made	No, it currently requires high upfront costs	IEA DHC (n.d.)
Other	Hydrogenation in steel production	Early	Moderate	Low (technology, infrastructure)	Coke, charcoal	Yes, for long-term carbon-neutral steel production	Yes, with research pilot projects	Hydrogen Europe Home (2024)
	Maritime industry (auxiliary power)	Early	Moderate	Low (technology, cost, infrastructure)	Diesel generators	Yes (cleaner operations)	With research pilot projects	Clean Shipping Coalition (2023), Fages et al. (2023)

2. Sectors where alternatives for hydrogen exist, but the decision depends on specific factors:
   - Building heating: While heat pumps are generally preferred for building electrification, hydrogen can be a viable option in areas with limited renewable resources or existing natural gas infrastructure. However, this depends on the economic viability of refitting the infrastructure to accommodate hydrogen.
   - Aviation: Batteries and sustainable fuels are also being developed for shorter and medium-distance flights. Hydrogen is more likely to be relevant for long-haul routes.
   - Maritime transport: Electric or hybrid solutions might be feasible for smaller vessels and shorter journeys. Hydrogen becomes more attractive for larger ships and longer voyages. Transporting hydrogen is usually in the form of ammonia or methanol. E-fuels and biofuels are also potential solutions.
   - Hydrogen can be used as a chemical feedstock in refineries and fertilizers, but its usefulness compared to other options depends on its cost, technological maturity, and infrastructure. Green hydrogen is a cleaner alternative for these industries, especially as an immediate solution, but its wider adoption depends on further developments and cost reductions. Alternatives like direct electrification and bio-based options could be good choices depending on the specific application and evolving landscape. Ultimately, the optimal solution will depend on the specific circumstances.
3. Sectors where alternatives currently offer a better solution for decarbonization:
   - Light-duty vehicles: Battery electric vehicles (BEVs) already offer a mature and cost-effective solution for personal transportation, exceeding hydrogen fuel cell electric vehicles (FCEVs) in most aspects.
   - Passenger trains: Electrification using overhead lines or battery-powered locomotives has lower infrastructure costs, is more efficient, and is generally the preferred option for reducing carbon emissions from train travel.
   - Most household appliances: Appliances like heating systems, cooking stoves, and water heaters can efficiently be electrified without needing hydrogen.

It is important to carefully analyze each application and consider existing alternatives and their limitations to ensure the effective and efficient deployment of hydrogen for decarbonization (IEA 2023; About the Council | Hydrogen Council 2023; Fuel Cell & Hydrogen Energy Association n.d.). Overall, prioritizing hydrogen depends on a complex interplay of factors: government policies, technological maturity, cost, infrastructure compatibility, and sector-specific challenges. For example, government incentives and regulations can create demand for clean hydrogen and encourage investment in the infrastructure needed to produce and use it. Technological advancements can also make producing, storing, and using hydrogen more efficient and affordable.

Hydrogen is a crucial solution for clean energy that can help combat pollution in various industries. Although the cost of producing green hydrogen is currently higher than traditional fossil fuels, the cost gap is quickly narrowing. Investing in research and development and providing targeted policy support can make hydrogen, especially green hydrogen, more competitive. This can be achieved by building a solid network of hydrogen production facilities, storage tanks, and refueling stations, which can accelerate its adoption and help create a cleaner and more sustainable future.

11.6 The Green Hydrogen Revolution

Green hydrogen was just a whisper in the wind a few years ago. Today, it is quickly evolving into a deafening roar, promising to reshape the global energy landscape and propel us toward a cleaner future. The current trend focuses on producing green hydrogen. This is created using electrolysis with renewable sources like solar or wind power without relying on fossil fuels or their related emissions. Green hydrogen is crucial due to its flexibility and adaptability, offering a diverse solution to global decarbonization.

Green hydrogen represents a critical piece of the decarbonization puzzle, particularly in sectors where direct electrification proves challenging. In heavy-duty transport, like trucks and ships, where battery limitations hinder range and efficiency, hydrogen fuel cells emerge as a powerful alternative, propelling vehicles without tailpipe emissions (Hydrogen Council's 2023). It plays a crucial role in industrial processes like steel and cement production, providing high-temperature heat and chemical reactions required for these often carbon-intensive sectors (GCCA Roadmap 2022; Swennenhuis et al. 2022). Additionally, green hydrogen is a potential alternative to grey hydrogen as a chemical feedstock in refineries and the fertilizer industry. Green hydrogen also stands as a potent long-term energy storage solution, capable of capturing excess renewable energy from wind and solar during peak production and feeding it back into the grid during periods of high demand, ensuring seamless grid stability and increased reliance on clean energy sources (IEA 2023).

Green hydrogen connects plentiful renewable energy sources with a variety of energy sectors that require clean and sustainable fuel. It enables us to capture and store excess renewable energy, overcoming its time-based limitations and making it available as a reliable energy source when needed. This unlocks the potential for wider use of renewable energy across sectors, speeding up the shift from fossil fuels to a genuinely sustainable energy future.

The green hydrogen revolution is gaining momentum, but there are still challenges to overcome. To compete with established fossil fuel options, reducing costs by advancing technology and increasing production capacity is essential. Developing a solid infrastructure, which includes deploying electrolyzers and building hydrogen transportation networks, is essential for large-scale adoption. Policy support through carbon pricing, subsidies, and research and development investments is critical to ensure a level playing field and accelerate the transition. This hydrogen journey will require the coordinated efforts of governments, industries, and research communities. However, the potential rewards, including a cleaner planet and reduced dependence on fossil fuels, make the green hydrogen revolution a significant undertaking worth embracing wholeheartedly.

11.7 Challenges and Opportunities in Adopting Green Hydrogen

The widespread adoption of green hydrogen demands a multifaceted approach, navigating the intricate terrain of infrastructure development, technological advancements, economic considerations, and social acceptance.

To establish a strong global supply chain, critical challenges such as transporting and distributing hydrogen over long distances must be addressed. This requires the development of pipelines, shipping vessels, and innovative transport methods, such as high-pressure pipelines (Sampson 2023). Additionally, efficient storage solutions are equally crucial and may involve exploring various options like underground salt caverns and innovative materials such as metal hydrides (IRENA n.d.). Encouraging public-private partnerships and innovative financing models is necessary to obtain the significant capital required for infrastructure development and technology deployment.

Technological advancements are driving the green hydrogen revolution forward. Researchers are working on new techniques and materials to increase electrolysis efficiency, leading to more hydrogen production per unit of electricity. They are also developing advanced catalysts and materials for electrolysis and other production processes to reduce costs and minimize the environmental impact (Lee et al. 2023). Integrating artificial intelligence and automation in hydrogen facilities improves production, monitoring, and safety. It is crucial to maintain investments in research and development to scale and reduce costs for making green hydrogen as cost-competitive as fossil fuels.

The green hydrogen revolution is not just about the positive impact on the environment but also a catalyst for socio-economic growth. The global green hydrogen market is expected to reach trillions of dollars in the coming decades, attracting significant investments and creating millions of new jobs across various sectors such as manufacturing, research and development, operation, and maintenance. This transition promotes sustainable economic growth by reducing reliance on fossil fuels, attracting renewable energy investments, and encouraging innovation in green technology.

Governments have a major role in shaping the future of green hydrogen by implementing supportive policies and regulations. National hydrogen strategies with incentives such as carbon pricing, subsidies, and feed-in tariffs encourage green hydrogen production and consumption. According to the World Bank report on policy frameworks for hydrogen, fostering international collaborations and standardized regulations can enable seamless hydrogen trade across borders while ensuring the safety and interoperability of infrastructure. Addressing challenges in policy implementation, such as harmonizing regulations across countries and mitigating potential job losses in fossil fuel industries, requires careful consideration and equitable solutions (World Bank 2022).

The adoption of green hydrogen faces not only technical hurdles but also social ones. Technical hurdles such as scaling up production, addressing material limitations, and establishing global standards for equipment and infrastructure need to be overcome. High upfront infrastructure and production facilities costs require innovative financing models and public-private partnerships. Addressing public concerns about safety and potential environmental impacts and ensuring equitable access to green hydrogen benefits are crucial for widespread social acceptance (Gallagher 2023).

Strategies to overcome barriers to the use of green hydrogen include:

- Demonstration projects: Displaying successful green hydrogen applications across various sectors can build trust and confidence in the public.
- Community interaction: Engaging local communities in decision-making processes ensures that benefits are distributed fairly and that concerns are addressed.
- Education and awareness campaigns: Raising public awareness about the potential of green hydrogen for economic growth and climate change mitigation can garner widespread support.

A future powered by green hydrogen presents a promising picture: energy systems that are diversified and resilient and integrate hydrogen with renewable energy sources. This provides long-term storage options and decarbonizes hard-to-abate sectors. By implementing regional hydrogen hubs, decentralized microgrids, and a global hydrogen trade network, adoption and accessibility can be ensured on a widespread level (IRENA 2021).

The potential of green hydrogen for sustainability and climate goals is enormous. It plays a crucial role in attaining net-zero emissions goals and addressing the impacts of climate change. The green hydrogen revolution is a collective responsibility, and by addressing challenges, harnessing opportunities, and fostering collaboration across governments, industries, and communities, we can unlock its boundless potential. This will illuminate a brighter path for future generations, leading to a cleaner, more equitable, and prosperous future for all.

11.8 Conclusion

This chapter explores the global shift toward renewable energy sources for decarbonization and the net-zero transition. The focus is on hydrogen as a potential decarbonization solution, particularly green hydrogen. Green hydrogen is produced through electrolysis using renewable electricity. The chapter provides a detailed analysis of the electrolysis process along with other production technologies.

Adopting hydrogen during the net-zero transition is a global effort, further emphasized by the growing number of national hydrogen strategies and international collaboration and partnerships. Although hydrogen is abundant and has potential as a clean energy carrier, its use case across sectors must be considered case by case. Research has shown that, in some cases, other solutions are more efficient, safer, and more economically viable. For hard-to-abate sectors and applications where direct electrification is impossible, hydrogen has the best chance of being used, for example, in steel and cement industries and heavy-duty transport. While grey hydrogen is currently used in refineries and fertilizers, green hydrogen is a cleaner alternative. Applications like light-duty vehicles, trains, and household heating do not require any form of hydrogen as a solution. End-use applications of hydrogen can also vary across regions based on the policy framework, infrastructure development, and available resources.

The shift toward a renewable energy future is becoming increasingly important, and green hydrogen and its derivatives are seen as an attractive solution for decarbonization. However, many challenges must be addressed for green hydrogen to be widely adopted. We need technological advancements, strategic roadmaps, and policy development to make it as competitive as grey hydrogen. Green hydrogen poses technical and social hurdles, such as scaling production, addressing material limitations, and establishing global standards. Additionally, high upfront costs require innovative financing models and public-private partnerships. It is also crucial to ensure safety and equitable access and address public concerns for social acceptance. Strategies to overcome barriers include demonstration projects, community engagement, and education campaigns.

The growing number of national hydrogen strategies and international collaboration and partnerships demonstrates the increasing recognition of the decarbonization potential of hydrogen and green hydrogen. Despite the challenges, the global push for renewable energy sources and the embrace of green hydrogen is a positive step toward a more sustainable future. Through ongoing innovation, collaboration, and strategic planning, we can conquer these obstacles, realize the full potential of hydrogen and green hydrogen, and make them essential drivers of the transition to net-zero emissions.

Competing Interest Declaration The authors have no conflicts of interest to declare that are relevant to the content of this book.

References

About the Council | Hydrogen Council (2023, October 30) Hydrogen Council. https://hydrogencouncil.com/en/about-the-council/

Ammonia Technology Roadmap – Analysis – IEA (n.d.) IEA. https://www.iea.org/reports/ammonia-technology-roadmap

Artrocker (n.d.) Methanol for hydrogen market size, transformative trends and future scenarios 2024–2031 – Artrocker. https://www.artrocker.tv/news/methanol-for-hydrogen-market-size-transformative-trends-and-future-scenarios-2024-2031/1656078/

Australia's National Hydrogen Strategy (n.d.). https://www.dcceew.gov.au/sites/default/files/documents/australias-national-hydrogen-strategy.pdf

Basic Hydrogen Strategy – 経済産業省 (2023). https://www.meti.go.jp/shingikai/enecho/shoene_shinene/suiso_seisaku/pdf/20230606_5.pdf

Bhandari R, Trudewind AC, Zapp P (2013) Life cycle assessment of hydrogen production via electrolysis e a review. J Clean Prod. https://doi.org/10.1016/j.jclepro.2013.07.048
CHFCC (n.d.) Hydrogen fuel cell light duty passenger vehicles. https://chfcc.org/fuel-cell-vehicles/
Clean Air Task Force (2023, November 21) Hydrogen for decarbonization: a realistic assessment &#8211. Clean air task force. https://www.catf.us/resource/hydrogen-for-decarbonization-a-realistic-assessment/
Clean Shipping Coalition (2023, March 9) Clean Shipping Coalition – clean shipping for people & planet. https://cleanshipping.org/
Clean Trucks and Buses (n.d.) environmental defense fund. https://www.edf.org/issue/clean-transportation/clean-trucks-and-buses
CSTEP (n.d.). Hydrogen applications segmentation and hierarchy. https://cstep.in/publications-details.php?id=2471
DCCEEW (n.d.). https://www.dcceew.gov.au/climate-change/international-climate-action/international-partnerships
EERE: Alternative Fuels Data Center Home Page (n.d.). https://afdc.energy.gov/
Energy.gov (2021) Bipartisan Infrastructure Law Implementation – Department of Energy. https://www.energy.gov/ceser/bipartisan-infrastructure-law-implementation
Energy.gov (n.d.-a). Regional clean hydrogen hubs selections for award negotiations. https://www.energy.gov/oced/regional-clean-hydrogen-hubs-selections-award-negotiations
Energy.gov (n.d.-b) FOTW #1282, March 20, 2023: The number of light-duty all-electric vehicle models nearly doubled from model year 2021 to 2022. https://www.energy.gov/eere/vehicles/articles/fotw-1282-march-20-2023-number-light-duty-all-electric-vehicle-models-nearly
Enking M (2023, March 16) North America' first hydrogen-powered train will debut this summer. Smithsonian Magazine. https://www.smithsonianmag.com/smart-news/north-america-first-hydrogen-powered-train-180981800/
European Clean Hydrogen Alliance (n.d.) Internal market, industry, entrepreneurship and SMEs. https://single-market-economy.ec.europa.eu/industry/strategy/industrial-alliances/european-clean-hydrogen-alliance_en
European Commission (2024, January 24) European Commission. https://commission.europa.eu/index_en
F & F (2019, June 16) Japan, E.U and U.S. Agree to hydrogen and fuel cells partnership. FuelCellsWorks. https://fuelcellsworks.com/news/japan-e-u-and-u-s-agree-to-hydrogen-and-fuel-cells-partnership/
Fages E, Weichenhain U, Schillaci G, Audouin C, Fages E, Weichenhain U, Schillaci G, Audouin C (2023, September 12) Ports and green hydrogen: match made in heaven? Roland Berger. https://www.rolandberger.com/en/Insights/Publications/Ports-and-green-hydrogen-Match-made-in-heaven.html
Fuel Cell & Hydrogen Energy Association (n.d.) Fuel Cell & Hydrogen Energy Association. https://www.fchea.org/
Gallagher J (2023) Emissions from natural hydrogen. Nat Energy 8(8):774–774. https://doi.org/10.1038/s41560-023-01344-2
GCCA Roadmap (2022). https://gccassociation.org/concretefuture/wp-content/uploads/2022/10/GCCA-Concrete-Future-Roadmap-Document-AW-2022.pdf
Home | Clean Aviation (n.d.). https://www.clean-aviation.eu/
Hydrogen (n.d.) Energy. https://energy.ec.europa.eu/topics/energy-systems-integration/hydrogen_en
Hydrogen Council's (2023) Path to hydrogen competitiveness. https://hydrogencouncil.com/wp-content/uploads/2020/01/Path-to-Hydrogen-Competitiveness_Full-Study-1.pdf
Hydrogen Europe Home (2024, January 26). Hydrogen Europe. https://hydrogeneurope.eu/
Hydrogen Newsletter (2023, December 20). https://www.hydrogennewsletter.com/
Hydrogen Production | Hydrogen Program (n.d.). https://www.hydrogen.energy.gov/program-areas/production

Hydrogen Production: Natural Gas Reforming (n.d.). Energy.gov. http://energy.gov/eere/fuelcells/hydrogen-production-natural-gas-reforming
ICDK (n.d.) Outlook on hydrogen economy & roadmap – ICDK. https://icdk.dk/-/media/websites/icdk/locations-reports/seoul/outlook-on-korean-hydrogen-economy-and-roadmap.ashx
IEA (2023) Global Hydrogen Review 2023, IEA, Paris. https://www.iea.org/reports/global-hydrogen-review-2023, License: CC BY 4.0
IEA (n.d.-a) Technology roadmap – hydrogen and fuel cells – analysis – IEA. https://www.iea.org/reports/technology-roadmap-hydrogen-and-fuel-cells
IEA (n.d.-b) The future of rail – analysis – IEA. https://www.iea.org/reports/the-future-of-rail
IEA DHC (n.d.). https://www.iea-dhc.org/home
International Council on Clean Transportation's "Global FCEV Market Guide 2023" (2023 November). https://www.edf.org/sites/default/files/2023-05/Electric%20Vehicle%20Market%20Update%20April%202023.pdf
IRENA (2020) Green hydrogen cost reduction: scaling up electrolysers to meet the 1.5°C climate goal. International Renewable Energy Agency, Abu Dhabi
IRENA (2021) Enabling measures roadmap for green hydrogen. https://www.irena.org/-/media/Files/IRENA/Agency/Collaborative-Frameworks/IRENA_Enabling_Measures_Roadmap_for_Green_H2_Jan22.pdf?la=en&hash=8FC3CDEB9128B1D23A90541B2E499C1F6DDEEFA6
IRENA (2022) World energy transitions outlook 1-5C pathway 2022 edition. https://www.irena.org/publications/2022/Mar/World-Energy-Transitions-Outlook-2022
IRENA (n.d.) IRENA report on hydrogen storage. https://www.irena.org/Energy-Transition/Technology/Hydrogen
IRS (2024) Inflation Reduction Act of 2022 | Internal Revenue Service. https://www.irs.gov/inflation-reduction-act-of-2022
Kukreja S, Narain M (2024, January 23) Green hydrogen BLUprint – Blume ventures. Blume Ventures. https://blume.vc/reports/green-hydrogen-bluprint
Lee GR, Kim J, Hong D, Kim YJ, Jang H, Han HJ, Hwang CK, Kim D, Kim JY, Jung YS (2023, September 5) Efficient and sustainable water electrolysis achieved by excess electron reservoir enabling charge replenishment to catalysts. Nat Commun 14(1). https://doi.org/10.1038/s41467-023-41102-2
Liebreich (n.d.-a) Source: Michael Liebreich/Liebreich Associates, Clean Hydrogen Ladder, Version 5.0, 2023. Concept credit: Adrian Hiel, Energy Cities. Image: Wenger (concept credit: Paul Martin). CC-BY 4.0
Liebreich (n.d.-b) Source: Michael Liebreich/Liebreich Associates, Clean Hydrogen Ladder, Version 5.0, 2023. Concept credit: Adrian Hiel, Energy Cities. CC-BY 4.0
Megía PJ, Vizcaíno AJ, Calles JA, Carrero A (2021) Hydrogen production technologies: from fossil fuels toward renewable sources. A mini review. Energy Fuel 35(20):16403–16415. https://doi.org/10.1021/acs.energyfuels.1c02501
Miller H, Bouzek K, Hnat J, Loos S, Bernäcker C, Weißgärber T, Röntzsch L, Meier-Haack J (2020) Green hydrogen from anion exchange membrane water electrolysis: a review of recent developments in critical materials and operating conditions. Sustain Energy Fuels 4:2114–2133. https://doi.org/10.1039/c9se01240k
Mission Innovation (n.d.). https://mission-innovation.net/wp-content/uploads/2022/09/Clean-Hydrogen-Mission-Action-Plan-Sept-22.pdf
National Green Hydrogen Mission | Ministry of New and Renewable Energy | India. (n.d.). https://mnre.gov.in/national-green-hydrogen-mission/
National Green Hydrogen Strategy Chile (n.d.). https://energia.gob.cl/sites/default/files/national_green_hydrogen_strategy_-_chile.pdf
National Hydrogen Strategy Update (n.d.). https://www.bmwk.de/Redaktion/EN/Publikationen/Energie/national-hydrogen-strategy-update.html

Publications Office of the EU (2016) Hydrogen roadmap Europe: a sustainable pathway for the European energy transition. https://op.europa.eu/en/publication-detail/-/publication/0817d60d-332f-11e9-8d04-01aa75ed71a1/language-en

REUTERS (n.d.). https://www.reuters.com/business/energy/germanys-energy-crisis-powers-hydrogen-switch-2022-12-29/

Sampson J (2023, May 23) Hydrogen Insights 2023. Hydrogen Council. https://hydrogencouncil.com/en/hydrogen-insights-2023/

Seddon D (2022, April 19) Hydrogen Economy, The: Fundamentals, Technology, Economics. World Scientific. http://books.google.ie/books?id=e2F2EAAAQBAJ&printsec=frontcover&dq=Hydrogen+economy&hl=&cd=8&source=gbs_api

State of Hydrogen (2022). https://www.dcceew.gov.au/sites/default/files/documents/state-of-hydrogen-2022.pdf

Swennenhuis F, de Gooyert V, de Coninck H (2022) Towards a CO2-neutral steel industry: justice aspects of CO2 capture and storage, biomass- and green hydrogen-based emission reductions. Energy Res Soc Sci 88:102598. https://doi.org/10.1016/j.erss.2022.102598

Técnicas Reunidas (n.d.) Hydrogen present and future. Part 2 – Técnicas Reunidas. https://www.tecnicasreunidas.es/articulo/hydrogen-present-and-future-part-2/

The Future of Hydrogen – Analysis – IEA (n.d.) IEA. https://www.iea.org/reports/the-future-of-hydrogen

U.S. National Clean Hydrogen Strategy and Roadmap (n.d.). https://www.hydrogen.energy.gov/library/roadmaps-vision/clean-hydrogen-strategy-roadmap

Whitehead J, Newman P, Whitehead J, Lim KL (2023) Striking the right balance: understanding the strategic applications of hydrogen in transitioning to a net zero emissions economy. Sustain Earth Rev 6(1). https://doi.org/10.1186/s42055-022-00049-w

World Bank (2022, November 15) World Bank group announces international low-carbon hydrogen partnership. https://www.worldbank.org/en/news/press-release/2022/11/15/hydrogen-for-development-partnership-h4d-launch

World Economic Forum (n.d.) Global Risks Report 2024 | World Economic Forum Global Risks Perception Survey 2023–24, World Economic Forum. https://www.weforum.org/publications/global-risks-report-2024/

Xu D, Dong L, Ren J (2023) Introduction of hydrogen routines. In: Hydrogen economy. Academic Press, pp 45–65

Chapter 12
Porous Organic Polymers for Renewable Energy Development

Shalinee Singh, Santosh Kumar, Dan Bahadur Pal, and Rajiv Ganguly

Abstract Renewable energy provides a critical assistance in sustainable, environmentally friendly, and economically viable electricity generation. Solar energy systems are pivotal among these sources, requiring high performance for optimal efficiency. Porous materials have emerged as effective and economical means to enhance heat transfer and energy efficiency within solar energy systems. Porous organic polymers (POPs) are an emerging class of porous materials made primarily from organic building blocks interconnected by strong covalent bonds. This structure creates rigid frameworks with specific void spaces, making them ideal for accommodating guest molecules. This chapter summarizes the applications, advantages, and limitations of porous materials in various types of solar energy systems. Additionally, it briefly summarises both energy conversion and energy storage components.

Keywords Renewable energy · Sustainable · Environmentally · Porous materials · Porous organic polymers

S. Singh · S. Kumar (✉)
Department of Chemistry, School of Basic and Applied Sciences,
Harcourt Butler Technical University, Kanpur, Uttar Pradesh, India
e-mail: santoshk@hbtu.ac.in

D. B. Pal
Department of Chemical Engineering, Harcourt Butler Technical University,
Kanpur, Uttar Pradesh, India

R. Ganguly
Department of Civil Engineering, Harcourt Butler Technical University,
Kanpur, Uttar Pradesh, India

S. Kumar, V. K. Singh (eds.), *Renewable Energy Development: Technology, Material and Sustainability*, Clean Energy Production Technologies,
https://doi.org/10.1007/978-981-97-9626-7_12

Abbreviations

BMIM-Cl-CB	1-butyl-3-methylimidazolium chloride conduction band
CMPs	conjugated microporous polymers
CNT	carbon nanotube
COFs	covalent organic frameworks
CON	covalent organic nanosheet
DFT	density functional theory
DSSC	dye-synthesized solar cell
ETL	electron-transporting layers
ETT	4,4′,4″,4″‘(ethane-1,1,2,2-tetrayl)tetranilino
FENG	ferroelectric nanogenerator
FF	fill factor
HTL	hole-transporting layers
ISC	short-circuit current
IUPA	International Union of Pure and Applied Chemistry
LED	light-emitting diode
M-PCP	molten/mullite-corundum porous ceramic preforms
OPV	organic photovoltaics
PBT-TT	benzodithiophene-TT copolymer
PCE	power conversion efficiency
PDMS	polydimethylsiloxane
PDI	perylene diimide
POPs	porous organic polymers
NPPC	nanoporous polycarbonate
PSC	perovskite solar cells
PSM	porous resin-supported
PV	photovoltaics
PVF-HFP	polyvinylidene fluoride-co-hexafluoropropylene
QSS	quasi-solid state
TNGs	triboelectric nanogenerators
TFPP	1,3,6,8-tetrakis(4-formylphenyl) pyrene
TW	terawatts
VOC	open-circuit voltage

12.1 Introduction

Over the past century, there has been a sharp rise in the need for energy due to population expansion and worldwide development. The world's energy demand was 16 TW in 2010; by 2030, it is expected to increase to 23 TW, and by 2050, it may reach 30 TW. According to recent figures, just 20.5% of the world's energy is consumed globally from renewable sources, which include hydropower, wind, biopower, and solar photovoltaics (PVs). The remaining 79.5% of energy comes from traditional sources including oil, coal, and natural gas (Zhu et al. 2020). As public awareness of environmental cleanliness rises, renewable energy is drawing more attention because of its ability to generate power with little to no emission of air pollutants and other ecologically favorable qualities. In addition to advancing sustainability, renewable energy is important economically. By producing energy from natural, renewable resources, it reduces the cost of producing electricity (Ang et al. 2022). Consequently, the development of advanced materials that can improve the effectiveness of energy conversion and energy storage systems is crucial. Porous materials, because of their large pore volume, high surface area-to-volume ratio, topological tunability of their underlying architecture, and controllable porosity offer optimal opportunities for energy conversion and storage. Porosity is an ancient concept, having existed in natural structures like rocks, biological tissues, and charcoal for millennia. The term "porous" typically describes a material with permanent voids that are often interconnected and allow the passage of liquids or gases, carrying a practical implication. In contemporary times, porous materials play a vital role in various scientific and technological fields. In the last two decades, there has been a significant increase in the development of new advanced porous functional materials. The International Union of Pure and Applied Chemistry (IUPAC) categorizes porous materials based on their pore diameter (d) sizes: (1) microporous ($d < 2$ nm), (2) mesoporous (2 nm $< d <$ 50 nm), and (3) macroporous ($d >$ 50 nm) represented in Fig. 12.1. There has been a significant increase in the investigation of porous materials recently. This includes a range of substances such as zeolites, activated carbon, mesoporous silica, metal-organic frameworks, porous organic polymers (POPs), and others. These materials are known for their controllable empty spaces at molecular and nanoscopic levels. They offer great potential in addressing various challenges in different areas, including separation processes, catalysis, adsorption, energy storage, and ion exchange (Lee and Cooper 2020). Since advent of the

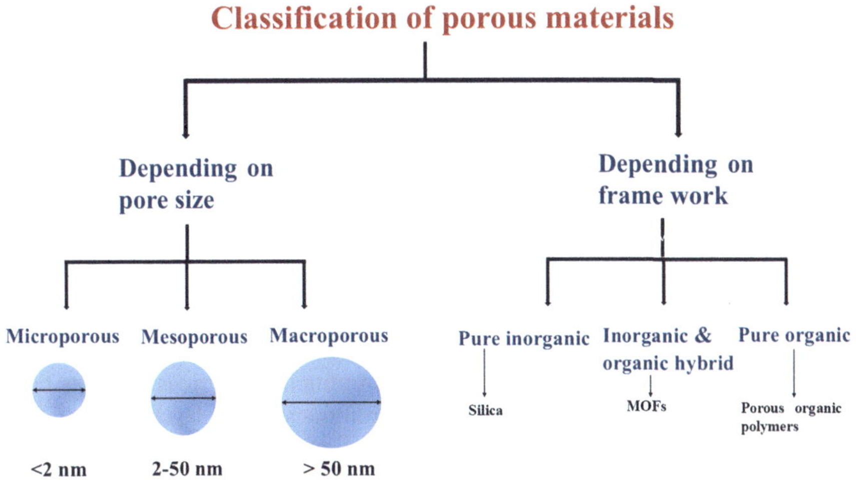

Fig. 12.1 Classification of porous material based on pore size and building framework

covalent POPs, extensive research has focused on leveraging their natural porosity and expansive surface, stemming from their distinctive interconnected structure. Researchers have also explored electrical capabilities, chemical and thermal stability (Kumar et al. 2019; Sarkar et al. 2021).

Achieving the required pore sizes and morphology is crucial for the effective functioning of porous organic polymers (POPs). Therefore, precise synthetic control is of paramount importance to tailor different frameworks and topologies according to the targeted polymer. Alongside numerous evolving post-synthetic modification methods, POPs offer a robust platform for creating both primary and high-order structures, as well as customizable end products. However, it's important to note that the structural rigidity of POPs significantly limits their solubility in common solvents, thereby restricting their processability (Bildirir et al. 2017; Choi et al. 2024; Kang et al. 2020; Zhang et al. 2022). In organic semiconductor materials, a π-conjugated system forms the electron-based platform for delocalization of charge. However, the spatial and geometric characteristics of POPs dictate the balance between rigidity and conjugation, posing a significant challenge for optimal design. To address this, various polymerization techniques have been employed to produce higher yield and topological-based fine molecular structures. These techniques include solvothermal (Lu et al. 2015), mechanochemical (Troschke et al. 2017; Zhang and Dai 2017), ion-thermal (Kuhn et al. 2008; Maschita et al. 2020), and other promising methods. The intriguing porosities, tunable backbones, and topological advantages of porous organic polymers (POPs) have sparked considerable research interest, particularly in exploring their versatility for photovoltaic applications. The reported efficient charge transport with long carrier lifetime in POPs emphasizes their potential for use in photovoltaic devices. PVs have demonstrated outstanding energy conversion capabilities, serving as a green energy resource by generating electricity through solar energy. The in-depth knowledge of photophysical characteristics of PVs are crucial for advancing solar cell technology that can achieve zero-carbon target by humanities.

In recent years, perovskite solar cells (PSCs) have emerged, achieving an efficiency of 25.20%, that is, nearing single-crystalline structured silicon. Researchers are also investigating alternative elements like nontoxic bismuth and antimony for the active layers in PSCs to promote a greener future. Meanwhile, organic solar cells (OSCs) are showing proficiency of 20%, as high luminescent fullerene acceptors are available. The ease of processing and tunability of organic materials give OSCs unique advantages, such as flexibility, affordable cost, and scalability, enhancing competition in the market. Dye-sensitized solar cells (DSSCs) have demonstrated exceptional performance under low and artificial light conditions, making them ideal for indoor applications. Since the initial demonstration of using porous organic polymers (POPs) in photovoltaics (PV) in 2013, significant efforts have been made to utilize POPs as the photoactive component. Additionally, various attempts have been made to employ POPs as the hole−/electron-transporting layers (HTL/ETL), further showcasing the versatility and potential of POPs in PV technologies. Figure 12.2 showed correlation chart of porous organic polymer with photovoltaics. This chapter aims to explore the potential of porous organic polymers (POPs) for photovoltaic (PV) applications, particularly within the sub-categories of organic

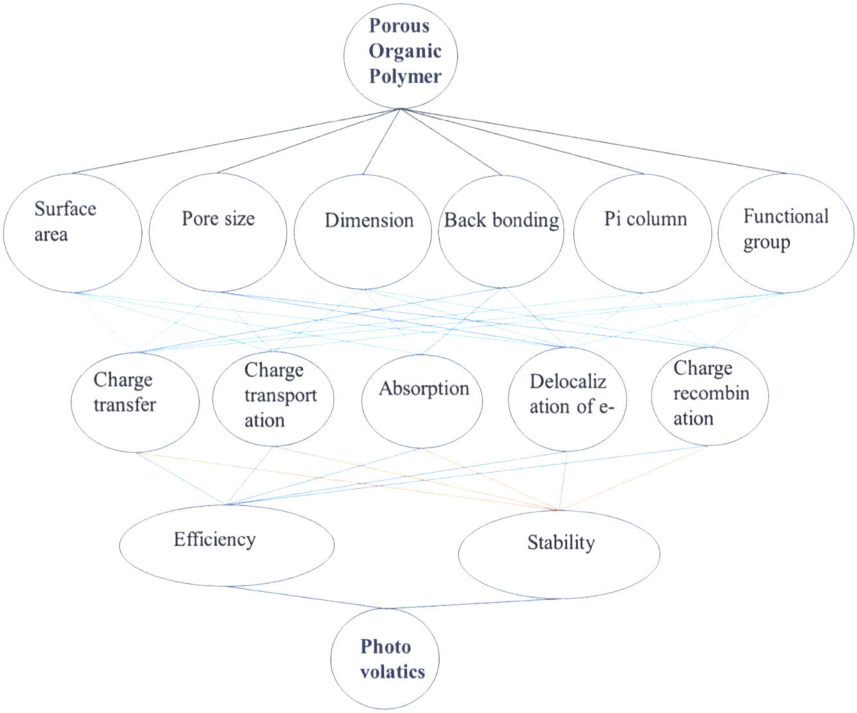

Fig. 12.2 Correlation chart of porous organic polymer with photovoltaics. Orange line represents future scope of work in PV

photovoltaics (OPVs), perovskite solar cells (PSCs), and dye-sensitized solar cells (DSSCs). To systematically enhance device performance, we will provide detailed analyses ranging from fundamental photo-physics to various device functionalities. Additionally, we will offer insights and perspectives on long-term stability, proposing alternative strategies to accelerate the commercialization of POP-incorporated solar cells. Building on recent progress, we intend to present a multifaceted engineering perspective that connects specific optoelectronic properties to key device performance factors. This approach will justify the gradual shift from focusing solely on the active layer to considering other functional components of the device and will inform design strategies for next-generation solar cells. These materials consist of a solid matrix with interconnected voids. Research shows that porous materials have improved thermal properties, like higher convective heat transfer coefficient and thermal conductivity, compared to base materials. It was also investigated by Rashidi et al. that fluid flow and forced convection heat transfer around a solid cylinder coated with a porous coating, recommending materials with higher permeability and higher thermal conductivity for better thermal performance (Rashidi et al. 2017).

Generally, porous materials can be utilized for various purposes in solar energy systems, including as adsorption materials, evaporation materials, insulation

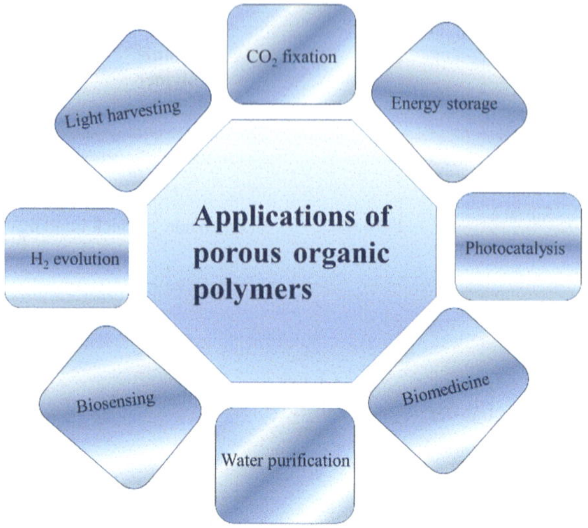

Fig. 12.3 Various applications of porous organic material

materials, thermal energy storage materials, and heat transfer augmentation materials as represented in Fig. 12.3. Metallic foams, porous metals, and hybrid mixture of activated alumina and zeolite-13X can be used as adsorption materials. For thermal energy storage in solar energy systems, CaSO4/PSM (calcium sulfate/porous-resin supported) and M-PCP/Na2SO4 (molten/mullite-corundum porous ceramic preforms) and the impregnation of porous material like diatom earth, expanded perlite, and gamma-alumina with erythritol, soil, and gravel are suitable. Fibrous porous materials, porous ceramics, porous silica material, and mineral wool (e.g., glass wool, stone wool, and ultimate) can be used as insulation material in solar energy systems. Porous materials with low thermal capacity such as blackened jute clothes, dark-painted sponge, and hay material thermal capacity are effective to be used as an evaporation material. Additionally, porous materials with high thermal conductivity, like metal foams, are recommended for heat transfer enhancement. The use of porous material has significantly enhanced the efficiency of solar systems. Current studies have been highly concentrated on meticulously constructing porous organic polymers (POPs) with regulated morphologies, such as nanosheets, nanoparticles, nanovesicles, and nanofibers. These precisely defined microstructures provide improved suitability for specific applications (Shen et al. 2021).

12.2 Generation of Renewable Energy Based on Porous Material

Renewable energy is increasingly used for power generation worldwide, driven by environmental concerns over conventional electricity generation method. Each of the five main types—bioenergy, geothermal energy, solar energy, hydro energy, and wind energy –has unique operational conditions and energy conversion efficiency, making their suitability dependent on location and condition. The process of producing charge flow via mechanical deformation, salinity gradients, and temperature changes or by using solar energy or wind energy is referred to as energy generation. For electrochemical sensors, measurements, piezoelectric, and monitoring systems, triboelectric and thermoelectric materials show the most promise among these techniques. Whereas thermoelectric technologies harness the waste heat from materials, piezoelectric materials turn mechanical deformation of particular materials into kinetic energy. Producing electricity from primary energy sources is known as energy generation. For example, electromechanical generators in power plants run on fuel combustion, nuclear fission, hydraulic forces, solar, geothermal, or wind energy to produce electricity. Large-scale renewable energy production is becoming the main focus of energy generation in an effort to lessen dependency on fossil fuels. The production of renewable energy has several benefits, including constant supply, low or no pollution, and safety (Sadasivuni et al. 2019). The primary objectives of employing RE sources for energy production in a forward-thinking manner are demand-driven energy savings, efficiency improvements, and decreased reliance on fossil fuels.

12.2.1 Piezoelectric Energy Generation

The mobility of electrochemical systems comes from their capacity to transform chemical energy into electrical energy, as demonstrated in batteries. Like solar cells, photovoltaic techniques produce energy by converting light energy into electrical energy. In electrically anisotropic materials, the piezoelectric effect uses mechanical strains to produce electricity, which is helpful in power harvesting applications. Thermocouples and thermopiles are two examples of devices that are used in the thermoelectric method to directly convert temperature differences into electrical energy (Sadasivuni et al. 2019). In order to collect energy from mechanical interactions, Fan et al. developed triboelectric nanogenerators (TNGs) utilizing a stretchy porous carbon nanotube (CNT)-based polydimethylsiloxane (PDMS) nanocomposite manufactured by a sacrificial template approach. The porous nanocomposite has an elongation of 243% and is made up of tightly packed cavities that have sizes between 200 and 400 μm. The partially exposed CNTs in the cavities function as micro-scale TNG units (Fan et al. 2016).

The production of mechanical energy from stress-voltage coupling method in an electrochemical system consisting of two lithium-alloyed silicon electrodes divided by an electrolyte was reported by Kim et al. 2016. When subjected to mechanical bending, one electrode experiences tension while the other suffers compression, yet both electrodes remain iso-potential in a stress-free state. The prototype devices' electrical measurements revealed an output peak power of around 0.53 $\mu W/cm^2$, an open-circuit voltage (Voc) of approximately 0.023 V, and a short-circuit current (Isc) of approximately 14.5 μA (Kim et al. 2016).

Vertically aligned ZnO nanowires, measuring about 12 μm in length and 184 nm in diameter, were fused directly with nanoporous polycarbonate template using a template-assisted electrodeposition mechanism. F.L. Boughey et al. investigated the vibrational energy-producing properties of these nanowires. Using a mechanical arm, the created ZnO-PC nanogenerators (NGs) were activated mechanically at a frequency of 5 Hz along the nanowire with an input energy of 3 μJ per cycle (Boughey et al. 2016).

Platinum nanoparticles embedded in a porous 79% porosity polyvinylidene fluoride-co-hexafluoropropylene (PVF-HFP) nanocomposite were created by Ghosh et al. to serve as a ferroelectric nanogenerators (FENGs). They looked into how well it could harvest energy. The FENG showed an open-circuit voltage (Voc) of around 18 V and an Isc of approximately 17.7 mA at a compressive normal stress of 4 MPa. The nanocomposite showed outstanding ferroelectric characteristics with a remanent polarization (Pr) of $\approx$61.7 μC/cm^2, an ultrahigh dielectric constant (εr) of 2678 at 1 kHz, and an improved piezoelectric charge coefficient (d33) of −686 pC/N and converse piezoelectric coefficient (g33) of −0.03 V·m/N. It was an effective material for piezoelectric devices because of these qualities. With an energy conversion efficiency of less than 0.2 percent, the developed FENG was able to power 55 blue LEDs and 25 green LEDs (Ghosh et al. 2015).

12.2.2 Porous Polymers in Solar Energy Generation

POP skeletons typically consist of knot and linker units with specific geometries. To achieve strong fluorescence and porosity, POPs are generally built from rigid knots containing π-electronic units or conjugated moieties, forming an extended π-conjugated linker through covalent bonding. Assembled π-conjugated systems and larger π-conjugated units serve as ideal building blocks for fluorescent POPs with high luminescence density. High electron delocalization enhances the transition from s0 → s1 and π* → π relaxation probabilities. Thus, incorporating fluorescent units into the POP skeletons through various polymerization reactions is an effective method for constructing fluorescent polymeric materials (Kumar et al. 2023; Zhang et al. 2022). To develop high-performance fluorescent matrixes for analytical applications, conjugated microporous polymers (CMPs) with high surface area, suitable pore structure, good physical and chemical stability, and high fluorescence intensity are highly desired. Due to their extended conjugated frameworks and optoelectronic properties, conjugated microporous polymers (CMPs), of which groundbreaking works were initially articled by Cooper et al., have opened a wide area for new possibilities as potent fluorescent materials. Typically, CMPs are made by various kinetically regulated couplings, and their chemical structure can be precisely regulated through design and specific polymerization technologies. In contrast to their linear counterpart, that is comparatively more prone for charge localization because of brittle nature as a result of geometric kinetics, POPs form strong covalent bonds that contribute to charge transport channel and good thermal stability. Isotropic multidimensional charge transport may be possible with higher-dimension POP networks. For advance PV applications through material characterization is essential; however, photophysical studies are constrained by POPs' insolubility and crystallinity problems. Thus, despite of their amorphous nature, research concentrates on CMPs because of their facile tunability and easy synthesis and on covalent organic frameworks (COFs) because of their crystallinity. Other categories of porous materials face challenges in optoelectronic applications due to stringent synthetic conditions limited crystallinity π-conjugation. For instance, covalent triazine frameworks (CTFs) have high nitrogen content but their triazine ring synthesis limits mass production and versatility. PAFs, made by Ullmann cross-coupling, struggles with large-scale, low-cost manufacturing. POPs are driving ongoing research to use them in electrolyte photoactive layer and ETL/HTL. The number of POP-specific PV reports is small compared other POP applications. However, combining POPs with perovskites has recently shown promise. Their hydrophobicity enhances stability, while their porous structure and tunable energy levels reduce charge recombination. In OPVs, POPs are increasingly used in HTL for better charge transport, and in DSSCs, replacing the liquid electrolyte prevents a short-circuit and improves durability. By integrating chemistry, physics, and materials science, our aim is to understand the interrelation between intrinsic and photophysical properties.

12.2.2.1 OPV Cell Based on Porous Organic Material

Over the last 5 years, integrated OPVs photoactive layer has not advanced as much as POPs in other areas. Research has increasingly focused on using POPs in the charge transport layer rather than as the photoactive component. A notable exception is the promising study by Xu et al., which simulated that quinazoline-based COF/ZnSe heterojunctions could potentially achieve a power conversion efficiency (PCE) of 22.32%. To fully leverage the direct bandgap and near-infrared (NIR) absorption of Q-COF, a ZnSe monolayer was selected as the acceptor to form a type II offset. Upon considering lattice strain, the PCE can be further enhanced. The low cleavage energy and higher mass energy density make Q-COFs a strong candidate as a donor material. However, since this report is entirely computational, it does not account for critical factors such as film formation, morphology, molecular orientation, and charge extraction (Xu et al. 2020). Further interesting development by Bildirir et al. demonstrated the PV application of an n-type solution-processable conjugated polymer network (PNT1). Synthesized via Stille cross-coupling reaction, PNT1 exhibits high thermal stability and sufficient photon absorption in the visible light spectrum. Using PNT1 as the acceptor component in a bulk heterojunction, the device was configured in an inverted structure of ITO/ZnO/PTB7-Th:PNT1/MoO_3/Ag. This configuration achieved a PCE of 1.18% and an EQE of 25%. In discussion, it is found that low fill factor (FF) could stem either from higher solubility of the donor and acceptor materials or low charge carrier mobility (Bildirir et al. 2017).

Zhang et al. utilized a hyper-branched HP-PDI in conjunction with PTB_7-Th. The hyper-branched structural arrangement and the affluence of terminal PDI units allowed for precise tuning of domain and balanced electron/hole drift velocity per unit of electrical field. This approach resulted in a device with a PCE of 2.15%, maintaining 91% of its natural efficacy after 6 days, demonstrating superior stability due to structural arrangements (Zhang et al. 2017). In contrast, Fu et al. proposed using a donor material soluble based on POP, specifically PDPP-C_{20}, which features lengthy branched alkyl chains ensuring good solubility in usual solvents and also a decent hole mobility. IT-4F paired combination achieved a remarkable Voc of 0.8 volts. Although, the unsatisfactory FF limited the PCE to just 2.46%. Instead of the lower PCE, the device demonstrated exceptional stability, operating steadily for up to 27 days under suitable environment while maintaining 90% of Jsc, Voc, and PCE, showcasing remarkable stability against external factors (Fu et al. 2021a, b). This postulation of a solution-processed conjugated polymer network is generating robust increase in fabrication and optoelectronic applications. Furthermore, acumens gained from usually studied organic blends can be transferred to precisely tuned nano-morphology and interfacial characteristics using various microfabrication methods.

12.2.2.2 Perovskite Solar Cells Based on Porous Organic Material

Perovskites of the classic ABX3 structure, particularly organic-inorganic lead halides like MAPbI3 and FAPbI3 (MA for methylammonium, FA for formamidinium), have garnered prominent attention in recent years in the field of PSC research (Ali et al. 2021; Kumar and Naidu 2021). Various strategies such as topological arrangements, microfabrication techniques, material combinations, and physics of device have been extensively explored to enhance their performance (Rong et al. 2018; Snaith 2013). The distinctive porosity and adjustable electrical properties of POPs hold potential for improving device performance. Unlike the trend observed in OPVs, POPs have been reported to enhance both HTL/ETL components and photoactive layer in perovskite devices, offering a new pathway for performance enhancement. 3D POPs, in particular, exhibit superior geometric features that facilitate charge transfer dynamics and photoelectric applications.

Lim et al. designed a soluble T-POP by polymer having melamine and alkyl-modified perylene motifs, which can be applied as a secondary electron transport layer on the PC70 BM layer. It was reported that Jsc has increased more than 6% increased, with increased FF value leading up to 13% power conversion capacity, owing to its strong π-π overlapping. This addition in charge transport in planar structured perovskite solar cells was produced by increased intermolecular interaction between the opto-active perylene units in back bones or perylene units in branches due to enhanced surface area of polymer compared to linear 1-D conjugated polymers (Lim et al. 2020).

Furthermore, Mohamed et al. have synthesized two highly crystalline and thermally stable covalent organic frameworks Car-ETT and TFPP-ETT leading to improved performance from 17.40% to ~19.80%. They coupled carbazole and pyrene units with ETT via [4 + 4] imine solvothermal condensations to get Car-ETT and TFPP-ETT COFs (Mohamed et al. 2020). In a study, Wu et al. investigated a spirobifluorene-based SP-3D-COF doped PSC (perovskite solar cell), resulting in improved PCE of up to 19% compared to undoped samples. By leveraging the charge transportation in multidimensions of SP-3D-COF in an inverted p-i-n configuration, they observed lower dark current density with satisfactory diode rectification, with great boosting in Jsc, Voc, and FF values. This extended leads to bathochromic shift in absorption conjugation due to doping in bulk. Researchers rationalized the performance enhancement through density functional theory (DFT) calculations, suggesting the formation of a shortly lived symmetry-breaking charge transfer state which facilitates the flow of excited electrons through a spirobifluorene species. Additionally, the spatial localization of HOMOs and LUMOs at the ends of each molecule effectively suppressed charge recombination. The larger area of the biphenyl-group in SP-3D-COF 2 improved interfacial interaction with $CH_3NH_3PbI_3$, enhancing morphology and promoting charge generation (Wu et al. 2021).

Mohamed et al. conducted a comprehensive study on 2D imine-linked Car-ETTA and TFPPy-ETTA COFs to delve deeper into the interfacial properties between POPs and perovskites. With incorporation of porous framework, they observed a hypsochromic shift (blue shift) in PL peak, indicating defect passivation in COF-modified ETTA HTL. This defect passivation led to heightened crystallinity of the perovskite cell through additional phase transition sites. The system (ITO)/PTAA/COFs/$(FAPbI_3)_{0.83}(MAPbBr_3)_{0.17}(CsPbI_3)_{0.05}$/PCBM/BCP/Ag achieved 19.79% of power conversion energy under forward scan conditions. Remarkably fast decay kinetics demonstrated by TCSPC indicated outstanding hole transfer properties for both TFPPy-ETTA and Car-ETTA COFs. The authors correlated a trend of photoactive current density with potent voltage, concluding that higher exciton generation and improved charge collection directly contributed to a higher Jsc level. This study underscored the structural impact of COFs on device performance, particularly in terms of defect passivation, nucleation site enhancement, and efficient charge transfer. Various POPs have demonstrated competent hole-transporting capabilities and deep energy alignment, making them compatible as HTLs (hole transport layers). In efforts to address the acidity issue associated with PEDOT/PSS in PSCs, POPs have been explored as alternatives. Wang et al. utilized electro-polymerization to deposit a PAF-86 film onto an ITO substrate as the HTL. Space charge limited current (SCLC) results indicated moderate hole mobility in the PAF-86 HTL. The device's working attainment improved notably without requiring encapsulation. This enhancement was attributed to improved surface interaction between PAF-86 and the perovskite layer. Subsequently, a stability test conducted on the device revealed that the retained PCE over 500 hours indicates the suitability of PAF-86 as a long-term stable component in PSCs. This mentioned effect is attributed to the hydrophobic nature of the porous film, which significantly retards humidity-related degradation processes. In addition to structural strategies, a novel approach involving thin-film materials as HTLs has been explored in PSCs (Wang et al. 2017).

In a recent study, Park et al. introduced a solvent-treatable covalent organic nanosheet (CON) into PSCs for efficient extraction of hole. Atomic force microscopy (AFM) revealed that the film quality of PEDOT/PSS depends heavily on the surface roughness of the CON substrate, which in turn increases the surface contact area and helps in charge transfer (CT), as indicated by strong photoluminescence quenching. The remarkable electron-blocking function and hole-transporting capability of CON, particularly CON-10, resulted in a PCE boost to 10.7%. Due to solvent-dependent morphology of CONs, it offers an easy method for tuning interlayer properties (Park et al. 2018). The rigorous processing conditions of POPs, stemming from their rigidity, have hindered their integration into modern solar cells. To address this, soluble groups have been introduced into the backbone of CON. Fu et al. synthesized a COF-like 2-D POP (PDPP-C_{20}) through incorporation of DPP derivative via side-chain engineering. This resulted in a dual role of charge extraction and passivation, achieving 21.92% high PCE. FTIR spectroscopy highlighted passivation effect, indicating contact between CO and under-coordinated Pb2+ deformities. Beyond photovoltaic parameter improvements, reduced hysteresis suggests increased reduction of ion migration in perovskites cells through POP incorporation (Fu et al. 2021a, b). Fu et al. made significant progress by utilizing a dopant-free 2DP-TDB to achieve a champion PSC efficiency of 22.17%. By incorporating bulky alkyl chains into the polymer, they drastically improved its solubility. The excellent crystallinity and strong face-on orientation observed in grazing incidence wide-angle X-ray diffraction (GIWAXS) data supported its high hole mobility and exceptional charge transport capability. X-ray photoelectron spectroscopy (XPS) further demonstrated the interaction between the thiophene units of 2DP-TDB and under-coordinated Pb2+ in the perovskite layer, effectively passivating surface defects with Lewis bases. This passivation prevented trap-assisted recombination and promoted charge transport. The synergistic effects of adopting 2DP-TDB, along with its deep HOMO level, positioned it as a competitive substitute for commonly used doped spiro-OMeTAD in PSCs. This achievement not only highlights the importance of material design but also underscores the potential of dopant-free strategies for enhancing device efficiency in photovoltaic applications (Fu et al. 2021a, b).

12.2.2.3 Dye-Sensitized Solar Cell Based on Porous Material

Dye-sensitized solar cells (DSSCs), which were first developed in 1988, have established a prominent position in photovoltaic industry. This is attributed to their unique characteristics, including flexibility, lightweight, affordable cost, and semi-transparency. These features make DSSCs appealing for wide application areas. Their cost-effectiveness and efficient functioning under lower light conditions attracted in renewable energy market (Shalini et al. 2016; Sugathan et al. 2015). In DSSCs, photoexcitation causes electrons to reach an excited state in the photosensitized dye. These excited electrons are then transported into the conduction band

(CB) of the adjacent semiconductor, usually TiO2. These excited electrons travel toward the direction of the contact areas and are gathered by the counter-electrode via the external circuit. The redox processes in the electrolyte, which often involve iodide and triiodide, compensate the loss of electrons in the dye. The redox cycle is completed when triiodide diffuses toward the counter-electrode for further reduction (Gong et al. 2012). However, DSSCs face challenges such as the adsorption of dyes, corrosion, removal of heavy metals, longer stability of the liquid electrolyte, and energy alignment. POPs, with their innate porosity, reasonable conductivity, and modifiable features, offer potential as competent material substitutes for various device components. Their ability to address these common issues makes POPs a promising avenue for enhancing the performance and stability of DSSCs. The highly porous structure and large surface area of the membrane provide more redox locations and increased ion loading capacity, resulting in higher ion conductivity and improved interfacial properties with the electrode (Shih et al. 2018). Thomas et al. have recently developed a quasi-solid-state (QSS) electrolyte using a porous membrane composed of PMMA/PIL/PIN in DSSCs (Thomas and Rajiv 2020). Incorporating PIL into the PMMA membrane caused the pores to shrink significantly due to phase inversion. SEM images of the final PMMA/PIL/PIN product revealed an increased pore numbers and greater pore volume. Due to the domineering thermal stability of PIN, this porous sheet shows great promise compared to thermally susceptible liquid electrolytes.

The excellent interconnectivity and higher porosity of the membrane contributes to a high electrolyte absorption, enhancing ionic conductivity and increasing the diffusion constant of redox species. The high Voc observed in the assembled device is attributed to suppressed charge recombination. The uniform surface coverage and intimate contact over TiO2, along with the imidazolium content of PIL, effectively passivate interfacial defects. Additionally, the chelation of electron-deficient Li + with PMMA significantly reduces its migration toward the interface, further suppressing recombination. Although the increased viscosity of the membrane impedes its penetration into TiO2, slightly compromising the Jsc, the device maintained 85% of its initial performance after 250 h, demonstrating excellent long-term stability compared to liquid electrolytes. As previously mentioned, replacing Pt electrodes is a heavily researched topic. Efforts have been made to incorporate microporous polyaniline (PANI) in DSSCs, but its instability remains a significant drawback for long-term efficient devices (Li et al. 2008). Ahmad et al. addressed this by integrating a nanoporous poly(3,4-propylenedioxythiophene) (PProDOT) layer into DSSCs, achieving a maximum PCE of over 9% (Ahmad et al. 2010). The large surface area and porosity of PProDOT resulted in higher catalytic activity and lower charge-transfer resistance. However, the incorporation of PProDOT led to a reduction in Voc, attributed to impurities introduced during synthesis. To further investigate the potential of PProDOT as a cathode material, the same group introduced dialkyl groups as side chains attached to the backbone, enhancing solubility (Ahmad et al. 2012). Although this modification resulted in a lower FF due to reduced conductivity, the modified PProDOT still outperformed the Pt electrode with higher Jsc and PCE values. The researchers suggested that the bulky side groups might create the hindrance in the mass diffusion mechanism, impacting overall performance.

12.3 Energy Storage Based on Porous Material

In order to control demand and enable flexible energy generation as needed, energy storage is crucial. By storing energy for future use or transforming it into a different form, it maximizes grid resources and energy creation in electrical systems. Energy storage provides a fossil fuel-free alternative power source (Krishnakumar et al. 2019; Lavate et al. 2023). Hao et al. (2014) cellulose containing bagasse was used to create hierarchical porous carbon aerogels, and their effectiveness as supercapacitor electrodes was evaluated. The specific capacitance that they were able to attain was roughly 142.1 F/g, with a discharge current density of roughly 0.5 A/g. The aerogels' uniquely crafted pore architectures were responsible for their capacitance retention of 93.9% over 5000 cycles (Hao et al. 2014). Pushparaj et al. created integrated hybrid energy storage devices, by using self-sustaining electrolyte BMIM-Cl (1-butyl-3-methylimidazolium chloride), a room-temperature ionic liquid (RTIL), and CNT-integrated cellulose nanocomposites as electrode spacers. By using thermal-chemical vapor deposition to manufacture MWCNTs vertically on a Si substrate and embedding porous cellulose dissolved in RTIL (BMIM-Cl), they created CNT-cellulose-RTIL paper. Supercapacitors, hybrid energy storage devices, and Li-ion batteries are configured using this CNT-cellulose-RTIL sheet (Pushparaj et al. 2007).

12.4 Future Perspective and Conclusion

Despite their strong light-harvesting capabilities and robust electron donating or electron withdrawing characteristics, using POPs as photoactive material presents significant adversaries in competing with other materials. Their insolubility and lower crystallinity have redirected efforts toward leveraging their special structural advantages to enhance charge transportation and interfacial properties. The POP-modified by hole transport layers (HTLs) have shown strong compatibility in perovskite solar cells (PSCs), achieving over 20% efficacy. There has also seen progress in understanding the excited state dynamics of POPs and exploring potential applications of POP-modified electrodes and electrolytes in dye-sensitized solar cells (DSSCs) in recent years. The donor-acceptor super-heterojunction design in covalent organic frameworks (COFs) and the rest of POPs significantly enhances electron de-localization, and shifts absorption to longer wavelengths. With specific stacking configurations, high mobility charges are fetched due to the spatially divided electron or hole transporting π-columns, which favor the charge extraction mechanism. However, the potential of POPs is hindered by singlet states, which impede exciton diffusion mechanism. Since exciton-exciton annihilation (EEA), follows long singlet states lifetime is highly desirable. High crystallinity and regulated domain sizes are needed to achieving this. Conjugated building blocks and post-synthetic modifications can be used for modification of the pore sizes and the polymer skeleton, which is interdependent. Research indicates that interfacial features are linked to crystallinity and solubility, necessitating a comprehension of planarity, bond rotations, and bond stretching.

Even though POPs typically boast excellent absorption extending into the near-infrared (NIR) region, their overall low external quantum efficiency (EQE) significantly compromises their power conversion efficiency (PCE). This indicates that charge dissociation plays a critical part. Revisiting the basic features of exciton-exciton annihilation (EEA), it is clear that energy transfer between two excited electrons requires consumption of energy, with only half of the excited electron advancing to the next charge dynamic phase. This shorter diffusion length in COFs exacerbates charge delocalization as well as self-annihilation process. Since EEA is sensitive to excited-state population, which makes COFs less viable for interior applications. Thus, utilizing COFs as the prime photoactive material presents significant adversaries for reaching desired goals. For photovoltaics that exploit the host-guest structural arrangement of COFs, respective essential issues must be addressed to achieve higher efficacy devices. Given that, the primary photophysical processes share many analogies with conventional donor-acceptor bulk (D-A) hetero-juction solar cell, the efficacyof dissociation can be engineered by adjusting the driving force. Further insights into excited-state dynamics and charge behavior, such as exciton splitting efficiency, are necessary. Multi objective modeling can help optimize energy levels and simultaneously improve overall device performance (Gasparini et al. 2021). However, incorporating the desired guest molecule into the COF framework is a tough challenge. Vapor deposition of C_{60} is the most commonly reported method, but incorporating highly promising polymers and emerging small molecules may require advanced techniques to retain the guests within the framework and adjust interfacial properties effectively.

The unique interplay of pore and skeleton design of POPs offers various building blocks and different topologies, providing unending applications in solar cells engineering. Despite the addition of exciting tectons, synthetic methods, and emerging materials, a corresponding boost in efficiency has left to covered. This underscores the demand of understanding the photophysical mechanism. There are only short number of literatures that precisely implement POPs as photoactive material correlate POP structure with observed charge transport and localization and energy losses. Characterizing common POPs is challenging due to their rigidity and insolubility, which significantly hampers the investigation of crystalline COFs. It took more than a decade to fully unravel the photophysical processes and underlying mechanisms since COFs were first invented, and a method to precisely probe charge yield through EEA has yet to be reported. Unlike traditional linear polymer characterization, where film investigations directly represent device component behavior, POPs rely more heavily on powder for experimental analysis. Upon polycondensation, several reports have noted structural variations in solid-state POP films due to stacking and steric hindrance. Thus, bridging the gap between experimental samples and photoactive components is urgent for effectively controlling the photophysical properties of POPs. Fortunately, with increased computing power and available molecular models, versatile density functional theory (DFT) calculations can help reconstruct intra- and intermolecular interactions at an atomic level. Leveraging existing knowledge about monomers and side groups, the actual molecular orientation and energetic behavior can be predicted. This molecular-level perspective can enhance the interpretation of characterization results and enable chemists to produce more customized and precise polymers with the desired electronic properties. Thus, bridging the gap between experimental samples and photoactive components is urgent for effectively controlling the photophysical properties of POPs. Fortunately, with increased computing power and available molecular models, versatile density functional theory (DFT) calculations can help reconstruct intra- and intermolecular interactions at an atomic level. Leveraging existing knowledge about monomers and side groups, the actual molecular orientation and energetic behavior can be predicted. This molecular-level perspective can enhance the interpretation of characterization results and enable chemists to produce more customized and precise polymers with the desired electronic properties. Regarding computer-aided approaches, multiple machine learning techniques have been employed to optimize perovskite material combinations. A similar approach can be extended to POPs to construct highly conjugated structures for effective charge delocalization. Additionally, advanced signal processing techniques may address the poor optical spectral features obtained from POPs. For example, specific transient absorption spectroscopy (TAS) features might be distinguishable, promoting more in-depth studies of excited dynamics. To enhance the commercial appeal of POPs, a different strategy should be adopted. Rather than focusing solely on improving power conversion efficiency (PCE), long-term stability and structural advantages should be further explored. Multiple studies have reported the positive encapsulation effect achieved through the incorporation of rigid POPs, highlighting the need for extensive degradation studies. Recent advances in understanding thermal/photodegradation involving ambient species (O_2/H_2O), thermodynamic-related degradation, and the impact of triplet states on stability have elucidated the molecular origins of these processes (Duan et al. 2020; Gasparini et al. 2021; Kawano et al. 2006; Zhang et al. 2022).

Precise control over molecule vibration, bond strain, twisting, and packaging is necessary for future molecular design. The amorphous nature POPs can be modified for certain applications, including increasing the catalytic activity of DSSC or perovskite crystallinity. It is possible to easily fabricate POPs that can be processed in a solution, according to recent reports. HTLs, ETLs, and primary photoactive components are examples of emerging usage. POPs are perfect for PSCs and DSSCs because of their superior surface passivation, charge transfer, hydrophobicity, and stability. To improve device implementation, more research is required to connect POP characteristics with charge dynamics. For amorphous POPs, flexible, affordable processing methods are crucial.

References

Ahmad S, Yum J-H, Butt H-J, Nazeeruddin MK, Grätzel M (2010) Efficient platinum-free counter electrodes for dye-sensitized solar cell applications. ChemPhysChem 11(13):2814–2819. https://doi.org/10.1002/cphc.201000612

Ahmad S, Bessho T, Kessler F, Baranoff E, Frey J, Yi C, Grätzel M, Nazeeruddin MK (2012) A new generation of platinum and iodine free efficient dye-sensitized solar cells. Phys Chem Chem Phys 14(30):10631–10639. https://doi.org/10.1039/C2CP41611E

Ali N, Shehzad N, Uddin S, Ahmed RS, Jabeen M, Kalam A, Al-Sehemi AG, Alrobei H, Kanoun MB, Khesro A, Goumri-Said S (2021) A review on perovskite materials with solar cell prospective. Int J Energy Res 45:19729–19745

Ang T-Z, Salem M, Kamarol M, Das HS, Nazari MA, Prabaharan N (2022) A comprehensive study of renewable energy sources: classifications, challenges and suggestions. Energ Strat Rev 43:100939. https://doi.org/10.1016/j.esr.2022.100939

Bildirir H, Gregoriou VG, Avgeropoulos A, Scherf U, Chochos CL (2017) Porous organic polymers as emerging new materials for organic photovoltaic applications: current status and future challenges. Mater Horiz 4(4):546–556. https://doi.org/10.1039/C6MH00570E

Boughey FL, Davies T, Datta A, Whiter RA, Sahonta SL, Kar-Narayan S (2016) Vertically aligned zinc oxide nanowires electrodeposited within porous polycarbonate templates for vibrational energy harvesting. Nanotechnology 27:28LT02

Choi Y, Byun H, Cho Y, Youm K, Hsan N, Kumar S, Koh J (2024) Effects of aromatic linker structures on carbon dioxide adsorption and conversion performance in melamine-based porous organic polymers. Fibers Polym. https://doi.org/10.1007/s12221-024-00600-w

Duan L, Guli M, Zhang Y, Yi H, Haque F, Uddin A (2020) The air effect in the burn-in thermal degradation of nonfullerene organic solar cells. Energ Technol 8(5):1901401. https://doi.org/10.1002/ente.201901401

Fan Y, Meng X, Li H, Kuang S, Zhang L, Wu Y, Wang Z (2016) Stretchable porous carbon nanotube elastomer hybrid nanocomposite for harvesting mechanical energy. Adv Mater 29. https://doi.org/10.1002/adma.201603115

Fu Q, Wang T, Sun Y, Zheng N, Xie Z, Lu D, Xu Z, Wan X, Zhang Y, Liu Y (2021a) A solution-processed nanoscale COF-like material towards optoelectronic applications. Sci China Chem 64(1):82–91. https://doi.org/10.1007/s11426-020-9865-3

Fu Q, Xu Z, Tang X, Liu T, Dong X, Zhang X, Zheng N, Xie Z, Liu Y (2021b) Multifunctional two-dimensional conjugated materials for dopant-free perovskite solar cells with efficiency exceeding 22%. ACS Energy Lett 6(4):1521–1532. https://doi.org/10.1021/acsenergylett.1c00385

Gasparini N, Camargo FVA, Frühwald S, Nagahara T, Classen A, Roland S, Wadsworth A, Gregoriou VG, Chochos CL, Neher D, Salvador M, Baran D, McCulloch I, Görling A, Lüer L, Cerullo G, Brabec CJ (2021) Adjusting the energy of interfacial states in organic photovoltaics for maximum efficiency. Nat Commun 12(1):1772. https://doi.org/10.1038/s41467-021-22032-3

Ghosh SK, Sinha TK, Mahanty B, Mandal D (2015) Self-poled efficient flexible "Ferroelectretic" nano generator: a new class of piezoelectric energy harvester. Energ Technol 3(12):1190–1197. https://doi.org/10.1002/ente.201500167

Gong J, Liang J, Sumathy K (2012) Review on dye-sensitized solar cells (DSSCs): fundamental concepts and novel materials. Renew Sust Energ Rev 16(8):5848–5860. https://doi.org/10.1016/j.rser.2012.04.044

Hao P, Zhao Z, Tian J, Li H, Sang Y, Yu G, Cai H, Liu H, Wong CP, Umar A (2014) Hierarchical porous carbon aerogel derived from bagasse for high performance supercapacitor electrode. Nanoscale 6(20):12120–12129. https://doi.org/10.1039/C4NR03574G

Kang DW, Kang M, Hong CS (2020) Post-synthetic modification of porous materials: superprotonic conductivities and membrane applications in fuel cells. J Mater Chem A 8(16):7474–7494. https://doi.org/10.1039/D0TA01733G

Kawano K, Pacios R, Poplavskyy D, Nelson J, Bradley DDC, Durrant JR (2006) Degradation of organic solar cells due to air exposure. Sol Energy Mater Sol Cells 90(20):3520–3530. https://doi.org/10.1016/j.solmat.2006.06.041

Kim S, Choi SJ, Zhao K, Yang H, Gobbi G, Zhang S, Li J (2016) Electrochemically driven mechanical energy harvesting. Nat Commun 7(1):10146. https://doi.org/10.1038/ncomms10146

Krishnakumar B, Kumar S, Gil JM, Mani D, Arivanandhan M, Sobral AJFN (2019) Synthesis and characterization of g/Ni–SiO2 composite for enhanced hydrogen storage applications. Int J Hydrog Energy 44(41):23249–23256. https://doi.org/10.1016/j.ijhydene.2019.07.073

Kuhn P, Antonietti M, Thomas A (2008) Porous, covalent Triazine-based frameworks prepared by Ionothermal synthesis. Angew Chem Int Ed 47(18):3450–3453. https://doi.org/10.1002/anie.200705710

Kumar NS, Naidu KCB (2021) A review on perovskite solar cells (PSCs), materials and applications. J Mater 7(5):940–956. https://doi.org/10.1016/j.jmat.2021.04.002

Kumar S, Kumar Mishra D, Sobral AJFN, Koh J (2019) CO2 adsorption and conversion of epoxides catalyzed by inexpensive and active mesoporous structured mixed-phase (anatase/brookite) TiO2. J CO2 Util 34:386–394. https://doi.org/10.1016/j.jcou.2019.07.019

Kumar S, Singh I, Hsan N, Swain BS, Koh J (2023) Synthesis of chitosan-based perylene dye material for photovoltaic solar-cell application. Int J Biol Macromol 253:126964. https://doi.org/10.1016/j.ijbiomac.2023.126964

Lavate S, Kumar S, Seena S, Srivastava R (2023) Chapter 11 – recent advances in environmental nanotechnology. In: Seena S, Rai A, Kumar S (eds) Nanoparticles and plant-microbe interactions. Academic Press, pp 293–318. https://doi.org/10.1016/B978-0-323-90619-7.00004-7

Lee J-SM, Cooper AI (2020) Advances in conjugated microporous polymers. Chem Rev 120(4):2171–2214. https://doi.org/10.1021/acs.chemrev.9b00399

Li Q, Wu J, Tang Q, Lan Z, Li P, Lin J, Fan L (2008) Application of microporous polyaniline counter electrode for dye-sensitized solar cells. Electrochem Commun 10(9):1299–1302. https://doi.org/10.1016/j.elecom.2008.06.029

Lim J, Kim M-S, Jang W, Wang DH, Park JK (2020) Solution-processable porous organic polymer for tailoring the charge transport property of planar perovskite solar cells. Dyes Pigments 178:108332. https://doi.org/10.1016/j.dyepig.2020.108332

Lu H, Wang C, Chen J, Ge R, Leng W, Dong B, Huang J, Gao Y (2015) A novel 3D covalent organic framework membrane grown on a porous α-Al2O3 substrate under solvothermal conditions. Chem Commun 51(85):15562–15565. https://doi.org/10.1039/C5CC06742A

Maschita J, Banerjee T, Savasci G, Haase F, Ochsenfeld C, Lotsch BV (2020) Ionothermal synthesis of imide-linked covalent organic frameworks. Angew Chem Int Ed 59(36):15750–15758. https://doi.org/10.1002/anie.202007372

Mohamed MG, Lee C-C, EL-Mahdy AFM, Lüder J, Yu M-H, Li Z, Zhu Z, Chueh C-C, Kuo S-W (2020) Exploitation of two-dimensional conjugated covalent organic frameworks based on tetraphenylethylene with bicarbazole and pyrene units and applications in perovskite solar cells. J Mater Chem A 8(22):11448–11459. https://doi.org/10.1039/D0TA02956D

Park S, Kim M-S, Jang W, Park JK, Wang DH (2018) Covalent organic nanosheets for effective charge transport layers in planar-type perovskite solar cells. Nanoscale 10(10):4708–4717. https://doi.org/10.1039/C7NR08797G

Pushparaj VL, Shaijumon MM, Kumar A, Murugesan S, Ci L, Vajtai R, Linhardt RJ, Nalamasu O, Ajayan PM (2007) Flexible energy storage devices based on nanocomposite paper. Proc Natl Acad Sci 104(34):13574–13577. https://doi.org/10.1073/pnas.0706508104

Rashidi S, Esfahani JA, Rashidi A (2017) A review on the applications of porous materials in solar energy systems. In: Renewable and sustainable energy reviews, vol 73. Elsevier Ltd., pp 1198–1210. https://doi.org/10.1016/j.rser.2017.02.028

Rong Y, Hu Y, Mei A, Tan H, Saidaminov MI, Seok SI, McGehee MD, Sargent EH, Han H (2018) Challenges for commercializing perovskite solar cells. Science 361(6408):eaat8235. https://doi.org/10.1126/science.aat8235

Sadasivuni KK, Cabibihan J-J, Deshmukh K, Goutham S, Abubasha MK, Gogoi JP, Klemenoks I, Sakale G, Sekhar BS, Rama Sreekanth PS, Rao KV, Knite M (2019) A review on porous polymer composite materials for multifunctional electronic applications. Polym-Plast Technol Mater 58(12):1253–1294. https://doi.org/10.1080/03602559.2018.1542729

Sarkar C, Shit SC, Das N, Mondal J (2021) Presenting porous–organic–polymers as next-generation invigorating materials for nanoreactors. Chem Commun 57(69):8550–8567. https://doi.org/10.1039/D1CC02616J

Shalini S, Balasundaraprabhu R, Thandalam S, Prabavathy N, Sundaram S, Prasanna S (2016) Status and outlook of sensitizers/dyes used in dye sensitized solar cells (DSSC): a review: sensitizers for DSSC. Int J Energy Res 40:1303. https://doi.org/10.1002/er.3538

Shen L, Shi Q, Zhang S, Gao J, Cheng DC, Yi M, Song R, Wang L, Jiang J, Karnik R, Zhang S (2021) Highly porous nanofiber-supported monolayer graphene membranes for ultrafast organic solvent nanofiltration. Sci Adv 7(37):eabg6263. https://doi.org/10.1126/sciadv.abg6263

Shih C-M, Wu Y-L, Wang Y-C, Kumar SR, Tung Y-L, Yang C-C, Lue SJ (2018) Ionic transport and interfacial interaction of iodide/iodine redox mechanism in agarose electrolyte containing colloidal titanium dioxide nanoparticles. J Photochem Photobiol A Chem 356:565–572. https://doi.org/10.1016/j.jphotochem.2018.01.034

Snaith HJ (2013) Perovskites: the emergence of a new era for low-cost, high-efficiency solar cells. J Phys Chem Lett 4(21):3623–3630. https://doi.org/10.1021/jz4020162

Sugathan V, John E, Sudhakar K (2015) Recent improvements in dye sensitized solar cells: a review. Renew Sust Energ Rev 52:54–64. https://doi.org/10.1016/j.rser.2015.07.076

Thomas M, Rajiv S (2020) Porous membrane of polyindole and polymeric ionic liquid incorporated PMMA for efficient quasi-solid state dye sensitized solar cell. J Photochem Photobiol A Chem 394:112464. https://doi.org/10.1016/j.jphotochem.2020.112464

Troschke E, Grätz S, Lübken T, Borchardt L (2017) Mechanochemical Friedel–crafts alkylation – a sustainable pathway towards porous organic polymers. Angew Chem Int Ed 56(24):6859–6863. https://doi.org/10.1002/anie.201702303

Wang Y, Zhang S, Wu J, Liu K, Li D, Meng Q, Zhu G (2017) Electropolymerization porous aromatic framework film as a hole-transport layer for inverted perovskite solar cells with superior stability. ACS Appl Mater Interfaces 9(50):43688–43695

Wu M, Shan Z, Wang J, Gu Z, Wu X, Xu B, Zhang G (2021) Three-dimensional covalent organic frameworks based on a π-conjugated tetrahedral node. Chem Commun 57(80):10379–10382. https://doi.org/10.1039/D1CC03219D

Xu B, Li S, Jiao H, Yin J, Liu Z, Zhong W (2020) A two-dimensional quinazoline based covalent organic framework with a suitable direct gap and superior optical absorption for photovoltaic applications. J Mater Chem A 8(7):3865–3871. https://doi.org/10.1039/C9TA12136F

Zhang P, Dai S (2017) Mechanochemical synthesis of porous organic materials. J Mater Chem A 5(31):16118–16127. https://doi.org/10.1039/C7TA04829G

Zhang J, Xie S, Zhang X, Lu Z, Xiao H, Li C, Li G, Xu X, Chen X, Bo Z (2017) Hyperbranched polymer as an acceptor for polymer solar cells. Chem Commun 53(3):537–540. https://doi.org/10.1039/C6CC07335B

Zhang T, Gregoriou VG, Gasparini N, Chochos CL (2022) Porous organic polymers in solar cells. Chem Soc Rev 51(11):4465–4483. https://doi.org/10.1039/d2cs00123c

Zhu J, Hu L, Zhao P, Wong KY (2020) Recent advances in electrocatalytic hydrogen evolution using nanoparticles. Chem Rev 120(2):851–918

Chapter 13
Advanced Bio-based Materials for Renewable Energy Development

Ira Singh, Santosh Kumar, Satyendra Singh, and G. L. Devnani

Abstract To address the growing environmental issues related with synthetic materials, it is imperative to produce sustainable and eco-friendly materials. Research on bio-based materials, such as biofibers, biopolymers, biofilms, and biocomposites, has exploded in the last few years. An extensive summary of the most recent developments in various fields is given in this book chapter. The remarkable mechanical, chemical, and physical qualities of biofibers sourced from plants and animals make them attractive options for use as reinforcing agents in polymer composites. PLA, PHA, and starch-based materials are examples of biopolymers that have drawn a lot of interest as possible substitutes for synthetic plastics. Biofilms derived from biopolymers have superior barrier qualities and antimicrobial efficacy and hold promise for intelligent packaging uses. Biofiber-reinforced biocomposites have improved mechanical and biodegradability characteristics. Furthermore, protein based biomaterials have special chances for the development of sustainable materials. This book chapter emphasizes the significance of looking into new sources of biorenewable materials and the requirement for additional study to address current issues. Bio-based materials are positioned to revolutionize a number of industries and help create a more environmentally friendly future due to the increasing demand for sustainable solutions.

Keywords Sustainable · Eco-friendly materials · Biocomposites · Biofibers · Biopolymers · Environmentally

I. Singh · S. Kumar (✉)
Department of Chemistry, School of Basic and Applied Sciences, Harcourt Butler Technical University, Kanpur, Uttar Pradesh, India
e-mail: santoshk@hbtu.ac.in

S. Singh · G. L. Devnani
Department of Chemical Engineering, Harcourt Butler Technical University, Kanpur, Uttar Pradesh, India

S. Kumar, V. K. Singh (eds.), *Renewable Energy Development: Technology, Material and Sustainability*, Clean Energy Production Technologies,
https://doi.org/10.1007/978-981-97-9626-7_13

13.1 Introduction

Environmental problems are currently the biggest challenge facing humans and other living things, aside from energy. Many synthetic materials and products have been developed over time to meet human needs. These artificial materials may have harmful effects on life and the environment on land, in the water, and in the air (Geyer et al. 2017; Stafford and Jones 2019). Potentially novel approaches that preserve the environment must be created in order to address these grave concerns. Researchers have been increasingly interested in bio-based materials, such as biofibers, biopolymers, and biocomposites, as a result of increased environmental awareness over the past few decades. These materials will eventually replace synthetic materials (Liu et al. 2019; Mohan and Kanny 2019). Natural fibers can be categorized as coming from plants, animals, or minerals, depending on whence they were extracted. These natural fibers are used as reinforcement in polymer matrices to form bio-based composites and polymer composites (Mazzanti et al. 2019; Sari et al. 2019). Researchers are becoming more interested in natural fibers because of their exceptional qualities, which include low density, low cost, easy availability, biodegradability, and ease of processing. In addition, it has strong fracture resistance and good mechanical, thermal, and acoustic qualities. Natural fibers are being used in place of synthetic materials in polymer composites and bio-based composites, according to numerous research. As a result, numerous sectors have moved to transform the application of bio-based products (Antunes et al. 2019; Battegazzore et al. 2019; Fombuena et al. 2019; Mohammad et al. 2019). Natural fibers, biopolymers, biofilms, and bio-based composites have all seen a discernible rise in use in a variety of adaptable applications in recent years, including home, automotive, and aerospace products. Thermal and acoustic insulators, food packages, interior constructions, pilot control panels, wing boxes, cabin panels, etc. are a few of the aerospace uses that have been investigated. Boeing put out a number of noteworthy initiatives to recycle spent biomaterials once an aircraft's useful life ended (Arockiam et al. 2018). Bio-based materials have a wide range of adaptable uses, including furniture, upholstery, train car interior panels, gardening supplies, packaging, building materials, and sporting equipment (Bakar et al. 2019; Kumar et al. 2019; Nagalakshmaiah et al. 2019; Sharma et al. 2019). Owing to some drawbacks, such as compatibility, hydrophilicity, and significant mechanical strength, bio-based composites are utilized in lightweight constructions and applications. Because bio-based products are being studied and used more frequently, there is a significant demand for bio-based raw materials. Many scientists were compelled by this urge to find other sources of biorenewable materials in order to stop the overuse of this resource. It increased the productivity of raw materials through animal and microorganism raising, agriculture, and plant cultivation. This offers a number of potential benefits, including lowering greenhouse gas emissions and global warming by increasing job prospects. The present study offers a comprehensive overview of the latest advancements in biofibers, biofilms, biopolymers, and biocomposites.

These developments aid industry and the engineering sector in creating sophisticated bio-based composites that may find use in agriculture (plant cultivation), animal husbandry, and microorganism rearing (Murawski et al. 2019; Payal 2019; Sanjay et al. 2018).

13.2 Sustainable Biomaterials

It is necessary to take into account the material's whole life cycle in order to comprehend what a sustainable biomaterial is. This covers the qualities of durable and desirable materials as well as the effects of manufacturing on the environment at large, encompassing all stages from initial manufacture to processing, deterioration, recycling, and final disposal at the end of the product's useful life (Lefèvre and Auger 2016; Biswal et al. 2020; Zhu et al. 2016). These characteristics allow for the classification of a sustainable biomaterial as one that is:

- Derived from renewable materials that are grown responsibly.
- Manufactured with eco-friendly techniques.
- Secure and not harmful to the surroundings.
- Repurposed and reused after fulfilling their initial function, such as fodder for composting or recycling.

Economic viability and societal advantages are crucial additional factors to take into account when determining sustainability standards. When evaluating a sustainable biomaterial's influence on the environment, for example, it should meet a legitimate societal demand so that the advantages it provides can outweigh the ecological footprint of its production (Lefèvre and Auger 2016). These social-economic concerns won't be covered in length in this book chapter, but it's important to emphasize how critical it is that they be integrated, as demonstrated in Fig. 13.1.

This will help achieve the 2030 Sustainable Development Goals set forth by the United Nations and enhance the sustainability of the next generation of biomaterials (United Nations Development Programme 2015).

13.3 Biofibers

Numerous natural sources, such as minerals, plants, and animals, include natural fibers. Numerous factors, including the fibers' geographical location, origin, extraction method, and processing method, impact their qualities. These fibers are employed in a variety of structural applications and polymer matrices as reinforcements. Plant fibers and animal fibers are two categories for biofibers that come from biological sources. In the sections that follow, biofibers are covered in great detail (Chen et al. 2019; Kluge et al. 2008).

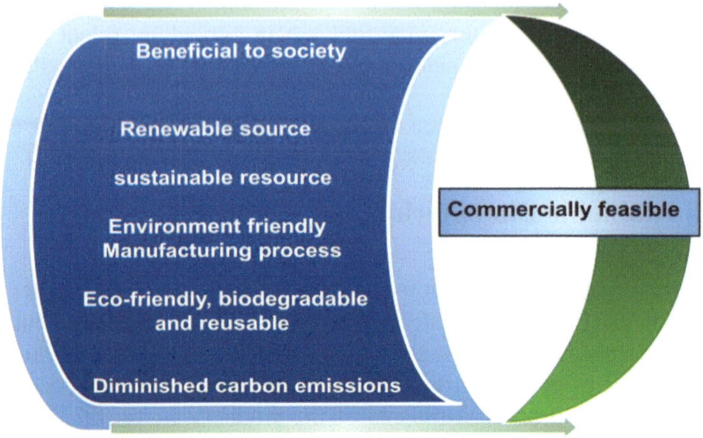

Fig. 13.1 Sustainability standards for materials include considerations of the environment, the economy, and social situations; these ideas are interrelated. (Lefèvre and Auger 2016)

13.3.1 Biofibers from Animals

Animal fibers are primarily divided into two categories based on their origin: hair and secretions. Animal fibers' exceptional mechanical, chemical, and physical qualities make them a possible reinforcing option for polymer matrices (Bharath et al. 2019; Shera et al. 2019). Animal fibers come from a variety of animal species; for example, wool is a type of hair that comes from alpaca, bison, sheep, angora, etc. Similarly, silk comes from a variety of insect sources, primarily from the approximately 14,000 species of butterflies and the 4000 species of spiders that are known to exist. Generally speaking, the textile industries mostly use wool and silk. Current wool research involves electronic uses, such as touch capacitors that light up LEDs by coating knitted wool materials with silver nanoparticles. Additional research on this technology may be used to develop biomedical devices that track motions in the joints and muscles (Gurarslan et al. 2019). Tissue engineering and surgical implants are two biological uses for silks. Because of its mechanical performance and endurance, research on spider silk has shown greater promise than that on other silk proteins. Synthetic spider silk-reinforced polymer composites may soon be utilized to create sustainable, high-strength composites as well as biological implants (Shehata et al. 2019). The primary component of chicken feathers, which are a waste byproduct from slaughterhouses, is keratin. Because of its absorption properties, keratin is used to make absorbent sponges for purification and in microbial corrosion-resistant applications. A recent study generated keratin sponges that have a higher propensity to absorb oil. Further research into keratin-based materials will enable large-scale fabrication of composite sponges that can be used to clean lenses, highly polished surfaces, and precision engineering industries (Sadeghi et al. 2019). Wool is a popular material for textiles. Wool's physicochemical characteristics vary depending on where it comes from. For instance, the fiber diameters of alpaca, angora, and qiviut wool are, correspondingly, 12–29, 12–16, and 15–20 mm (Ramamoorthy et al. 2015). Keratin, an animal protein, is the primary component of wool fibers, animal hairs, and chicken feathers. In addition to being used in textile applications as a sustainable waste management technique, wool waste was used to create composite fibers with a tensile strength of 157 MPa, which are based on keratin and graphene oxide. In polymer composites, this composite fiber has the potential to replace natural fibers like coir, banana, etc. (Li et al. 2019). Insect-secreted silk is extensively utilized in the biomedical and textile industries. Chitin, a highly organized protein found in silk, gives the material its excellent mechanical strength and strong chemical resistance. Different species have different protein structures. Throughout its life, a spider produces a silk protein called dragline silk (*Nephila*). It is the strongest silk among the *Nephila* species, with a tensile strength of 1.1 GPa, which is equivalent to high tensile steel (1.5 GPa) (Yazawa et al. 2019). Fibrin and sericin make up the two primary components of mulberry silk, the type of silk that is most frequently employed. There are 5263 heavy chain and 262 light chain amino acids in fibroin. While it is utilized below 170° C, it is extremely stable at temperatures as high as 250 °C. Its tensile strength of around 600 MPa is greater than most of the plant fibers indicated in Table 13.1. As a result, it is employed in the creation of medical scaffolds, tissue engineering, and low- and high-density polyethylene (LDPE) reinforcements, among other applications (Ramamoorthy et al. 2015).

13.3.2 Biofibers from Plants

Fibers extracted from plants and trees primarily consist of cellulose, hemicellulose, lignin, pectin, wax, etc. The mechanical, morphological, chemical, and physical properties of plant fibers are all influenced by the environment, the part of the plant or trees from which the fibers are gathered, and the extraction technique. The term "ligino-cellulose fibers" refers to plant-derived fibers that can take the shape of thick fibers like coir, hair like kapok, cotton, etc., or bast fibers like jute, kenaf, kudzu, linden, etc. Primary and secondary utility fibers from plants are separated based on the plant or tree's intended use. The main fiber-producing plants that are grown for their fiber are sisal, jute, hemp, cotton, kenaf, etc. Secondary utility fibers are a byproduct of oil palm, banana, pineapple, and coir plants, among others (Chen et al. 2019; Faruk et al. 2012). Plant fibers are further categorized according to their physiological characteristics and place of origin. The fibers derived from the stem include fruit fibers (coir, luffa, etc.), grass fibers (bamboo, wheat straw, bagasse, etc.), wood fibers (teak wood, rosewood), bast fibers (flax, hemp, kenaf, jute, isora, etc.), and seed fibers (cotton, soya, kapok, curaua, palm, etc.). (Chen et al. 2019; Adebayo et al. 2021; Navaneethakrishnan et al. 2020; Teixeira et al. 2019). Figure 13.2 illustrates the many plant fiber varieties. Numerous novel plant fibers were discovered in 2019 that could be used as reinforcement for polymer matrices. These fibers comprised *Ficus racemosa*, *Carica papaya* bark, *Tridax procumbens*, *Ficus halicababum*, *Impomoea pescaprae*, *Catharanthus roseus*, *Parthenium hysterophorus*, *Saccharum bengalense* grass, *Cereus hildmannianus*, *Morus alba*

Table 13.1 A comparison of the thermal and chemical characteristics of newly discovered plant fibers with those that are sold commercially (Vinod et al. 2020)

S. no	Natural cellulose fibers	Chemical properties				Crystalline properties		Thermal properties	References
		a-Cellulose (wt %)	Hemi cellulose (wt %)	Lignin (wt %)	Moisture (wt %)	Crystalline index (%)	Crystal size (nm)	Thermal stability	
1.	*Coccinia grandis* L.	62.35	13.42	15.61	5.6	52.17	13.38	213.4	Senthamaraikannan and Kathiresan (2018)
2.	*Leucas aspera*	50.7	13.2	9.7	11.31	20.23	6.7	203	Vijay et al. (2021)
3.	*Dracaena reflexa*	70.32	11.02	11.35	5.19	57.32	19.01	232.32	Manimaran et al. (2019b)
4.	*Cereus hildmannianus*	58.40	17.14	10.36	8.86	40.19	28.27	285.9	Subramanian et al. (2021)
5.	*Parthenium hysterophorus*	51.5	9.7	14.3	8.6	40.68	–	255	Vijay et al. (2019d)
6.	Thespesia populnea	70.27	12.64	16.34	10.83	48.17	–	320	Kathirselvam et al. (2019a, b)
7.	*Catharanthus roseus*	47.3	9.1	15.1	8.3	25.9	–	203	Vinod et al. (2019)
8.	*Tridax procumbens*	32	6.8	3	11.2	34.46	25.04	270	Vijay et al. (2019b)

Fig. 13.2 Varieties of plant fibers

L. stem, and *Coccinia grandis* stem (Jebadurai et al. 2019; Saravana Kumaar et al. 2019; Manimaran et al. 2019a; Subramanian et al. 2021; Vijay et al. 2019a, b, c, d; Vinod et al. 2019, 2021). Plant fiber applications shouldn't be restricted to known species. Due to the cultivation of crops that follow a similar trend, it may result in the extinction of species and an ecological imbalance. Thus, in order to meet the need for raw materials, it is imperative to discover new ligino-cellulose fibers. This would incentivise scientists and researchers to develop a range of commercial products based on biotechnology.

13.4 Biopolymers

Synthetic polymer waste is being generated at an alarming rate. Less than 10% of synthetic plastics manufactured are recycled, according to research. This raises questions about synthetic polymer synthesis. The use of biopolymers as a synthetic

Table 13.2 The properties and uses of biopolymers (Vinod et al. 2020)

S. no	Biopolymer	Property	Application	References
1.	Acetate of cellulose	Excellent foam formation and easily spun as thread.	Air filters, photo films, textiles, and thin frames	Huang and Dean (2020)
2.	Microcrystalline cellulose	Excellent absorption, nontoxic, and easily pliable	Emulsifiers and stabilizers in the food sector	Ahsan et al. (2020)
3.	Methylcellulose	Tensile strength is increased by the additive nature	Materials for construction, adhesives, and cosmetics	Schmidt et al. (2020)
4.	Carboxymethyl cellulose	Stabilizer and alters viscosity, increases tensile strength	Biomedical, food, textile, and paper	Miyashiro et al. (2020)
5.	Hemicellulose and acrylic Acid copolymer	Excellent heavy metal adsorbent for Pb, cd, and Zn	Purification of solutions	Chen et al. (2020)
6.	Carboxymethyl chitosan hemicellulose	–	Used in water purification	Gopi et al. (2019a, b)
7.	Polyaniline-lignosulfonate/epoxy Starch-sodium lignosulfonate	Antioxidant and resistant to corrosion	The food, coatings, and cosmetics industries	Yao et al. (2020)
8.	Lignopolyurethanic-sodium lignosulfonate	Strong mechanical properties	Boosts the flexural modulus	da Silva et al. (2019)

polymer substitute has grown recently due to environmental concerns. Numerous biodegradable polymers have been created to far, including gluten, polylactic acid (PLA), poly-hydroxy-alkanoates (PHA), poly-3-hydroxybutyrate (PHB), polyhydroxyvalerate (PHV), and poly-hydroxyhexanoate (PHH) derived from sustainable resources. Biopolymers have worse thermomechanical characteristics compared to synthetic polymers. Biopolymers are often employed extensively in the food, pharmaceutical, and biomedical industries. A selection of the most recently investigated biopolymers and their prospective uses is shown in Table 13.2.

13.4.1 Classification of Biopolymers

Agro-resources (plants) and microorganisms (fungi) are the origins of biopolymers, which are made from bioderived monomers (Vinayagamoorthy and Venkatakoteswararao 2020). Sources include bacteria, plants, and animals. The biopolymers PLA, PHA, PHB, PHV, and PHH, as well as proteins, are generated from these sources. Carbohydrates, often called polysaccharides, are agro-based polymers. $Cx(H_2O)y$ is the universal chemical formula. Polysaccharides are created

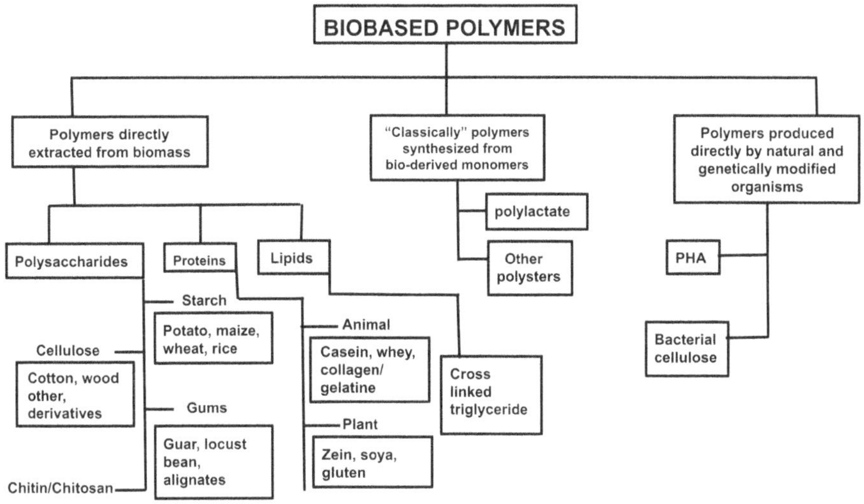

Fig. 13.3 Categorization of biopolymers. (Gurunathan et al. 2015)

when a lot of monosaccharides, such as ketones or hydroxyaldehydes, mix. Carrageenan, cellulose, chitin, alginate, xanthan gum, and dextrin are among the significant polysaccharides (Luzi et al. 2019). Because of its appealing qualities of availability, thermoplastic behavior, recyclability, and renewability, starch is seen as a promising material (Nevoralová et al. 2019). There is a list of the biopolymers' classifications in Fig. 13.3.

Biopolymer degradation is influenced by a number of variables, including composition, environment, polymer type, and chemical interactions between polymers. This is how the deterioration process is categorized:

- Biodegradable: Microorganisms are responsible for the degradation process.
- Hydro-biodegradable: Breakdown takes place when water and microorganisms are present.
- Delinking of molecules in the presence of light is known as photodegradation.
- Bioerodable: Erosion brought on by abrasion from the environment.
- Compostable: Bacterial action leads to degradation, which enhances soil quality.

13.5 Biofilms

Biofilms are made from biopolymers such as PLA, PHA, PHBV, starch, cellulose, proteins, gluten, PEEK, etc. Studies show that these biopolymers have a high film-forming capacity and can be used in a range of industrial contexts, such as the

Table 13.3 Natural plasticizers utilized in biomass products (polysaccharide, protein, and lipid-based films) and other films made by extracting microorganisms to create biodegradable films (Vieira et al. 2011)

Type	System of application	Plasticizer	References
Polysaccharide-based films	Rice starch modified with citric acid (CARS) and pea starch modified with citric acid (CAPS)	GLY	Ma et al. (2009)
	c-carrageenan edible films	GLY and water	Karbowiak et al. (2006)
	eatable flicks GLY together with water (Miyashiro et al. 2020)	GLY and EG	Smits et al. (2003)
Protein-, lipid-based films	Zein	Acids oleic and linoleic	Budi Santosa and Padua (1999)
	Caseinate-pullulan	Sorbitol and water	Kristo an Biliaderis (2006)
	Whey protein	Sorbitol and GLY	Kim and Ustunol (2001)
Other films (from microbial sources)	*Amaranthus cruentus* flour edible films	GLY	Colla et al. (2006)
	Poly(3-hydroxybutyrate) (PHB)	Trilaurin, lauric acid, tributyrin, and dodecanol	Yoshie et al. (2000)
	Poly (3-hydroxybutyrate-co-3-hydroxyvalerate) (PHBV)	Triethyl citrate (TEC), epoxidized soybean oil (ESO), dibutyl phthalate (DBP), and soybean oil (SO)	Choi and Park (2004)

additional considerations (Liu et al. 2019; Ferreira et al. 2019). Because the matrix is hydrophobic and the biofiber is hydrophilic, it is imperative to chemically alter the fibers' surfaces to improve adhesion between the fiber and matrix. By dissolving the hydroxyl and carbonyl groups on the fiber's surface, chemical treatment reduces the hydrophilic character of plant fiber. According to several research, biofibers can be employed as reinforcement in a variety of matrices, including polyester, epoxy, polypropylene, polystyrene, and polylactic acid (Madhu et al. 2019; Teklu et al. 2019).

Since each material's properties vary depending on the processing environment, biocompatibility and selecting the right processing window are two of the main issues encountered in this respect. Creating biocomposites using the right processing methods is crucial to achieving better mechanical, thermal, and physical qualities.

13.6.2 Properties of Composites

The mechanical properties of biocomposite are significantly influenced by temperature, applied force, amount and rate of deformation, and heating and cooling rates. Important variables that affect these conditions include the kind of polymer, crystallinity, molecular weight, morphology, chemical composition, cross linking, plasticization, copolymerization, molecular orientation, concentration, and type of reinforcing. The production environment, type, orientation, and polymer all have an impact on a composite's tensile strength. A few examples of the many mechanical properties are tensile strength, tensile modulus, flexural strength, flexural modulus, yield strength, yield modulus, compressive strength, impact strength, hardness, rate of elongation, storage modulus, loss modulus, Poisson's ratio, creep, micromechanical analysis, etc. Research has indicated that the tensile strength of composites can be increased by chemically treating natural fibers to decrease their hydrophilic nature, which in turn improves the matrix fiber interaction. A similar pattern may be seen in Bio-Polyethylene-Based Composites Reinforced with Alkali and Palmitoyl Chloride-Treated Coffee Silverskin (Dominici et al. 2019). Jute fibers that are oriented longitudinally are found to function better under tensile and flexural loads. Nonetheless, the orientation of the reinforcement is chosen based on the composite's loading application (Laranjeira et al. 2006). Numerous studies have reported a significant increase in hardness, impact strength, interlaminar shear strength, and flexural strength when natural fibers are added to different polymer matrices. In biomedical applications, the biocompatibility of biofiller-reinforced biopolymer is increasingly important; yet, there is a dearth of research on bioimplants, and few are undergoing clinical trials. Future biocompatible implant development is crucial to assisting patients in avoiding the need for subsequent surgery to remove their implants.

13.7 Protein-Based Biomaterials

Many different kinds of protein-based biomaterials are used in various industries, but to meet sustainability requirements, the protein source needs to be renewable, and the whole production process—including the creation of biomaterials—has to be low-energy and environmentally friendly (Zuin and Ramin 2018). In this situation, using proteins from food side streams or agricultural wastes offers a dependable, affordable, and sustainable feedstock (Sözer et al. 2017). Promising avenues for the development of eco-friendly biomaterials include the direct insertion of different fibrous proteins, such as keratin, silk, resilin, and elastin, into biodegradable hierarchical biomaterials (Abascal and Regan 2018; Costa et al. 2018). Recent years have seen the emergence of several proteins as novel sustainable protein biomaterials because of their biocompatibility, accessibility, and simple production methods.

Table 13.4 Proteins utilized in the creation of biomaterials, as well as the sources of these proteins that are readily available and often used (Lefèvre and Auger 2016)

Protein	Traditional source	Alternative source	References
Keratin	Wool fibers, butcher-derived horns, nails, and feathers	Biomimetic recombinant hagfish thread keratin	Shavandi et al. (2017), Grigsby et al. (2020), Girotti et al. (2020), Perotto et al. (2019), Fu et al. (2015)
Collagen	Type I collagen is found in mammalian skin and tendon tissues (of porcine, bovine, and ovine origin); type II collagen is found in cartilaginous tissues of cattle, pigs, and chickens	Recombinant human collagen in different prokaryotic	Sorushanova et al. (2019), Grønlien et al. (2019), Davison-Kotler et al. (2019)
Silk	*Nephila clavipes* and *Araneus diadematus* are two species of spiders; domestic silkworms are *Bombyx mori* (*B. mori*); wild silkworms are *Antheraea pernyi* and *Samia cynthia ricini*	Recombinant silk proteins in different host systems	Aigner et al. (2018), Chambre et al. (2020), Nguyen et al. (2019), Peng et al. (2016)
Resilin and elastin	Tropoelastin-obtained from animals, recombinant manufacture	Protein	Costa et al. (2018), Su et al. (2019)

Nevertheless, there are drawbacks to getting these proteins from natural sources, including variation from batch to batch, impurities, the quantity that is available, the potential for quick biodegradation, and the danger of spreading illness (Aigner et al. 2018; Gauza-Włodarczyk et al. 2017). Biotechnological techniques have been created to manufacture more protein with improved biological safety and consistency in larger quantities (Wang et al. 2019; Chambre et al. 2020; Huang et al. 2018; Werten et al. 2019). Table 13.4 provides a partial list of the different proteins that are used to make biomaterials as well as details on additional and often used sources for each protein.

13.8 Perspectives on Sustainable Biomaterials

One of the most important steps in solving the environmental problems that synthetic materials present is the creation of sustainable biomaterials. Biofibers, biopolymers, biocomposites, and biofilms have surfaced as viable substitutes, presenting an extensive array of appealing attributes and uses. But in order for them to reach their full potential, a few issues must be resolved. These include strengthening the mechanical qualities and long-term durability of biopolymers, optimizing the processing conditions for biocomposites, and promoting the compatibility between biofibers and polymer matrices. Novel sources of biorenewable materials, such as underused plant species and agricultural waste streams, require more investigation. Furthermore, the development of sophisticated processing methods combined with the application of nanotechnology can result in the production of high-performing bio-based materials with customized characteristics. Particularly protein-based biomaterials have a great deal of promise for the development of sustainable materials, but further study is needed to address issues with batch-to-batch variability and possible disease transmission. The complete life cycle of bio-based materials must be taken into account, from obtaining renewable feedstocks to managing end-of-life issues, in order to achieve real sustainability. This entails creating effective recycling and composting plans and evaluating the social and economic advantages and environmental effects of bio-based materials.

13.9 Conclusion

The book chapter provides a thorough analysis of developments in the field of sustainable biomaterials, emphasizing the value of biologically derived materials such as biofibers, biopolymers, biofilms, and biocomposites in addressing environmental problems. It focuses on biomaterials that are made with the intention of being recycled and reused, derived from renewable resources, and produced using eco-friendly processes. Along with reducing negative environmental effects, these materials also have positive social and economic effects, which aligns with the 2030 Sustainable Development Goals. The advantages of biofibers—which come from plants, animals, and minerals—include their low density, low cost, biodegradability, and exceptional mechanical qualities. Because of their advantages over synthetic materials, they are used in many industries, such as the automotive, aerospace, and construction sectors. In particular, the potential of plant and animal fibers—such as jute and hemp—to reinforce polymer matrices and provide environmentally friendly substitutes for traditional materials is being investigated. Despite having somewhat worse thermo-mechanical characteristics than synthetic polymers, biopolymers—which are derived from sustainable resources—are promoted as potential alternatives. They are widely used in the biological, pharmaceutical, and food industries. The paper also explores biofilms derived from biopolymers, highlighting their use in industrial contexts for food packaging, medicine delivery, and wound healing. A detailed examination of their application in various matrices is presented, along with an exploration of the potential of biocomposites reinforced with biofibers to enhance the mechanical properties of composites. In order to improve compatibility with hydrophobic matrices, the document emphasizes the necessity of surface-modifying biofibers. In order to optimize the mechanical, thermal, and physical properties of biocomposites, it also highlights how important it is to select the appropriate processing methods. The book chapter concludes by outlining a vision for sustainable biomaterials' future and highlighting their vital role in promoting a more sustainable and ecologically friendly approach to material science. In order to lessen the effects on the environment and promote a sustainable future, it promotes the ongoing research and development of bio-based substitutes for conventional materials. The thorough analysis emphasizes how sustainable biomaterials have the ability to revolutionize a variety of industries, including aerospace and agriculture, and calls for cooperation and innovation to fully realize this potential.

References

Abascal NC, Regan L (2018) The past, present and future of protein-based materials. R Soc Open Biol 8(10):180113

Adebayo GO, Hassan A, Yahya R, Sarih NM, Odesanya KO (2021) Impact of water saturation on the tensile and thermal properties of heat-treated mangrove/high-density polyethylene composites. J Thermoplast Compos Mater 34(4):508–522

Ahsan HM, Zhang X, Liu Y, Wang Y, Li Y, Li B et al (2020) Stable cellular foams and oil powders derived from methylated microcrystalline cellulose stabilized pickering emulsions. Food Hydrocoll 104:105742

Aigner TB, DeSimone E, Scheibel T (2018) Biomedical applications of recombinant silk-based materials. Adv Mater 30(19):1704636

Alves D, Marques A, Milho C, Costa MJ, Pastrana LM, Cerqueira MA, Sillankorva SM (2019) Bacteriophage ϕIBB-PF7A loaded on sodium alginate-based films to prevent microbial meat spoilage. Int J Food Microbiol 291:121–127

Antunes A, Faria P, Silva V, Brás A (2019) Rice husk-earth based composites: a novel bio-based panel for buildings refurbishment. Constr Build Mater 221:99–108

Arantes ACC, Silva LE, Wood DF, das Graças Almeida C, Tonoli GHD, de Oliveira JE et al (2019) Bio-based thin films of cellulose nanofibrils and magnetite for potential application in green electronics. Carbohydr Polym 207:100–107

Arockiam NJ, Jawaid M, Saba N (2018) Sustainable bio composites for aircraft components. In: Sustainable composites for aerospace applications. Woodhead Publishing, pp 109–123

Bakar NA, Sultan MTH, Azni ME, Ariffin AH (2019) Investigation of the mechanical properties of Napier-grass-reinforced composites for the aerospace industry: a review. In: Durability and life prediction in biocomposites, fibre-reinforced composites and hybrid composites, pp 321–334

Battegazzore D, Abt T, Maspoch ML, Frache A (2019) Multilayer cotton fabric bio-composites based on PLA and PHB copolymer for industrial load carrying applications. Compos Part B 163:761–768

Bharath KN, Manjunatha GB, Santhosh K (2019) Failure analysis and the optimal toughness design of sheep–wool reinforced epoxy composites. In: Failure analysis in biocomposites, fibre-reinforced composites and hybrid composites. Woodhead Publishing, pp 97–107

Biswal T, BadJena SK, Pradhan D (2020) Sustainable biomaterials and their applications: a short review. Mater Today: Proc 30:274–282

Budi Santosa FX, Padua GW (1999) Tensile properties and water absorption of zein sheets plasticized with oleic and linoleic acids. J Agric Food Chem 47(5):2070–2074

Chambre L, Martín-Moldes Z, Parker RN, Kaplan DL (2020) Bioengineered elastin-and silk-biomaterials for drug and gene delivery. Adv Drug Deliv Rev 160:186–198

Chen H, Lin Z, Tan C (2019) Classification of different animal fibers by near infrared spectroscopy and chemometric models. Microchem J 144:489–494

Chen Y, Li Q, Li Y, Zhang Q, Huang J, Wu Q, Wang S (2020) Fabrication of cellulose nanocrystal-g-poly (acrylic acid-co-acrylamide) aerogels for efficient Pb (II) removal. Polymers 12(2):333

Choi JS, Park WH (2004) Effect of biodegradable plasticizers on thermal and mechanical properties of poly (3-hydroxybutyrate). Polym Test 23(4):455–460

Colla E, do Amaral Sobral PJ, Menegalli FC (2006) Amaranthus cruentus flour edible films: influence of stearic acid addition, plasticizer concentration, and emulsion stirring speed on water vapor permeability and mechanical properties. J Agric Food Chem 54(18):6645–6653

Costa F, Silva R, Boccaccini AR (2018) Fibrous protein-based biomaterials (silk, keratin, elastin, and resilin proteins) for tissue regeneration and repair. In: Peptides and proteins as biomaterials for tissue regeneration and repair, pp 175–204

da Silva CG, de Oliveira F, Frollini E (2019) Sugarcane bagasse fibers treated and untreated: performance as reinforcement in phenolic-type matrices based on lignosulfonates. Waste Biomass Valori 10:3515–3524

Davison-Kotler E, Marshall WS, García-Gareta E (2019) Sources of collagen for biomaterials in skin wound healing. Bioengineering 6(3):56

Dominici F, Garcia Garcia D, Fombuena V, Luzi F, Puglia D, Torre L, Balart R (2019) Bio-polyethylene-based composites reinforced with alkali and palmitoyl chloride-treated coffee silverskin. Molecules 24(17):3113

Ezati P, Rhim JW (2020) pH-responsive chitosan-based film incorporated with alizarin for intelligent packaging applications. Food Hydrocoll 102:105629

Faruk O, Bledzki AK, Fink HP, Sain M (2012) Biocomposites reinforced with natural fibers: 2000–2010. Prog Polym Sci 37(11):1552–1596

Ferreira DP, Cruz J, Fangueiro R (2019) Surface modification of natural fibers in polymer composites. In: Green composites for automotive applications. Woodhead Publishing, pp 3–41

Fombuena V, Petrucci R, Dominici F, Jordá-Vilaplana A, Montanes N, Torre L (2019) Maleinized linseed oil as epoxy resin hardener for composites with high bio content obtained from linen byproducts. Polymers 11(2):301

Fu J, Guerette PA, Miserez A (2015) Self-assembly of recombinant hagfish thread keratins amenable to a strain-induced α-helix to β-sheet transition. Biomacromolecules 16(8):2327–2339

Gaspar M, Benkő Z, Dogossy G, Reczey K, Czigany T (2005) Reducing water absorption in compostable starch-based plastics. Polym Degrad Stab 90(3):563–569

Gauza-Włodarczyk M, Kubisz L, Mielcarek S, Włodarczyk D (2017) Comparison of thermal properties of fish collagen and bovine collagen in the temperature range 298–670 K. Mater Sci Eng C 80:468–471

Geyer R, Jambeck JR, Law KL (2017) Production, use, and fate of all plastics ever made. Sci Adv 3(7):e1700782

Girotti A, Gonzalez-Valdivieso J, Santos M, Martin L, Arias FJ (2020) Functional characterization of an enzymatically degradable multi-bioactive elastin-like recombinamer. Int J Biol Macromol 164:1640–1648

Gopi S, Balakrishnan P, Chandradhara D, Poovathankandy D, Thomas S (2019a) General scenarios of cellulose and its use in the biomedical field. Mater Today Chem 13:59–78

Gopi S, Pius A, Kargl R, Kleinschek KS, Thomas S (2019b) Fabrication of cellulose acetate/chitosan blend films as efficient adsorbent for anionic water pollutants. Polym Bull 76:1557–1571

Grigsby WJ, Scott SM, Plowman-Holmes MI, Middlewood PG, Recabar K (2020) Combination and processing keratin with lignin as biocomposite materials for additive manufacturing technology. Acta Biomater 104:95–103

Grønlien KG, Pedersen ME, Sanden KW, Høst V, Karlsen J, Tønnesen HH (2019) Collagen from Turkey (Meleagris gallopavo) tendon: a promising sustainable biomaterial for pharmaceutical use. Sustain Chem Pharm 13:100166

Gurarslan A, Özdemir B, Bayat İH, Yelten MB, Karabulut Kurt G (2019) Silver nanowire coated knitted wool fabrics for wearable electronic applications. J Eng Fiber Fabr 14:1558925019856222

Gurunathan T, Mohanty S, Nayak SK (2015) A review of the recent developments in biocomposites based on natural fibres and their application perspectives. Compos A: Appl Sci Manuf 77:1–25

Hirvikorpi T, Vähä-Nissi M, Nikkola J, Harlin A, Karppinen M (2011) Thin Al2O3 barrier coatings onto temperature-sensitive packaging materials by atomic layer deposition. Surf Coat Technol 205(21–22):5088–5092

Huang H, Dean D (2020) 3-D printed porous cellulose acetate tissue scaffolds for additive manufacturing. Addit Manuf 31:100927
Huang W, Ling S, Li C, Omenetto FG, Kaplan DL (2018) Silkworm silk-based materials and devices generated using bio-nanotechnology. Chem Soc Rev 47(17):6486–6504
Jebadurai SG, Raj RE, Sreenivasan VS, Binoj JS (2019) Comprehensive characterization of natural cellulosic fiber from Coccinia grandis stem. Carbohydr Polym 207:675–683
Karbowiak T, Hervet H, Léger L, Champion D, Debeaufort F, Voilley A (2006) Effect of plasticizers (water and glycerol) on the diffusion of a small molecule in iota-carrageenan biopolymer films for edible coating application. Biomacromolecules 7(6):2011–2019
Kathirselvam M, Kumaravel A, Arthanarieswaran VP, Saravanakumar SS (2019a) Isolation and characterization of cellulose fibers from Thespesia populnea barks: a study on physicochemical and structural properties. Int J Biol Macromol 129:396–406
Kathirselvam M, Kumaravel A, Arthanarieswaran VP, Saravanakumar SS (2019b) Assessment of cellulose in bark fibers of Thespesia populnea: influence of stem maturity on fiber characterization. Carbohydr Polym 212:439–449
Kim SJ, Ustunol Z (2001) Solubility and moisture sorption isotherms of whey-protein-based edible films as influenced by lipid and plasticizer incorporation. J Agric Food Chem 49(9):4388–4391
Kluge JA, Rabotyagova O, Leisk GG, Kaplan DL (2008) Spider silks and their applications. Trends Biotechnol 26(5):244–251
Kristo E, Biliaderis CG (2006) Water sorption and thermo-mechanical properties of water/sorbitol-plasticized composite biopolymer films: Caseinate–pullulan bilayers and blends. Food Hydrocoll 20(7):1057–1071
Kumar R, Ul Haq MI, Raina A, Anand A (2019) Industrial applications of natural fibre-reinforced polymer composites–challenges and opportunities. Int J Sustain Eng 12(3):212–220
Laranjeira E, De Carvalho LH, De L. Silva SM, d'Almeida JRM (2006) Influence of fiber orientation on the mechanical properties of polyester/jute composites. J Reinf Plast Compos 25(12):1269–1278
Lavery LA, Bhavan K, Wukich DK (2019) Biofilm and diabetic foot ulcer healing: all hat and no cattle. Ann Transl Med 7(7):159
Lefèvre T, Auger M (2016) Spider silk inspired materials and sustainability: perspective. Mater Technol 31(7):384–399
Li YB, Liu HH, Wang XC, Zhang XX (2019) Fabrication and performance of wool keratin–functionalized graphene oxide composite fibers. Mater Today Sustain 3:100006
Liu W, Chen T, Fei ME, Qiu R, Yu D, Fu T, Qiu J (2019) Properties of natural fiber-reinforced biobased thermoset biocomposites: effects of fiber type and resin composition. Compos Part B 171:87–95
Luzi F, Puglia D, Torre L (2019) Natural fiber biodegradable composites and nanocomposites: a biomedical application. In: Biomass, biopolymer-based materials, and bioenergy. Woodhead Publishing, pp 179–201
Ma X, Chang PR, Yu J, Stumborg M (2009) Properties of biodegradable citric acid-modified granular starch/thermoplastic pea starch composites. Carbohydr Polym 75(1):1–8
Macha IJ, Karacan I, Ben-Nissan B, Cazalbou S, Müller WH (2019) Development of antimicrobial composite coatings for drug release in dental, orthopaedic and neural prostheses applications. SN Appl Sci 1:1–10
Madhu P, Sanjay MR, Pradeep S, Bhat KS, Yogesha B, Siengchin S (2019) Characterization of cellulosic fibre from Phoenix pusilla leaves as potential reinforcement for polymeric composites. J Mater Res Technol 8(3):2597–2604
Manimaran P, Saravanan SP, Prithiviraj M (2019a) Investigation of physico chemical properties and characterization of new natural cellulosic fibers from the bark of Ficus Racemosa. J Nat Fibers 18(2):274–284
Manimaran P, Saravanan SP, Sanjay MR, Siengchin S, Jawaid M, Khan A (2019b) Characterization of new cellulosic fiber: Dracaena reflexa as a reinforcement for polymer composite structures. J Mater Res Technol 8(2):1952–1963

Mazzanti V, Pariante R, Bonanno A, de Ballesteros OR, Mollica F, Filippone G (2019) Reinforcing mechanisms of natural fibers in green composites: role of fibers morphology in a PLA/hemp model system. Compos Sci Technol 180:51–59

Miyashiro D, Hamano R, Umemura K (2020) A review of applications using mixed materials of cellulose, nanocellulose and carbon nanotubes. Nano 10(2):186

Mohammad F, Arfin T, Bwatanglang IB, Al-lohedan HA (2019) Starch-based nanocomposites: types and industrial applications. In: Bio-based polymers and nanocomposites: preparation, processing, properties & performance, pp 157–181

Mohan TP, Kanny K (2019) Compressive characteristics of unmodified and nanoclay treated banana fiber reinforced epoxy composite cylinders. Compos Part B 169:118–125

Mohd Aris ZF, Bavishi V, Sharma R, Nagarajan R (2019) Barrier properties and abrasion resistance of biopolymer-based coatings on biodegradable poly (lactic acid) films. Polym Eng Sci 59(9):1874–1881

Murawski A, Diaz R, Inglesby S, Delabar K, Quirino RL (2019) Synthesis of bio-based polymer composites: fabrication, fillers, properties, and challenges. In: Polymer nanocomposites in biomedical engineering, pp 29–55

Nagalakshmaiah M, Afrin S, Malladi RP, Elkoun S, Robert M, Ansari MA et al (2019) Biocomposites: present trends and challenges for the future. In: Green composites for automotive applications. Elsevier Science, Cambridge, pp 197–215

Navaneethakrishnan G, Karthikeyan T, Saravanan S, Selvam V, Parkunam N, Sathishkumar G, Jayakrishnan S (2020) Structural analysis of natural fiber reinforced polymer matrix composite. Mater Today: Proc 21:7–9

Nevoralová M, Koutný M, Ujčić A, Horák P, Kredatusová J, Šerá J et al (2019) Controlled biodegradability of functionalized thermoplastic starch based materials. Polym Degrad Stab 170:108995

Nguyen TP, Nguyen QV, Nguyen VH, Le TH, Huynh VQN, Vo DVN et al (2019) Silk fibroin-based biomaterials for biomedical applications: a review. Polymers 11(12):1933

Payal R (2019) Reliable natural-fibre augmented biodegraded polymer composites. In: Sustainable polymer composites and nanocomposites, pp 961–975

Peng Q, Zhang Y, Lu L, Shao H, Qin K, Hu X, Xia X (2016) Recombinant spider silk from aqueous solutions via a bio-inspired microfluidic chip. Sci Rep 6(1):36473

Perotto G, Sandri G, Pignatelli C, Milanesi G, Athanassiou A (2019) Water-based synthesis of keratin micro-and nanoparticles with tunable mucoadhesive properties for drug delivery. J Mater Chem B 7(28):4385–4392

Ramamoorthy SK, Skrifvars M, Persson A (2015) A review of natural fibers used in biocomposites: plant, animal and regenerated cellulose fibers. Polym Rev 55(1):107–162

Sabbah M, Di Pierro P, Cammarota M, Dell'Olmo E, Arciello A, Porta R (2019) Development and properties of new chitosan-based films plasticized with spermidine and/or glycerol. Food Hydrocoll 87:245–252

Sadeghi S, Dadashian F, Eslahi N (2019) Recycling chicken feathers to produce adsorbent porous keratin-based sponge. Int J Environ Sci Technol 16:1119–1128

Sanjay MR, Madhu P, Jawaid M, Senthamaraikannan P, Senthil S, Pradeep S (2018) Characterization and properties of natural fiber polymer composites: a comprehensive review. J Clean Prod 172:566–581

Saravana Kumaar A, Senthilkumar A, Sornakumar T, Saravanakumar SS, Arthanariesewaran VP (2019) Physicochemical properties of new cellulosic fiber extracted from Carica papaya bark. J Nat Fibers 16(2):175–184

Sari NH, Sanjay MR, Arpitha GR, Pruncu CI, Siengchin S (2019) Synthesis and properties of pandanwangi fiber reinforced polyethylene composites: evaluation of dicumyl peroxide (DCP) effect. Compos Commun 15:53–57

Schmidt PW, Morozova S, Ertem SP, Coughlin ML, Davidovich I, Talmon Y et al (2020) Internal structure of methylcellulose fibrils. Macromolecules 53(1):398–405

Senthamaraikannan P, Kathiresan M (2018) Characterization of raw and alkali treated new natural cellulosic fiber from Coccinia grandis. L. Carbohydr Polym 186:332–343

Sharma A, Thakur M, Bhattacharya M, Mandal T, Goswami S (2019) Commercial application of cellulose nano-composites–a review. Biotechnol Rep 21:e00316

Shavandi A, Silva TH, Bekhit AA, Bekhit AEDA (2017) Keratin: dissolution, extraction and biomedical application. Biomater Sci 5(9):1699–1735

Shehata N, Hassounah I, Sobolciak P, Krupa I, Lewis R, Kandas I (2019) Spider silk fibers: synthesis, characterization, and related biomedical applications. In: Materials for biomedical engineering. Elsevier, pp 289–307

Shera SS, Kulhar N, Banik RM (2019) Silk and silk fibroin-based biopolymeric composites and their biomedical applications. In: Materials for biomedical engineering. Elsevier, pp 339–374

Smits ALM, Kruiskamp PH, Van Soest JJG, Vliegenthart JFG (2003) Interaction between dry starch and plasticisers glycerol or ethylene glycol, measured by differential scanning calorimetry and solid state NMR spectroscopy. Carbohydr Polym 53(4):409–416

Sorushanova A, Delgado LM, Wu Z, Shologu N, Kshirsagar A, Raghunath R et al (2019) The collagen suprafamily: from biosynthesis to advanced biomaterial development. Adv Mater 31(1):1801651

Sözer N, Nordlund E, Ercili-Cura D, Poutanen K (2017) Cereal side-streams as alternative protein sources. Cereal Foods World 62(4):132–137

Stafford R, Jones PJ (2019) Viewpoint–Ocean plastic pollution: a convenient but distracting truth? Mar Policy 103:187–191

Su RSC, Gill EE, Kim Y, Liu JC (2019) Characterization of resilin-like proteins with tunable mechanical properties. J Mech Behav Biomed Mater 91:68–75

Subramanian SG, Rajkumar R, Ramkumar T (2021) Characterization of natural cellulosic fiber from cereus hildmannianus. Journal of Natural Fibers 18(3):343–354

Teixeira RS, Santos SFD, Christoforo AL, Payá J, Savastano H Jr, Lahr FR (2019) Impact of content and length of curauá fibers on mechanical behavior of extruded cementitious composites: analysis of variance. Cem Concr Compos 102:134–144

Teklu T, Wangatia LM, Alemayehu E (2019) Effect of surface modification of sisal fibers on water absorption and mechanical properties of polyaniline composite. Polym Compos 40(S1):E46–E52

United Nations Development Programme (2015) Evaluation of the role of {UNDP} in supporting national achievement of the millennium development goals. United Nations, Thematic Development Reports

Vaidya AA, Collet C, Gaugler M, Lloyd-Jones G (2019) Integrating softwood biorefinery lignin into polyhydroxybutyrate composites and application in 3D printing. Mater Today Commun 19:286–296

Vieira MGA, Da Silva MA, Dos Santos LO, Beppu MM (2011) Natural-based plasticizers and biopolymer films: a review. Eur Polym J 47(3):254–263

Vijay R, Manoharan S, Vinod A, Singaravelu DL, Sanjay MR, Siengchin S (2019a) Characterization of raw and benzoyl chloride treated Impomea pes-caprae fibers and its epoxy composites. Mater Res Exp 6(9):095307

Vijay R, Singaravelu DL, Vinod A, Sanjay MR, Siengchin S, Jawaid M et al (2019b) Characterization of raw and alkali treated new natural cellulosic fibers from Tridax procumbens. Int J Biol Macromol 125:99–108

Vijay R, Singaravelu DL, Vinod A, Raj IFP, Sanjay MR, Siengchin S (2019c) Characterization of novel natural fiber from saccharum bengalense grass (Sarkanda). J Nat Fibers 17(12):1739–1747

Vijay R, Singaravelu DL, Vinod A, Sanjay MR, Siengchin S (2019d) Characterization of alkali-treated and untreated natural fibers from the stem of parthenium hysterophorus. J Nat Fibers 18(1):80–90

Vijay R, Manoharan S, Arjun S, Vinod A, Singaravelu DL (2021) Characterization of silane-treated and untreated natural fibers from stem of leucas aspera. J Nat Fibers 18(12):1957–1973

Vinayagamoorthy R, Venkatakoteswararao G (2020) Synthesis and property analysis of green resin-based composites. J Thermoplast Compos Mater 33(10):1429–1445
Vinod A, Vijay R, Singaravelu DL, Sanjay MR, Siengchin S, Moure MM (2019) Characterization of untreated and alkali treated natural fibers extracted from the stem of Catharanthus roseus. Mater Res Exp 6(8):085406
Vinod A, Sanjay MR, Suchart S, Jyotishkumar P (2020) Renewable and sustainable biobased materials: an assessment on biofibers, biofilms, biopolymers and biocomposites. J Clean Prod 258:120978
Vinod A, Vijay R, Singaravelu DL, Sanjay MR, Siengchin S, Yagnaraj Y, Khan S (2021) Extraction and characterization of natural fiber from stem of cardiospermum halicababum. J Nat Fibers 18(6):898–908
Wang Y, Katyal P, Montclare JK (2019) Protein-engineered functional materials. Adv Healthc Mater 8(11):1801374
Werten MW, Eggink G, Stuart MAC, de Wolf FA (2019) Production of protein-based polymers in Pichia pastoris. Biotechnol Adv 37(5):642–666
Wu J, Sun Q, Huang H, Duan Y, Xiao G, Le T (2019) Enhanced physico-mechanical, barrier and antifungal properties of soy protein isolate film by incorporating both plant-sourced cinnamaldehyde and facile synthesized zinc oxide nanosheets. Colloids Surf B: Biointerfaces 180:31–38
Wu C, Li Y, Sun J, Lu Y, Tong C, Wang L et al (2020) Novel konjac glucomannan films with oxidized chitin nanocrystals immobilized red cabbage anthocyanins for intelligent food packaging. Food Hydrocoll 98:105245
Yao Y, Sun H, Zhang Y, Yin Z (2020) Corrosion protection of epoxy coatings containing 2-hydroxyphosphonocarboxylic acid doped polyaniline nanofibers. Prog Org Coat 139:105470
Yazawa K, Malay AD, Masunaga H, Numata K (2019) Role of skin layers on mechanical properties and supercontraction of spider dragline silk fiber. Macromol Biosci 19(3):1800220
Yoshie N, Nakasato K, Fujiwara M, Kasuya K, Abe H, Doi Y, Inoue Y (2000) Effect of low molecular weight additives on enzymatic degradation of poly (3-hydroxybutyrate). Polymer 41(9):3227–3234
Zhao Y, Saldaña MD (2019) Use of potato by-products and gallic acid for development of bioactive film packaging by subcritical water technology. J Supercrit Fluids 143:97–106
Zhou X, Yang R, Wang B, Chen K (2019) Development and characterization of bilayer films based on pea starch/polylactic acid and use in the cherry tomatoes packaging. Carbohydr Polym 222:114912
Zhu Y, Romain C, Williams CK (2016) Sustainable polymers from renewable resources. Nature 540(7633):354–362
Zuin VG, Ramin LZ (2018) Green and sustainable separation of natural products from agro-industrial waste: challenges, potentialities, and perspectives on emerging approaches. In: Chemistry and chemical technologies in waste valorization, pp 229–282

Chapter 14
Achieving Net-Zero Emission: A Sustainable Future with Renewable Energy

Nirupama, Sweta Swaraj, Shoaib Akhtar, Swadha Shree, and Dan Bahadur Pal

Abstract The rapid growth of worldwide industrialization and excessive reliance on fossil fuels have resulted in the emission of greenhouse gases, intensifying the rise in global temperatures and raising environmental concerns. However, because of favorable regulations and falling costs, there has been a significant increase in the capacity of renewable energy in recent years. Even while renewable energy has come a long way, it still has a small impact on industries like heating and transportation. The push for the use of renewable energy comes mostly from efforts to mitigate climate change, such as the 2015 Paris Climate Agreement, which aims to keep global warming to 1.5 °C. Greenhouse gas emissions must be reduced to half during the next 10 years and reach "net-zero carbon emmision" by 2050 in order to meet this lofty target. This chapter summarizes the advancements, chances, and difficulties encountered during this revolutionary voyage toward decarbonization, in which renewable energy is essential. Energy efficiency, behavioral modifications, electrification, renewable energy, hydrogen and hydrogen-based fuels, bioenergy, and carbon capture, use, and storage (CCUS) are important decarbonization pillars. Achieving a low-carbon future requires a wide range of policy alternatives, supported by international cooperation, including carbon price, regulations, investments in green technologies, and sustainable land-use practices. Future generations can live in a world that is more sustainable and climate-resilient if we establish a well-thought-out policy framework now.

Keywords Energy efficiency · Behavioral modifications · Electrification · Renewable energy · Hydrogen and hydrogen-based fuels · Bioenergy

---

Nirupama (✉) · S. Swaraj · S. Akhtar · S. Shree
Department of Chemical Engineering, Birsa Institute of Technology, Sindri, Jharkhand, India

D. B. Pal
Department of Chemical Engineering, Harcourt Butler Technical University, Kanpur, Uttar Pradesh, India

S. Kumar, V. K. Singh (eds.), *Renewable Energy Development: Technology, Material and Sustainability*, Clean Energy Production Technologies,
https://doi.org/10.1007/978-981-97-9626-7_14

14.1 Introduction

The over-exploitation of fossil fuels and growing global industrialization have left our world in a state of existential crisis: dangerous greenhouse gases (GHGs) are being released into the atmosphere by human activity, leading to climate change (Wang et al. 2021). Carbon dioxide (CO_2) emissions are primarily responsible for climate change. CO_2 is released into the atmosphere through various human activities, the primary being the burning of fossil fuels for energy production, transportation, and industrial processes. The average worldwide atmospheric CO_2 concentration rose significantly from 285 to 419 ppm between 1850 and 2022, according to Chen's report from 2021. It causes the average global temperature to rise to 0.97 to 1.21 °C, and by 2050, it's predicted that greenhouse gas emissions will have increased to 50% worldwide. The main cause of greenhouse gas emissions is the use of nonrenewable energy, specifically CO2 emissions (Rabaey and Ragauskas 2014). As a result, regulating CO_2 emissions into the atmosphere is crucial. Most scientists concur that we need to dramatically reduce emissions and reach net-zero carbon emissions in order to prevent catastrophic outcomes and ensure a sustainable future.

Rising global temperatures are causing thermal expansion of oceans, and melting of glaciers leads to rise in sea level. This threatens the communities around coastal areas with flood, displacing populations and disrupting infrastructure. Additionally, climate change is expected to intensify extreme weather conditions like heat waves, droughts, floods, and storms. These events can cause widespread damage to infrastructure, property, and agriculture, leading to economic losses and humanitarian crises. Climate change disrupts natural ecosystems, altering temperature regimes, precipitation patterns, and ocean acidification. These changes threaten the survival of plant and animal species, potentially leading to mass extinctions and disrupting ecological food webs. The loss of biodiversity has pouring consequences, impacting human well-being through reduced ecosystem services such as pollination, water purification, flood control, etc. Climate change poses significant risks to global food and water security. Changes in precipitation patterns and increased heat stress can lead to agricultural failures, impacting food production and potentially causing food shortages and price hikes. Additionally, melting glaciers and altered precipitation patterns can disrupt freshwater availability, threatening water security for human consumption and irrigation purposes. Climate change is expected to inflame existing public health challenges and introduce new ones. Heat waves can cause heatstroke, respiratory problems, and mortality, particularly among vulnerable populations. Increased flooding can increase the risk of waterborne diseases. Additionally, changes in climate can alter the geographic distribution of zoonotic diseases such as malaria and dengue fever.

In response to rising global temperatures and greenhouse gas emissions, 197 member gatherings of the United Nations Framework Convention on Climate Change (UNFCCC) solidly accepted the Paris Agreement at the Paris Climate Change Conference (PCCC) on December 12, 2015. Plans for worldwide climate change action beyond 2020 are outlined in the accord (Berndes et al. 2016). Every nation agreed to limit the increase in global temperature to less than 2 °C and strive toward less than 1.5 °C as part of the 2015 Paris Agreement. In February 2021, a pledge was made by 124 nations to achieve net-zero carbon emissions by 2050 or 2060 and to become carbon neutral (Chen 2021). Net-zero carbon emissions must be achieved by a combination of social, economic, environmental, and technological techniques in order to meet the targets outlined in the Paris Agreement and support sustainable development. Some of the action needed to achieve net-zero emission is presented in Fig. 14.1.

Net-zero carbon emissions is a state of delicate equilibrium. It signifies a point where the emissions of greenhouse gases, primarily carbon dioxide, are balanced by their removal from the atmosphere. The only way to stop global warming's unstoppable progress and its disastrous impacts is to achieve this ambitious aim, despite its seeming impossibility. But reaching net-zero requires a multidimensional marathon rather than a single sprint. It requires a multifaceted strategy, with each component properly designed and put into action. This net-zero journey calls for a diversified strategy. Subsequently, it is imperative that we scrutinize the current worldwide initiatives and the varied array of approaches implemented in pivotal domains—ranging from remaking industry, agriculture, and land use to reforming energy production and transportation (Beckmann and Feige 2023; Gangadhari et al. 2023; van der Spek et al. 2022). But achieving net-zero involves more than just environmental concerns. It is intricately woven into the socioeconomic realities that collectively make up our global community. Ensuring a "just transition" is crucial as we manage this shift toward a net-zero economy. This idea promotes inclusivity and guarantees

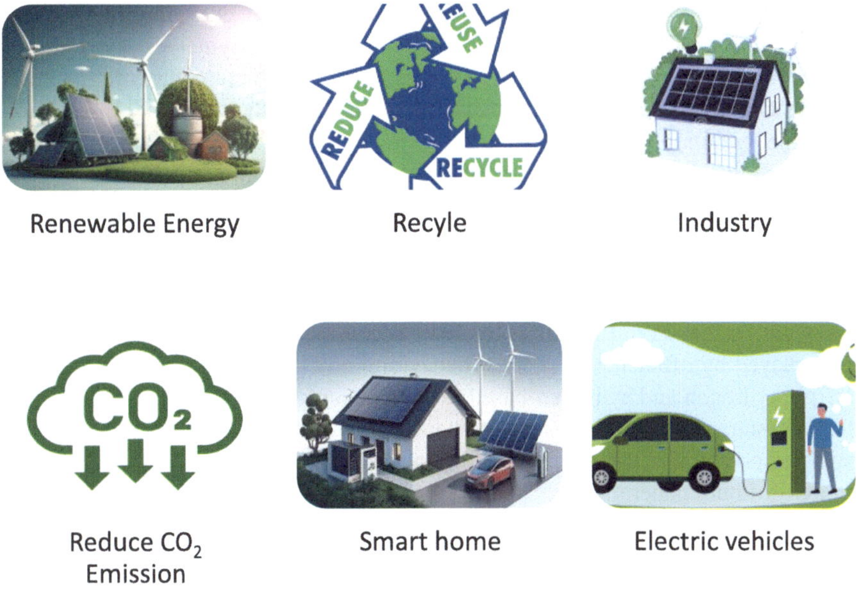

Fig. 14.1 Action needed on net-zero carbon emission

that everyone, not just a chosen few, gains from the voyage (Zohuri and Balgehshiri 2023; Gözlügöl 2022). It demands the establishment of new possibilities, especially for communities that are most at risk from the disruptions brought on by climate change. Furthermore, it is essential for developing international collaboration. In this cooperative endeavor, no country—developed or developing—can afford to fall behind. Since we are all occupants of this planet, there is an inherent connection between our fates.

In the electrical industry, neither nuclear energy nor carbon capture and storage (CCS) has developed at a rate that is comparable. In recent years, there has been a significant advancement in the social, political, technical, and economic viability of wind, solar, and electricity storage technologies. Emissions from energy-intensive enterprises would need to be drastically reduced in order to keep global warming to 1.5 °C. To do this, strategies like hydrogen (H_2), bio-based feedstock, electrification, replacement, and, in certain situations, CO_2 capture, use, and storage (CCUS) would be employed. However, institutional, technological, and financial barriers have made these opportunities unattainable, increasing the financial risk for numerous established companies. In the industrial sector, energy conservation is more financially feasible and advantageous to the industrial system. In the near future, locally produced low- and zero-carbon fuels can be an option for energy security which can lessen dependency on fossil fuel supply in addition to having less detrimental impacts on the environment. Considering that it is imperative that low- and zero-carbon fuels be included into technology, industry, and energy systems as soon as possible. Indeed, careful market design is crucial for ensuring a smooth transition from current energy schemes to ones that integrate various fuels, particularly as we move toward low- and zero-carbon options. Considering the risks related to supply security, the challenging issue of network transition, and the changing supply adaptation of low- and zero-carbon fuels, it is crucial for policymakers and industry stakeholders to collaborate on developing resilient infrastructure, investing in innovative technologies, and ensuring regulatory frameworks that support a smooth and equitable transition to sustainable energy systems (Buckingham et al. 2022). The pursuit of net-zero actualization by 2050 and the requirement to maintain a continuous, uninterrupted supply chain necessitate the use of a wider range of fossil fuel substitutes that are reasonably priced, widely applicable, long-storable, and easily accessible locally, even though zero-carbon fuels—especially hydrogen—are ideal.

In this chapter, we provide a detailed examination of the energy sector's change based on our Net-Zero Emissions by 2050 Scenario. We also illustrate the critical roles played by efficiency measures, behavioral change, electrification, renewable energy, hydrogen and hydrogen-based fuels, bioenergy, and carbon capture, utilization, and storage (CCUS). This comprehensive exploration aims to shed light on the path and potential of achieving net-zero, a goal crucial for securing a sustainable future for future generations.

14.2 Net-Zero Carbon Emission

The current status of CO_2 emissions, their effect on climate change, and the necessity of mitigation techniques will all be covered in this section. After rising continuously for decades, global CO_2 emissions reached a record high of 36.3 billion tons in 2021. For the first time in human history, atmospheric CO_2 concentrations have surpassed 420 parts per million (ppm) due to this steady growth, crossing a vital threshold that hasn't been seen in millions of years. The atmosphere's global temperatures rise as a result of the increased CO_2 concentration acting as a blanket to trap heat (Bouckaert et al. 2021). The chemical industry confronts particular difficulties since the majority of chemical products include carbon, making it very impossible to decarbonize. About 5% of CO_2 emissions worldwide are caused by the chemical sector, which is essential to reaching net-zero emission goals. However, because of the widespread use of carbon-rich raw materials, decarbonizing carbon-intensive industries will require a concerted effort involving technological innovation, policy support, and collaboration across the public and private sectors. While the challenges are significant, the potential benefits in terms of reducing greenhouse gas emissions, enhancing energy security, and fostering sustainable economic growth make the pursuit of decarbonization imperative. Cumulative emission shares to global warming by selected countries are presented in Table 14.1.

Absolutely, achieving net-zero greenhouse gas (GHG) emissions is crucial for limiting global warming to below the 1.5 °C threshold set by the Paris Agreement. This involves not only reducing emissions from human activities but also ensuring that any remaining emissions are balanced by removing an equivalent amount of GHGs from the atmosphere. By around the middle of the century, anthropogenic CO_2 emissions must be nil (Rogelj et al. 2018). This necessitates not solely curbing emissions from human endeavors but also guaranteeing that any residual emissions are offset by removing an equivalent quantity of GHGs from the atmosphere (Masson-Delmotte et al. 2021).

Global CO_2 releases from energy-related and industrial processes are predicted to drop to zero by 2050 and to about 21 Gt CO_2 by 2030 (IEA 2021). Developed and emerging economies will have almost zero CO_2 emissions by 2045, and by 2050, these countries will have taken out 0.2 gigatons of CO_2 from the atmosphere collectively (Iyer et al. 2021). Even though these growing economies will all have

Table 14.1 Cumulative emission shares to global warming by selected countries (Climate Equality Monitor n.d.)

S. no	Country	Share (%)
1.	The United States	22
2.	India	4.5
3.	China	13.4
4.	Germany	4.5
5.	Russia	6.9
6.	Brazil	1.8
7.	others	46.9

net-zero emissions well before 2050, their combined emissions of carbon dioxide will still be roughly 0.2 gigatons by that year. In industrialized economies, these are lessened by sequestering carbon dioxide, leading to net-zero global CO_2 emissions. Significant cuts in CO_2 emissions must be made immediately in order to prevent the worst effects of climate change. This calls for a diversified strategy that includes change to energy sources that are renewable. Emissions can be greatly decreased by switching from fossil fuels to solar, wind, geothermal, and other renewable energy sources (RESs). Energy demand and related emissions can be greatly reduced by putting energy-efficient technologies and practices into buildings, transportation, and industry. Technology that stores CO_2 emissions underground from factories and power stations can be a vital adjunct to other strategies. Reducing CO_2 emissions is not only necessary for the environment; it is also a vital task to protect the welfare of current and future generations. By recognizing the immediacy of the challenge and employing efficient mitigation strategies, we must strive for advancement toward a future that is both sustainable and resilient to climate impacts.

By electrifying and creating carbon-free (C-free) electricity, it could be comparatively easy to decarbonize some energy services, like electricity, light-duty transportation, and home heating and cooling.

Electrification alone is not enough to address emissions in hard-to-abate industries like chemical, aviation, cement, iron, and steel, which collectively contribute over one-third of global carbon emissions. These industries mostly depend on high-temperature heat and carbon as a material feedstock. Therefore, achieving net-zero emissions in these sectors necessitates a combination of carbon-neutral carbon feedstock and carbon-free energy sources (Davis et al. 2018). Various technological options exist to produce chemicals with net-zero CO_2 emissions, including biomass utilization, CO_2 consumption, recycling, and carbon capture and storage. However, the feasibility of these routes may be constrained by the availability of local energy sources and natural resources such as land and water, as noted by Gabrielli (2023). Even if a number of studies assess the viability of a net-zero chemical industry in terms of global resources, there are still unresolved regional differences. Because of this, given each country's distinct demands and resources, it is still unknown if the net-zero chemical industry will be technically, ecologically, and physiologically sustainable (Meng et al. 2023; Bachmann et al. 2023).

14.3 Ambitious Climate Action to Reach Net-Zero Emissions

In order to achieve net-zero emissions, countries, corporations, and other organizations must make ambitious efforts to cut their greenhouse gas emissions to the point where any leftover emissions are balanced by removing an equal amount of greenhouse gases from the environment. It's common to refer to this condition as reaching "net-zero emissions."

Some key components of ambitious climate action to reach net-zero emissions includes the following:

Energy System Transformation: Decarbonizing the energy sector is paramount. A rapid transition from fossil fuels to renewable energy sources such as hydropower, tidal power, solar, wind, and geothermal is essential. Scaling up renewable energy deployment, coupled with energy efficiency improvements and smart grid modernization, will be crucial for creating a clean and reliable energy system (Change 2007). Additionally, research and development efforts toward innovative technologies like most advanced energy storage solutions further enhance the integration of renewable into the grid.

Sustainable Transportation: The transportation sector is a major contributor to greenhouse gas emissions. Promoting a shift toward electric vehicles (EVs) powered by clean energy sources is a key strategy. To do this, it is necessary to construct a reliable infrastructure for charging electric vehicles, to inspire the acceptance of electric vehicles through legislative frameworks, and to promote the growth of public transportation and bicycle infrastructure in order to lessen dependency on private automobiles (Chakravarty and Ghosh 2024).

Industry Decarbonization: Transforming industrial processes is critical for achieving net-zero. This involves utilizing renewable energy sources for industrial operations, adopting energy-efficient technologies, and exploring carbon capture, utilization, and storage (CCUS) technologies to mitigate emissions from hard-to-abate sectors like steel, cement, and chemicals (Tsioumani 2020). Moreover, advocating for circular economy principles, which prioritize resource efficiency and the reuse of materials, can lead to substantial emission reductions within the industrial sector.

Nature-Based Solutions: Conserving and restoring natural ecosystem plays an important role in climate change mitigation. Forests, mangroves, and peatlands act as vital carbon sinks, absorbing significant amounts of atmospheric CO_2. Protecting and restoring these ecosystems while promoting sustainable land management practices will enhance their carbon sequestration capacity.

Global Collaboration and Collective Action: The challenge of climate change requires a global response. International cooperation is essential to ensure ambitious emission reduction targets are set, technology transfer and knowledge sharing are facilitated between developed and developing nations, and financial resources are mobilized to support climate action initiatives in vulnerable regions. The path toward net-zero necessitates overcoming significant challenges, including overcoming vested interests in fossil fuel-based industries, ensuring a just transition for workers impacted by decarbonization efforts, and mobilizing the necessary financial resources. However, the potential benefits are undeniable—mitigating climate change risks, fostering innovation, creating jobs in green sectors, and building a better future for all. Achieving net-zero emissions requires a joint effort from governments, businesses, and individuals. By embracing ambitious climate action through a multifaceted approach, we can chart a course toward a more sustainable future for our planet.

14.4 Key Pillars of Decarbonization

In the past, the term "decarbonization" used to refer to the decreasing average carbon intensity of primary energy output over time, or, more broadly, to the decrease in carbon emissions from energy and industrial activities (Clarke et al. 2014; Fisher et al. 2007).

Full Decarbonization Global CO_2 emissions from energy or manufacturing operations will cease to exist once the economy is completely decarbonized. When CO_2 emissions from energy and industrial processes are not offset by storing CO_2 in the ground or by carbon capture, they are referred to as unabated CO_2 emissions. This term appeared in the 2015 G7 summit statement. It is possible to view the G7's pledge to decarbonize "over the course of the century" as a step toward complete decarbonization by the end of the century.

Carbon Neutrality The concept of carbon neutrality in the global economy entails the absence of annual CO_2 emissions on a global scale. This encompasses emissions from energy, industry, and land use, as they are all anthropogenic in nature. Additionally, "carbon neutrality" denotes a scenario where for every ton of CO_2 released into the atmosphere due to human activities, exactly 1 ton of CO_2 is actively removed from the atmosphere through other human activities. This term serves as a scientific equivalent to "net-zero carbon emissions." Climate neutrality, on the other hand, can be understood in various ways. It was initially introduced about 20 years ago, and UNEP continued to promote it through outreach efforts in 2008 and 2011. UNEP (2008) defines climate neutrality as "living in a way which produces no net greenhouse gas (GHG) emissions" on a global scale. This implies the possibility of achieving net-zero greenhouse gas emissions from a scientific standpoint. According to scientific literature (UNEP 2014), achieving net-zero worldwide GHG emissions means that all global Kyoto-GHG emissions (methods) reach net-zero. This suggests that negative CO_2 emissions offset any remaining CO_2 and non-CO_2 emissions (such as nitrous oxide or methane, measured in units of CO_2 equivalent).

Achieving zero greenhouse gas (GHG) emissions in relation to carbon emissions would mean that there are no human-induced GHG emissions at all, with no "net" offsetting actions. This situation is unlikely, though, as it has been determined that some industries, including agriculture, livestock, and grazing, have low technological mitigation potential (Smith and Sagar 2014).

Zero Carbon Emissions Net emissions of CO_2 from human activities—which include land use and agriculture in addition to industrial output and energy services—must approach zero in order to stabilize the global mean temperature.

Decarbonizing energy services like light-duty transportation, heating, cooling, and lighting should be comparatively simple if we electrify and generate electricity from dispatchable ("on-demand") nonrenewable sources (like nuclear energy and fossil fuels with carbon capture and storage) and variable renewable energy sources (like wind and solar).

14.5 Pathways and Challenges Toward Decarbonization and Carbon Neutrality

It is difficult to picture the chemical industry functioning without primary chemicals and feedstock based on carbon, given that carbon is present in most chemical end products. Indeed, when discussing emission reductions within the chemical industry, it may be more appropriate to focus on achieving net-zero emissions rather than solely decarbonization. In practice, fossil fuels serve primarily as raw materials to provide carbon and/or hydrogen for finished products, constituting approximately half of the energy input required by the chemical industry. Therefore, efforts to achieve net-zero emissions in the chemical sector should address not only the energy sources used but also the emissions associated with raw material inputs and production processes. Carbon and hydrogen are abundant elements found in nature, existing within various complex molecules such as water, CO_2, biomass, and fossil fuels. However, fossil fuels have historically been favored by the chemical industry due to their inherent energy content, which is embedded within them rather than needing to be supplied separately. This characteristic has made fossil fuels a preferred choice for providing both carbon and energy required in chemical production processes. Reducing greenhouse gas emissions, especially CO_2 emissions, significantly is essential to the fight against climate change. This means that carbon reduction pathways, which are detailed policies outlining steps toward a low-carbon future, must be investigated and put into action. Undertakings such as technical developments, legislative shifts, social upheavals, and economic viability must all be taken into account in order to determine the most efficient and realistic routes.

Energy Efficiency Improving energy efficiency is a fundamental, workable, and affordable strategy for decarbonization. It can result in quick emission reductions and, for the most part, does not necessitate significant modifications to industrial operations. The goals of the collaboration are to promote new and inventive technologies, encourage organizational change, and hasten the adoption of more energy-efficient practices. Helping industrial firms set and achieve aggressive goals for reducing energy use, water use, waste, and carbon emissions is the goal of the Better Plants Initiative. Through industry partnerships, the Better Plants program seeks to assist top manufacturers in reducing their carbon footprint through energy efficiency improvements, as well as enhancing resilience, efficiency, and economic competitiveness. With the program, DOE provides support to 3600 facilities, or 14% of the manufacturing base in the United States. These partners have claimed saving USD

10.6 billion and 2.2 QBtu (2.3 exajoule (EJ)) of energy collectively (US, DOE Better Plants Program 2022). This equates to a reduction of 131 MMT in CO2 emissions. In 2050, energy efficiency plus heat recovery alone could cut emissions by 8 MMT CO2, or 13% of total emissions. The refining (43%) and pulp and paper (41%) industries are the primary contributors to emission reductions, with the food and beverage (36%) business coming in second (WSP 2015). Similarly, waste heat recovery applications are anticipated to be the primary driver of the 25% reduction in industrial energy consumption in the European Union between 2015 and 2050 brought about by energy efficiency advances. The European Commission projects significant reductions in energy-related CO2 emissions across various industrial sectors by 2050. Specifically, they estimate reductions of 22% in the iron and steel sector, 22% in the chemical sector, 35% in the nonmetallic minerals sector (including cement and lime), 15% in the nonferrous metals sector (such as aluminum, copper, and lead), and 32% in refineries. These cuts are contrasted with a baseline scenario, which is based on current and agreed-upon energy and climate policies and represents the EU's decarbonization trajectory from 2015 to 2050 (European Commission, "A Clean Planet for All," 2018). Climate Works Australia (2014) estimates that by 2050, Australia's industrial sector's energy efficiency gains could lead to a 40% decrease in energy intensity when compared to 2010 levels. This indicates a significant potential for reducing energy consumption and associated emissions within the industrial sector over the coming decades.

Behavioral Change Without the enthusiastic and active participation of citizens, the NZE's vision of a completely transformed energy sector will not be realized. People are ultimately responsible for driving demand for energy-related goods and services, and guiding the energy system toward sustainability will be greatly influenced by society norms and individual decisions. A little less than 40% of the NZE's emission reductions come from the use of low-carbon technologies, such as those used in steel and energy production, which demand significant investment and regulatory support but minimal active participation from the public. To achieve a further 55% reduction in emissions, a combination of low-carbon technology deployment and active citizen and consumer engagement is needed, such as installing solar water heaters or purchasing electric vehicles. The remaining 8% of emission reductions come from improvements in material efficiency and behavioral modifications that lower energy demand, such as reducing business travel by plane. Public image-conscious businesses can also be influenced by consumer perceptions while making investment decisions (Bouckaert et al. 2021). Traditional mitigation strategies often focus on technological advancements and infrastructure changes, but individual and collective behavioral shifts can significantly contribute to carbon reduction efforts. Promoting sustainable practices in everyday activities, such as reducing energy consumption in homes and workplaces, choosing low-carbon transportation options, and adopting mindful consumption habits, can collectively lead to substantial reductions in emissions. Encouraging shifts toward more plant-based diets, reducing food waste, and minimizing resource consumption further contribute to this effort. Effective behavior changes interventions involve understanding underlying motivations and barriers. Utilizing targeted communication strategies, social norm nudges, and fostering a sense of collective responsibility are crucial for influencing individual behavior. Additionally, promoting education and awareness campaigns can empower individuals to make informed choices and act as responsible steward of the environment. While technological innovation remains essential, harnessing the power of behavioral change offers a complementary and accessible pathway toward a low-carbon future.

Electrification The global demand for a sustainable, dependable, and ecologically friendly energy system seeks to integrate various technologies as best they can to enable a seamless and financially sustainable transition to electrification. Because energy is needed for heating, cooling, and electricity at the same time and is typically produced by burning fossil fuels, small, medium, and big industrial activities are pertinent contributors to the creation of CO2 emissions worldwide. As a result, the power sector is undergoing a dramatic change as it searches for novel approaches to satisfy the world's expanding energy needs while reducing its emissions of greenhouse gases (Gambini et al. 2023).

One of the primary factors influencing the NZE's emission reductions is the direct substitution of low-emission power for fossil fuels, which accounts for around 20% of the total decrease in emissions. Between 2020 and 2050, the world's demand for energy will more than double. Of the end-use sectors, industry will have the most absolute increase in power use, accounting for about 11,000 TWh. In the NZE, the percentage of electricity used in transportation rises from less than 2% in 2020 to almost 45% in 2050. By 2030, EVs will account for almost 60% of all passenger car sales worldwide (up from 5% in 2020), and by 2050, practically all cars will be electrified (the remaining vehicles will be hydrogen-powered). Decarbonization of the electric power sector is essential for overall emission control, as it is the greatest energy user and carbon emitter globally, contributing approximately 40% of CO2 emissions (Bouckaert et al. 2021). If power were produced by low-carbon sources, electrification in the commercial, residential, transportation, and industrial sectors might reduce carbon emissions from the direct use of fossil fuels significantly.

The decarbonization of electricity has made significant strides, partly driven by the increasing adoption of renewable energy sources such as solar and wind. While discussions around renewable energy often focus on its impact on the electrical grid, household appliances, and transportation, it's important to recognize that the commodity chemical sector also plays a substantial role in energy consumption and greenhouse gas emissions globally. Nonetheless, the commodity chemical sector uses a lot of energy and produces a significant amount of greenhouse gas emissions worldwide. Reducing the carbon footprint of chemicals might be achieved by electrifying the sector using renewable energy sources, which would have a substantial influence on global carbon dioxide emissions (Schiffer and Manthiram 2017). By shifting to electrically powered systems, such as electric vehicles, electric heating and cooling systems, and electric appliances, we can significantly reduce greenhouse gas emissions associated with transportation, heating, and industrial processes. Through the replacement of fossil fuel-powered facilities with renewable energy sources including solar, wind, geothermal, and hydropower, electrification can directly reduce emissions in the power sector. This change eliminates the associated worries about air pollution from fossil fuels in addition to reducing emissions from the production of power. The global electric power grid has been rapidly decarbonizing due to incentive programs and falling prices of renewable energy. Between 2010 and 2020, the amount of electricity generated by wind and solar energy increased by a factor of roughly 5.5. In 2020, nonfossil energy sources accounted for around 38.6% of the world's electricity production; these sources included 16.2% hydropower, 10.1% nuclear, 6.0% wind, 3.1% solar, and 3.2% other (Yin et al. 2023).

Renewables Future global issues will heavily rely on renewable energy sources. Renewable energy sources, also known as alternative energy sources, are those that can be utilized repeatedly to produce energy, such as solar, wind, biomass, geothermal, and other energy sources. Energy services with nearly zero emissions of greenhouse gases and air pollutants could be provided by renewable energy sources that satisfy home energy needs. The creation and execution of renewable energy projects in rural areas can reduce migration to metropolitan areas by generating work opportunities. One way to reliably, cheaply, and ecologically friendly address the energy needs of rural and small-scale communities is to harvest renewable energy in a decentralized manner (Rathore and Panwar 2007; Zakhidov 2008; Bergmann et al. 2008).

One-fourth of the world's energy demand is met by renewable energy sources (RESs) (UNDP 2015). Biomass, hydropower, geothermal, solar, wind, and marine energies are all included in RES. Primary, clean, home, and endless energy sources are considered renewable. 20% of the electricity produced worldwide comes from large-scale hydropower. A promising energy source is wind power in coastal and other windy areas. Alternative energy sources are another name for RESs. It is anticipated that the percentage of RESs will rise dramatically in 2100, from 30% to 80% (Fridleifsson 2001). Concern over the energy problem and rising fossil fuel usage to fulfill existing demands have led to a renewed focus on promoting renewable energy options to meet the expanding energy needs of the developing globe (Hiemstra-van der Horst and Hovorka 2009). Carbon dioxide from the overuse of fossil fuels has contributed to global warming; hence, there is a pressing need to promote clean, renewable energy sources. An agreement was reached with the Kyoto Protocol's goals and overall pollution prevention aims in order to track the release of these greenhouse emissions (Holm-Nielsen et al. 2009).

Over the past three decades, the successful advancement of solar and other renewable energy technologies can be attributed to three primary factors. Firstly, advancements in the technologies themselves, including their maturation, reliability, and cost-effectiveness, have played a crucial role. Secondly, the implementation of policies at local, regional, and national levels that facilitate the adoption of renewable energy has been instrumental. Thirdly, the availability of low-cost financing, particularly from the private sector with increasingly favorable terms, has further supported their development. In addition to these interconnected factors, there has been a noticeable increase in public acceptance and endorsement of these technologies. The drive to mitigate the impacts of climate change and improve air quality in urban areas is a significant motivation propelling the advancement of these technologies (Renné 2022).

Persistent worries regarding global warming, the imperative to fully decarbonize the energy sector, and broader issues such as fossil fuel depletion, local air quality deterioration, and disparities in energy access and utilization have spurred the concept of transitioning to 100% renewable energy. The Coalition for Action Working Group "Towards 100% Renewable Energy" of the International Renewable Energy Agency (IRENA) has officially defined 100% renewable energy (IRENA 2020). "100% Renewables" refers to a scenario wherein all energy sources, encompassing bioenergy, geothermal, hydropower, ocean, solar, and wind energy, are utilized continuously, every day of the year, within a specific location, region, or nation to fulfill all end-use energy requirements, including power, transportation, heating, and cooling. To meet these energy demands, renewable energy can be generated locally or sourced from outside the area through the utilization of infrastructure and supportive policies, such as hydrogen production facilities, electrical networks, or hot water supplies. Moreover, any storage facilities employed to maintain the balance of the energy supply must exclusively utilize energy derived from renewable resources (Renné 2022).

As the urgency to mitigate climate change intensifies, the shift toward renewable energy sources emerges as a fundamental pillar of global carbon reduction endeavors. These renewable natural resources, which include hydropower, geothermal, wind, and solar energy, provide a sustainable and clean substitute for fossil fuels, which are the main source of greenhouse gas emissions. Their potential is multifaceted, encompassing several key aspects:

Emission Reduction: Compared to fossil fuels, renewable energy sources pump out significantly fewer, if any, greenhouse gases while running. This significantly reduces the carbon footprint of electricity generation, which is crucial in the fight against climate change. By displacing fossil fuels, renewable energy directly contributes to the decarbonization of the energy sector, resulting in reduced environmental harm.

Sustainable Resource Utilization: In contradistinction to finite resources like coal and oil, renewable energy sources exhibit the remarkable characteristic of perpetual replenishment through the agency of natural processes. This inherent characteristic guarantees their longitudinal availability, minimizing contingency upon resources with a finite supply. Consequently, this fosters energy security and bolsters the principles of sustainability.

Technological Advancements and Cost Reductions: Rapid technological breakthroughs in the field of renewable energy are resulting in cost savings and better efficiency. Once thought to be too expensive, solar power and wind power are now more competitive with conventional fossil fuels, making them a more appealing and practical choice for large-scale energy generation.

Despite the undeniable allure of renewable energy, several challenges persist. One prominent concern revolves around grid integration due to the inherent intermittency of certain renewable sources. This variability necessitates the development of efficient and cost-effective energy storage solutions to mitigate these fluctuations and ensure a reliable energy supply. However, ongoing research and development endeavors offer promising prospects for overcoming these hurdles. Advancements in energy storage technology coupled with grid modernization initiatives are leading the way for a more resilient and adaptable energy infrastructure, capable of seamlessly integrating a significantly higher proportion of renewable energy sources. However, embracing renewable energy is not just a technological shift; it's a paradigm shift toward a sustainable future. By harnessing the potential of these clean and abundant resources, we can significantly reduce our carbon footprint and ensure energy security.

Sources of renewables: Combating climate change demands a multifaceted approach, and the harnessing of renewable energy sources stands as a cornerstone of this effort. These naturally replenishing resources, such as solar, wind, geothermal, hydropower, and biomass offer or stand as a cleaner and more sustainable alternative option to fossil fuels, the primary contributor to greenhouse gas emissions. Their potential lies in several key aspects like emission reduction, diverse portfolio, technological advancements and cost reductions, sustainable resource utilization, etc.

Biomass: The energy needed to synthesize chemical compounds is found in biomass, together with the atoms of hydrogen and carbon. But compared to fossil fuels, biomass feedstocks have a less advantageous chemical structure, for instance, having a lower energy content and a higher water content. In the biomass pathway, there are no net CO2 emissions because CO2 is absorbed from the atmosphere during biomass growth and released throughout the synthesis and end-of-life stages of the biomass-based product (Holmgren et al. 2014; Holmgren et al. 2012).

Solar Energy Solar thermal energy is the most plentiful renewable energy source available, and it can be obtained directly or indirectly. About 1.8×10^{14} kW of the 3.8×10^{23} kW of energy that the Sun emits is intercepted by the Earth. There is a great deal of potential to use solar energy for thermal applications, like crop drying, water heating, and cooking. The most practical and direct use of solar energy is in solar cooking. A viable alternative that has the potential to rank among the best energy sources for cooking is solar energy. A study conducted in Costa Rica and globally compared the advantages and disadvantages of solar ovens with traditional firewood and electric stoves. It revealed that even with utilization for 6–8 months per year, a typical hot box-type solar oven has a payback period of 12 to 14 months. Additionally, it can potentially reduce carbon dioxide emissions by 38.4 million tons annually and save approximately 16.8 million tons of firewood (Thirugnanasambandam et al. 2010).

Because of these benefits, solar energy is crucial and essential source of energy. The world is plenty of isolated places without access to power, but since these places receive a lot of solar radiation, it is still feasible to use solar energy to generate electricity there. An apparatus that uses solar radiation to generate electricity is known as a solar thermal electricity-generating system. Essentially, solar energy is gathered and transformed into electrical energy by means of a heat-to-electricity conversion converter. As fossil fuels are limited and rapidly exhausting, electrical energy is the cornerstone of all developmental efforts in both developed and developing countries. In recent decades, energy-related issues have grown in importance and include resource optimization, environmental impact from pollutant emissions, and conventional energy resource consumption (Demirbas 2007).

Traditionally, photovoltaic (PV) cells have been used for direct solar energy conversion to electricity by harnessing the photovoltaic (PV) effect. Perhaps the most basic and popular type of solar cell for usage in space and on Earth is silicon solar cells. Even with the application of current technologies, the PV system is a promising source of electricity generation for energy resource conservation and CO2 emission reduction. Energy requirements and greenhouse gas emissions will decrease with further advancements in solar cell efficiency, material content, and system design for optimal utilization of recycled materials. With an average solar radiation of 5.5 kWh/m^2, the annual net CO2 emission mitigation potential of a 1.8 kWp solar photovoltaic pump is approximately 2085 kg for diesel-operated pumps and about 1860 kg for petrol-operated pumps. Substituting diesel for gasoline presents a greater opportunity to mitigate CO2 emissions, primarily due to the lower fuel utilization efficiency of diesel engine pumps (Panwar et al. 2011).

While solar energy holds immense potential as a clean and sustainable energy source, it's not without its challenges. The primary hurdle lies in its intermittent nature, as sunlight availability fluctuates. This necessitates the integration of energy storage solutions and smart grid management to maintain grid stability and reliability. Additionally, the initial costs of large-scale solar projects can be prohibitive, although they are experiencing a downward trend. However, the future of solar energy remains bright. Technological advancements, coupled with supportive policy frameworks and declining costs, are paving the way for wider adoption. Continuous research and development efforts focused on efficient storage solutions, grid modernization, and cost reduction further bolster the potential of solar energy as a key pillar in a foreseeable sustainable energy future. Leveraging its inherent abundance and clean nature, solar energy presents itself as a powerful solution to the pressing issues of climate change and energy security.

Wind Energy Among the diverse renewable energy sources, wind energy has emerged as a powerful, clean, and cost-effective solution for combating climate change and pave the way for a more sustainable energy future. Wind is a widely available and naturally replenishing resource, offering immense potential across various geographical locations. Unlike fossil fuels, wind energy generation produces minimal to no greenhouse gas emissions during operation, significantly contributing to decarbonizing the energy sector. Wind turbine technology has experienced significant advancements, leading to increased efficiency and cost reductions. Larger and more powerful turbines have been developed, maximizing energy capture and making wind power increasingly competitive with traditional fossil fuels. Wind energy can be integrated effectively into existing electricity grids, contributing to a diverse and flexible energy mix. Additionally, advancements in grid management and energy storage technologies are further unlocking the potential of wind energy by addressing concerns related to intermittency. The wind energy sector creates jobs in manufacturing, installation, operation, and maintenance, part-taking in the economic growth and diversification of local communities (Bouckaert et al. 2021). Wind power can indeed be beneficial for addressing small power needs in remote areas. However, to ensure optimal flexibility and reliability, it is often best utilized in combination with other power generation technologies. Research on wind energy potential indicates an abundance of wind resources worldwide, with estimates suggesting a global wind energy potential of 26,000 TWh/year, of which 9000 TWh/year could be harnessed for commercial and other uses. Wind power presents a feasible option for electricity generation in developing nations facing power shortages. Its quick installation and transportability make it particularly suitable for remote and challenging-to-access locations. Wind energy produces an endless supply of electricity that never runs out or becomes more expensive. By producing power, these devices may be able to avoid using several billion barrels of oil and millions of tons of carbon and other pollutants. At an average wind speed of 4.5 m/s, the projected net annual CO2 emission mitigation potential is highest (7401 kg) for the SICO model and lowest (2874 kg) for the GM-II model when substituting diesel. Similarly, Panwar et al. (2011) provided estimates for the same models at identical wind speeds, reporting values of 2194 and 5713 kg, respectively.

While wind energy offers significant benefits, challenges remain. Concerns regarding visual and noise impacts on nearby communities need to be addressed through careful planning and responsible siting practices. Additionally, integrating large-scale wind farms into the grid and managing intermittency require ongoing research and development efforts in energy storage and smart grid technologies.

Hydrogen and Hydrogen-Based Fuels Hydrogen (H_2) has developed as a potential game-changer in the pursuit of clean energy, offering a versatile and clean-burning fuel with the potential to decarbonize various sectors. The primary appeal of hydrogen lies in its clean-burning characteristic. Unlike fossil fuels, its combustion produces only water vapor, thereby significantly reducing greenhouse gas emissions and mitigating their impact to climate change. This clean-burning property makes hydrogen a promising candidate for applications in sectors with limited low-carbon alternatives, such as heavy-duty transportation, steel production, and industrial heating. The versatility of hydrogen extends beyond its clean-burning characteristics. It can be used in various forms, including gaseous H_2, liquid hydrogen (LH_2), and ammonia (NH_3), catering to diverse applications across different sectors. Fuel cell electric vehicles (FCEVs) powered by hydrogen offer a zero-tailpipe emission alternative to internal combustion engine vehicles, particularly in long-haul transportation and heavy-duty applications where batteries face limitations. Additionally, hydrogen can be injected into natural gas pipelines, offering a potential low-carbon blending option for existing infrastructure (Bouckaert et al. 2021).

However, its journey toward widespread adoption is fraught with challenges, demanding a nuanced understanding of its potential and limitations. Despite its potential, the large-scale utilization of hydrogen faces several significant challenges. The current production of hydrogen primarily relies on fossil fuels (grey hydrogen) through a process called steam methane reforming, resulting in significant emissions. Transitioning toward "green hydrogen" produced through electrolysis using renewable energy is crucial to achieve true carbon neutrality. However, this process currently requires significant energy input and faces challenges in terms of cost competitiveness.

The widespread adoption of hydrogen necessitates the development of a robust hydrogen infrastructure, including production facilities, transportation networks, and refueling stations. This infrastructure development requires substantial investment and presents a significant hurdle in the near term. Additionally, safety concerns regarding the storage and transportation of hydrogen need to be addressed to ensure the responsible and safe handling of this energy carrier.

By 2030, the share of low-carbon hydrogen will increase from 10% in 2020 to 70%. By 2030, electrolysis will account for around half of the low-carbon hydrogen produced worldwide, with the remaining portion coming from coal and natural gas with CCUS, though regional variations may greatly affect this ratio. In gas networks, hydrogen is also mixed with natural gas. By 2030, the average worldwide blend will contain 15% of hydrogen by volume, which will result in a 6% decrease in CO2 emissions from gas use. In the 2020s, there will also be a significant surge in the installation of hydrogen end-use equipment, with over 15 million hydrogen fuel cell cars on the road by 2030. The usage of low-carbon hydrogen in all NZE sectors is expected to grow quickly after 2030. Hydrogen and hydrogen-based fuels offer a significant low-carbon source of system flexibility in the electricity sector. This is primarily achieved by converting some existing gas-fired capacity to co-fire with hydrogen and some coal-fired power plants to co-fire with ammonia. By 2050, about 60% of all fuel used in shipping will come from hydrogen-based sources. Ammonia and other synthetic liquids and gases, along with hydrogen-based fuels, account for over 30% of the low-carbon hydrogen utilized in 2050. Electrolyzers are responsible for a growing portion of hydrogen generation; by 2050, they will produce 60% of all hydrogen produced (Bouckaert et al. 2021).

Hydrogen presents itself as a compelling and versatile energy carrier with the potential to decarbonize various sectors. However, embracing its full potential necessitates acknowledging and addressing the challenges associated with its production methods, infrastructure development, and safety considerations. This holistic approach is critical for ensuring responsible and sustainable integration of hydrogen into the future energy landscape. Ongoing research and development endeavors focused on cost-effective green hydrogen production, robust infrastructure development, and enhanced safety protocols are paramount to solving the full utilization of hydrogen. By addressing these challenges through collaborative efforts and rigorous research, we can pave the way for the sustainable integration of hydrogen into the future energy mix, ultimately contributing to a cleaner and more sustainable future for all. The risks associated with climate change are vast and multifaceted, spanning environmental, social, and economic domains. Addressing these risks requires immediate and collaborative action to reduce greenhouse gas emissions and adapt to the changing climate.

Bioenergy: Biogas, Biodiesel, Biomass Gasifier, Gasifier Based Poor Generation System Bioenergy, encompassing diverse pathways like biogas, biodiesel, and biomass gasification, is emerging as a game-changer in the pursuit of clean and sustainable energy solutions. These pathways utilize organic matter like crops, agricultural residues, and waste materials, offering several key advantages over traditional fossil fuels:

Fossil fuels, which run out and pollute the air, and bioenergy sources like plants are constantly being renewed and can be used in a way that doesn't harm the environment in the long run. This is because as these plants grow, they soak up carbon dioxide, a gas that contributes to climate change. When we burn these plants for energy, the carbon dioxide they release is balanced out by the amount they absorb while growing, making it a more sustainable option.

Bioenergy offers a versatile portfolio of options, catering to various energy needs. Biogas, produced from the anaerobic digestion of organic matter, can be used for electricity generation and heat production or converted to biomethane for transportation fuel. Biodiesel, a renewable alternative to diesel, powers vehicles in the transportation sector, while biomass gasification produces syngas, a versatile fuel utilized for power generation, heat production, or conversion into valuable products like synthetic natural gas (SNG). Additionally, bioenergy contributes to sustainable waste management by transforming organic waste streams like agricultural residues or municipal solid waste into valuable energy resources, promoting a circular economy approach.

While promising, bioenergy faces certain challenges. Ensuring feedstock sustainability, minimizing land-use impacts, and optimizing conversion processes for minimal environmental footprints are crucial considerations. Additionally, some bioenergy pathways require further research and development to enhance economic viability and compete effectively with fossil fuels.

Bioenergy, with its diverse pathways, potential for carbon neutrality, and waste management capabilities, presents a fascinating solution for a sustainable energy. By addressing sustainability concerns, optimizing technologies, and fostering responsible practices, bioenergy can play a pivotal role in the transformation toward a cleaner and more secure energy (Bouckaert et al. 2021).

Carbon Capture and Utilization (CCU) By addressing emissions from current assets, providing a solution for emissions from some of the most challenging sectors, providing an affordable means of rapidly scaling up the production of low-carbon hydrogen, and facilitating the removal of CO_2 from the atmosphere, CCUS has the potential to expedite the changeover to net-zero CO_2 emissions. Policies in the NZE support a range of initiatives that aim to develop markets for CCUS investment and encourage companies that produce hydrogen and biofuels, manage industrial hubs, and update coal-fired power plants to use shared CO_2 transport and storage infrastructure (Bouckaert et al. 2021).

Shifting the carbon source for carbon-based chemicals from high-energy reduced fossil carbon. To oxidized low-energy carbon sourced from captured CO_2—whether from point-source emitters via point-source capture (PSC) or directly from the atmosphere through direct air capture (DAC)—the chemical industry can achieve net-zero carbon emissions through the process of carbon capture and utilization (CCU). To enable CCU, it is essential to develop new chemical processes, organic chemistry techniques, and catalysts capable of transforming CO_2 into the desired carbon-based products. Alternatively, CO_2 can be converted into CO-rich synthesis gas (syngas)—a mixture primarily composed of H_2, CO, and CO_2-methane, or dimethyl ether. The conventional approach (BAU) can then be employed to convert this gas into the desired carbon products. Since low-carbon hydrogen and energy can no longer be sourced from fossil hydrocarbons, CCU relies on them as inputs for product synthesis in any scenario. Furthermore, it is important to emphasize that CO_2 capture is essential for both carbon capture and storage (CCS) and CCU, even though permanent CO_2 storage is a distinct component of CCS (Centi et al. 2019; Huo et al. 2023).

14.5.1 Tackling Global Carbon Emissions: Effective Policy Options for Mitigation

Combating climate change demands a multipronged approach, with effective policy measures playing a crucial role in driving down global carbon emissions (Bouckaert et al. 2021). This section explores various policy options for mitigating emissions across different sectors:

Carbon Pricing: Implementing a carbon price mechanism through carbon taxes systems which incentivizes polluters to reduce emissions. By putting a cost on carbon, these policies create economic motivation for businesses and individuals to adopt cleaner technologies and lower-carbon practices (Lilliestam et al. 2021). Revenue generated from carbon pricing can be used to fund clean energy investments, support vulnerable communities impacted by climate change, and promote innovation in low-carbon technologies.

Regulatory Measures: Regulatory frameworks play an important role in getting emission standards and mandating cleaner technologies. Stringent regulations for vehicle fuel efficiency standards, building energy codes, and industrial emissions can significantly contribute to emission reductions. These regulations provide clear guidelines and stimulate market adoption of low-carbon technologies.

Investment in Green Technologies: Government investment and focus in research, development, and demonstration of clean energy technologies, such as renewable energy, energy efficiency solutions, and carbon capture and storage (CCS), are crucial for accelerating innovation and reducing their costs. Additionally, public investments can help build essential infrastructure for renewable energy deployment, such as charging stations for electricity-based vehicles and smart grids.

Land-Use Policies and Sustainable Practices: Sustainable land-use policies and practices play a vital role in mitigating emissions and enhancing carbon sequestration. Promoting REDD+ (Reducing Emissions from Deforestation and Forest Degradation) initiatives that incentivize forest conservation and restoration can significantly contribute to carbon capture. Additionally, encouraging sustainable agricultural practices like reduced tillage and cover cropping can minimize soil carbon emissions and enhance soil health (Kelly 2017).

International Cooperation and Collaboration: Addressing climate change necessitates a global response. Strong international cooperation is crucial for sharing best practices, facilitating technology transfer between developed and developing countries, and mobilizing financial resources to support climate action initiatives in vulnerable regions. Agreements like the Paris Agreement provide a crucial foundation for international collaboration in reducing greenhouse gas emissions (Moloney et al. 2017).

Implementing effective mitigation policies requires very cautious consideration of several economic, social, and political factors. Balancing emission reduction goals with ensuring energy security, affordability, and a just transition for workers impacted by decarbonization is crucial. Additionally, addressing concerns about equity and ensuring fair carbon pricing mechanisms that don't disproportionately burden vulnerable populations is essential. A diverse set of policy options exists to effectively tackle global carbon emissions. Implementing a combination of carbon pricing, regulations, investments in green technologies, and sustainable land-use practices, coupled with strong international cooperation, is vital for achieving a low-carbon future.

International Cooperation and Frameworks: International frameworks like the Paris Agreement provide a vital platform for fostering global collaboration and setting collective goals for emission reductions. These frameworks establish a shared vision, encourage transparency and accountability, and create a platform for facilitating financial and technological support for developing countries.

Achieving global participation for net-zero presents multiple challenges. Overcoming political and economic differences between nations, ensuring equitable burden sharing, and mobilizing sufficient financial resources are just some of the hurdles that need to be addressed. However, the potential benefits of a united global response are immense—mitigating climate change risks, fostering sustainable development, and creating a secure and prosperous future for all. Global participation is not merely an option but a necessity in the quest of net-zero emissions. By embracing a shared responsibility, prioritizing equity, fostering collaboration, and building capacity, the global community can collectively achieving a sustainable future for all. Stopping climate change is a giant job, and everyone on Earth needs to help. Big countries that pollute a lot and tiny islands that feel the effects the most all have to work together. Making things fair is important too. Poor countries shouldn't have to give up progress to go green, and rich countries should help them out with money and ideas. By sharing ideas and working together, like swapping tips on clean energy and planting trees, we can all move faster. Learning how to do things right is key too, so every country has the know-how to make these changes stick. If we all work together, share responsibility, and help each other out, we can build a healthy planet for everyone, now and in the future.

14.6 Conclusion

The rapid expansion of global industrialization and the utilization of fossil fuels have hastened the emission of greenhouse gases, thereby exacerbating the ascent of global temperatures and the deterioration of the environment. This chapter provides a concise overview of the path toward achieving net-zero emissions while embracing renewable energy for a sustainable future. It examines progress made, opportunities identified, and challenges encountered in this transformative journey. Emphasizing the pivotal role of renewable energy, the narrative highlights its potential to revolutionize global energy systems. Key elements of decarbonization include energy efficiency, behavioral shifts, electrification, renewables, hydrogen and hydrogen-based fuels, bioenergy, and carbon capture, utilization, and storage (CCUS). Combating climate change demands a multifaceted approach, and the harnessing of renewable energy sources stands as a cornerstone of this effort. These naturally replenishing resources, such as solar, wind, geothermal, hydropower, and biomass, offer or stand as a cleaner and more sustainable alternative option to fossil fuels, the primary contributor to greenhouse gas emissions. Crafting effective mitigation policies requires careful consideration of economic, social, and political factors, balancing emission reduction with energy security, affordability, and fair transitions for affected workers. Addressing equity concerns and implementing equitable carbon pricing mechanisms are crucial. A range of policy options, including carbon pricing, regulations, green technology investments, and sustainable land-use practices, underpinned by international collaboration, are essential for achieving a low-carbon future. A comprehensive and well-designed policy framework holds the potential to pave the way toward a sustainable and climate-resilient world for future generations.

References

Agreement P (2015, December) Paris agreement. In report of the conference of the parties to the United Nations framework convention on climate change (21st session, 2015: Paris). Retrieved December (Vol. 4, No. 2017, p. 2). Getzville, NY, USA: HeinOnline

Bachmann M, Zibunas C, Hartmann J, Tulus V, Suh S, Guillén-Gosálbez G, Bardow A (2023) Towards circular plastics within planetary boundaries. Nat Sustain 6(5):599–610

Beckmann M, Feige I (2023) The road to net zero and beyond: looking back, taking stock, and moving forward. In: Road to net zero: strategic pathways for sustainability-driven business transformation. Springer International Publishing, Cham, pp 265–285

Bergmann A, Colombo S, Hanley N (2008) Rural versus urban preferences for renewable energy developments. Ecol Econ 65(3):616–625

Berndes G, Abt B, Asikainen A, Cowie A, Dale V, Egnell G, Lindner M, Marelli L, Paré D, Pingoud K, Yeh S (2016) Forest biomass, carbon neutrality and climate change mitigation. From Sci Poli 3(7):1–27

Bouckaert S, Pales AF, McGlade C, Remme U, Wanner B, Varro L, D'Ambrosio D, Spencer T (2021) Net zero by 2050: A roadmap for the global energy sector

Buckingham J, Reina TR, Duyar MS (2022) Recent advances in carbon dioxide capture for process intensification. Carbon Capture Sci Technol 2:100031

Centi G, Iaquaniello G, Perathoner S (2019) Chemical engineering role in the use of renewable energy and alternative carbon sources in chemical production. BMC Chem Eng 1(1):1–16

Chakravarty T, Ghosh P (2024) SDG 13 and climate change in India. In: Sustainability: science, policy, and practice in India: challenges and opportunities. Springer International Publishing, Cham, pp 175–186

Change OC (2007) Intergovernmental panel on climate change. World Meteorological Organization, vol 52, p 1–43

Chen JM (2021) Carbon neutrality: toward a sustainable future. Innovation 2(3):100127

Clarke L, Jiang K, Akimoto K, Babiker M, Blanford G, Fisher-Vanden K, Hourcade JC, Krey V, Kriegler E, Löschel A., McCollum D (2014) Assessing transformation pathways

Climate Equity Monitor (n.d.) Online Dashboard. M.S. Swaminathan Research Foundation and National Institute for Advanced Studies. https://climateequitymonitor.in/index.html

Climateworks Australia (2014) Pathways to Deep Decarbonisation in 2050. 2014. Available online: https://www.climateworkscentre.org/resource/pathways-to-deep-decarbonisation-in-2050-how-australia-can-prosper-in-a-low-carbon-world/

Davis SJ, Lewis NS, Shaner M, Aggarwal S, Arent D, Azevedo IL, Benson SM, Bradley T, Brouwer J, Chiang YM, Clack CT (2018) Net-zero emissions energy systems. Science 360(6396):eaas9793

Demirbas MF (2007) Electricity production using solar energy. Energy Sour Part A Recov Util Environ Eff 29(6):563–569

Fisher RA, Williams M, Da Costa AL, Malhi Y, Da Costa RF, Almeida S, Meir P (2007) The response of an Eastern Amazonian rain forest to drought stress: results and modelling analyses from a throughfall exclusion experiment. Glob Chang Biol 13(11):2361–2378

Fridleifsson IB (2001) Geothermal energy for the benefit of the people. Renew Sust Energ Rev 5(3):299–312

Gabrielli P, Rosa L, Gazzani M, Meys R, Bardow A, Mazzotti M, Sansavini G (2023) Net-zero emissions chemical industry in a world of limited resources. One Earth 6:682

Gambini M, Mazzoni S, Vellini M (2023) The role of cogeneration in the electrification pathways towards decarbonization. Energies 16(15):5606

Gangadhari RK, Karadayi-Usta S, Lim WM (2023) Breaking barriers toward a net-zero economy. In: Natural resources forum. Blackwell Publishing Ltd., Oxford

Gözlügöl AA (2022) The clash of 'E'and 'S'of ESG: just transition on the path to net zero and the implications for sustainable corporate governance and finance. J World Energy Law Bus 15(1):1–21

Hiemstra-van der Horst G, Hovorka AJ (2009) Fuelwood: the "other" renewable energy source for Africa? Biomass Bioenergy 33(11):1605–1616

Holmgren KM, Berntsson T, Andersson E, Rydberg T (2012) System aspects of biomass gasification with methanol synthesis–process concepts and energy analysis. Energy 45(1):817–828

Holmgren KM, Andersson E, Berntsson T, Rydberg T (2014) Gasification-based methanol production from biomass in industrial clusters: Characterisation of energy balances and greenhouse gas emissions. Energy 69:622–637

Holm-Nielsen JB, Al Seadi T, Oleskowicz-Popiel P (2009) The future of anaerobic digestion and biogas utilization. Bioresour Technol 100(22):5478–5484

Huo J, Wang Z, Oberschelp C, Guillén-Gosálbez G, Hellweg S (2023) Net-zero transition of the global chemical industry with CO 2-feedstock by 2050: feasible yet challenging. Green Chem 25(1):415–430

IRENA (2020) Coalition for Action, 2020: Towards 100% renewable energy: utilities in transition

Iyer G, Clarke L, Edmonds J, Fawcett A, Fuhrman J, McJeon H, Waldhoff S (2021) The role of carbon dioxide removal in net-zero emissions pledges. Energy Climate Change 2:100043

Kelly AC (2017) Improving REDD+ (Reducing Emissions from Deforestation and Forest Degradation) Programs (Doctoral dissertation)

Lilliestam J, Patt A, Bersalli G (2021) The effect of carbon pricing on technological change for full energy decarbonization: a review of empirical ex-post evidence. Wiley Interdiscip Rev Clim Chang 12(1):e681

Masson-Delmotte VP, Zhai P, Pirani SL, Connors C, Péan S, Berger N, Caud Y, Chen L, Goldfarb MI, Scheel Monteiro PM (2021) Ipcc, 2021: Summary for policymakers. In: Climate change 2021: the physical science basis contribution of working group I to the sixth assessment report of the intergovernmental panel on climate change

Meng F, Wagner A, Kremer AB, Kanazawa D, Leung JJ, Goult P, Guan M, Herrmann S, Speelman E, Sauter P, Lingeswaran S (2023) Planet-compatible pathways for transitioning the chemical industry. Proc Natl Acad Sci 120(8):e2218294120

Moloney S, Fünfgeld H, Granberg M (2017) Climate change responses from the global to local scale: an overview. In: Local action on climate change, pp 1–16

Panwar NL, Kaushik SC, Kothari S (2011) Role of renewable energy sources in environmental protection: a review. Renew Sust Energ Rev 15(3):1513–1524

Rabaey K, Ragauskas AJ (2014) Editorial overview: energy biotechnology. Curr Opin Biotechnol 27:V–VI

Rathore NS, Panwar NL (2007) Renewable energy sources for sustainable development. New India Publishing

Renné DS (2022) Progress, opportunities and challenges of achieving net-zero emissions and 100% renewables. Solar Compass 1:100007

Rogelj J, Popp A, Calvin KV, Luderer G, Emmerling J, Gernaat D, Fujimori S, Strefler J, Hasegawa T, Marangoni G, Krey V (2018) Scenarios towards limiting global mean temperature increase below 1.5 C. Nat Clim Chang 8(4):325–332

Schiffer ZJ, Manthiram K (2017) Electrification and decarbonization of the chemical industry. Joule 1(1):10–14

Smith KR, Sagar A (2014) Making the clean available: escaping India's Chulha Trap. Energy Policy 75:410–414

Thirugnanasambandam M, Iniyan S, Goic R (2010) A review of solar thermal technologies. Renew Sust Energ Rev 14(1):312–322

Tsioumani E (2020) Convention on biological diversity: a review of the post-2020 global biodiversity framework working group negotiations. Envtl Pol'y & L 50:55

U.S. DOE Better Plants Program (2022) Better Plants 2022 Progress Update: Working with Industrial Partners to Advance Energy Efficiency & Decarbonization. Available online: https://betterbuildingssolutioncenter.energy.gov/resources/2022-better-plants-progress-update

UNDP (2015) World energy assessment: energy and the challenge of sustainability.

UNEP (2014) Stockholm convention on persistent organic pollutants

UNEP Sustainable Building and Construction Initiative (2008) The Kyoto protocol, the clean development mechanism, and the building and construction sector: a report for the UNEP sustainable buildings and construction initiative. UNEP/Earthprint
van der Spek M, Banet C, Bauer C, Gabrielli P, Goldthorpe W, Mazzotti M, Munkejord ST, Røkke NA, Shah N, Sunny N, Sutter D (2022) Perspective on the hydrogen economy as a pathway to reach net-zero CO 2 emissions in Europe. Energy Environ Sci 15(3):1034–1077
Wang F, Harindintwali JD, Yuan Z, Wang M, Wang F, Li S, Yin Z, Huang L, Fu Y, Li L, Chang SX (2021) Technologies and perspectives for achieving carbon neutrality. Innovation 2(4):100180
WSP; Parsons Brinckerhoff; DNV-GL (2015) Industrial decarbonisation and energy efficiency roadmaps to 2050: cross-sector summary; U.K. Department of Energy and Climate Change and Department for Business, Innovation and Skills. Available online: https://www.gov.uk/government/publications/industrial-decarbonisation-and-energy-efficiency-roadmaps-to-2050
Yin Z, Lu X, Chen S, Wang J, Wang J, Urpelainen J, Fleming RM, Wu Y, He K (2023) Implication of electrification and power decarbonization in low-carbon transition pathways for China, the US and the EU. Renew Sust Energ Rev 183:113493
Zakhidov RA (2008) Central Asian countries energy system and role of renewable energy sources. Applied Solar Energy 44:218–223
Zohuri B, Balgehshiri SKM (2023) Navigating net zero: the crucial role of nuclear power and innovations. Sci Set J Econ Res 2(3):01–09

Chapter 15
Renewable Energy in Focus: Development Trends, Challenges, and Policy Responses

Vivek Kumar Singh and Gyanendra Singh Sisodia

Abstract Renewable energy development is essential for sustainable growth, addressing climate change, and reducing dependency on fossil fuels. Key renewable sources include solar, wind, hydro, and biomass, each offering unique benefits and challenges. Technological advancements, such as improved photovoltaic cells, efficient wind turbines, and smart grid innovations, are critical for maximizing energy output and reliability. Globally, countries are increasingly adopting renewable energy, with varying degrees of success and investment. Challenges comprise steep initial costs, intermittency issues, and need in large-scale storage solutions. Robust policies and initiatives, such as subsidies, tax incentives, feed-in tariffs, renewable portfolio standards, and investment in research and development, are crucial in fostering growth and overcoming barriers. International collaboration between developing and developed countries enhances global renewable energy deployment. Developed countries can provide technological expertise and financial support, while developing nations can offer vast, untapped resources and potential for large-scale projects. This cooperation promotes technological exchange, capacity building, and unified efforts against climate change, ensuring a collective movement towards a sustainable energy future.

Keywords Renewable energy · Climate change · Fossil fuels · Photovoltaic cells · Collaboration · Policy response

V. K. Singh (✉)
Energy for Sustainability Initiative, University of Coimbra, Coimbra, Portugal

Emeraldgears Initiative Foundation (EGIF) Global, Greater Noida, Uttar Pradesh, India
e-mail: Vivekkumarsingh22@gmail.com

G. S. Sisodia
Ajman University of Science and Technology, College of Business Administration, Ajman University, Ajman, UAE

S. Kumar, V. K. Singh (eds.), *Renewable Energy Development: Technology, Material and Sustainability*, Clean Energy Production Technologies,
https://doi.org/10.1007/978-981-97-9626-7_15

15.1 Introduction

In an era overshadowed by the far-reaching impacts of fossil fuel dependence—manifesting in climate change, resource depletion, and geopolitical tensions—renewable energy emerges as both a beacon of hope and a practical solution (O'sullivan and Sandalow 2017). The global shift towards renewable resources including solar, hydropower, geothermal, wind, and biomass represents a pivotal pathway towards achieving both global energy security and sustainability (Hassan et al. 2024). As global energy demand continues its upward trajectory driven by population growth, industrialization, and technological advancements, the dwindling reserves of traditional fossil fuels underscore an urgent need for diversification. This imperative aims to ensure stability, affordability, and environmental stewardship in energy production. At the heart of the renewable energy campaign lays a crucial imperative: mitigating climate change. The combustion of fossil fuels releases greenhouse gases (GHGs) into the atmosphere, contributing significantly to global warming and its concomitant impacts—ranging from ascending sea levels and extreme weather upshots to disruptions in ecosystems and livelihoods. In stark contrast, renewable energy offers a clean and sustainable alternative, operating with minimal to no emissions and substantially reducing the carbon footprint associated with energy generation (Jaiswal et al. 2022).

Over recent decades, technological innovation has been crucial in the evolution of renewable energy. For example, solar photovoltaic (PV) cells have become more competent and cost-efficient, boosting the competitiveness of solar energy in the global market (Dada and Popoola 2023). Similarly, advancements in wind turbine technology have increased their size and efficiency, enabling them to harness stronger winds and reduce the cost per kilowatt hour of electricity produced. Innovations in energy storage options, such as batteries and pumped hydro storage, covers the intermittency challenge of renewables, facilitating their greater integration into existing energy grids. Despite significant progress, renewable energy faces considerable challenges. The primary challenge is intermittency—the variability in energy production from sources like solar and wind due to varying weather conditions. This intermittency necessitates sophisticated grid management and robust energy storage solutions to ensure the reliability and stability of energy supply. Economic challenges, including high initial costs and fluctuating fossil fuel prices, can deter investment in renewable projects. Social factors such as public acceptance, land-use conflicts, and cultural considerations also significantly influence the deployment of renewable energy infrastructure (Batel 2020).

Governments worldwide shape the renewable energy landscape through policy frameworks and incentives by playing crucial and seminal roles. These range from ambitious renewable energy targets and feed-in tariffs to tax credits and subsidies that encourage investment and adoption. International agreements, such as the Paris Agreement, set ambitious targets for reducing greenhouse gas emissions, compelling nations to prioritize renewable energy integration within their energy portfolios. Local regulations and planning permissions further influence the development of renewable energy projects, balancing environmental considerations with economic benefits and societal impacts. Looking ahead, the future of renewable energy is both promising and complex. Technological advancements in green hydrogen are driving down costs and improving efficiency, making renewables increasingly competitive with fossil fuels. The urgent need to combat climate change and achieve sustainable development goals underscores the importance of accelerating the global adoption of renewable energy solutions. However, achieving widespread adoption requires overcoming technical, economic, and social challenges, while navigating the intricate landscapes of policy and regulation.

This chapter offers a comprehensive review of global renewable energy development, challenges, and policies. By examining technological advancements, barriers to adoption, and the evolving policy frameworks shaping the renewable energy landscape, we aim to provide insights for decision-makers, stakeholders, and the public.

15.2 Exploring Key Renewable Energy Sources

Renewable energy technologies are transforming the global energy landscape, offering sustainable and environmentally friendly alternatives to fossil fuels. Among these technologies, solar energy stands out as one of the most well-known and widely adopted forms. Solar energy harnesses the power of the sun, an abundant and inexhaustible resource, through two primary technologies: photovoltaic (PV) panels and solar thermal systems. PV panels are the most common method of converting sunlight into electricity (Nikolaidis 2023; Tyagi et al. 2012). They utilize semiconductor materials, typically silicon, to absorb sunlight and generate an electric current. When photons from the sun strike the PV cells, they knock electrons loose from their atoms, creating a flow of electricity. These panels' installations can be on the rooftops, integrated into material of construction of the building, or arrayed in large solar farms. Significant advancements in PV technology have led to increased efficiency and reduced costs, making solar energy more accessible to both residential and commercial users.

In contrast, solar thermal systems exploit mirrors or lenses to focus sunlight onto a small area, producing heat. This heat subsequently can be used to generate steam that pushes turbines to generate electricity, or it can be used directly for heating purposes (Tyagi et al. 2012). Concentrated solar power (CSP) plants are large-scale installations that typically utilize solar thermal technology. They have the advantage of being able to store thermal energy in materials like molten salt, allowing for electricity generation even when the sun is not shining. This ability to store energy makes CSP a valuable component of a diversified renewable energy portfolio.

Wind energy is another major renewable energy source, harnessing the kinetic energy of wind through turbines. These turbines transform the mechanical energy of the rotating blades into electrical energy via a generator. Wind farms can be situated onshore or offshore, where wind speeds are generally higher and more steady

(Enevoldsen and Valentine 2016). Onshore wind farms are situated on land, typically in rural areas with strong and steady winds. They are relatively easy to construct and maintain compared to offshore wind farms. Onshore turbines vary in size, with some capable of generating several megawatts of power. They have become a common sight in many countries, conducing significantly to the renewable energy flux. Offshore seaward wind farms, built in bodies of water, usually in the ocean, benefit from stronger and more consistent wind speeds than those on land. Offshore turbines are larger and can produce more power than their onshore counterparts. However, the construction and maintenance of offshore wind farms are more complex and costly. Despite this, advancements in technology and economies of scale are cutting costs, leading to a surge in offshore wind projects worldwide.

Hydropower, or hydroelectric power, is one of the oldest and most widely used forms of renewable energy. It generates electricity by utilizing the verve of flowing water, typically from rivers or dams. Large-scale hydropower plants often use dams to create reservoirs. Water stored in the reservoir is released through turbines, converting the kinetic energy of flowing water into mechanical energy and then into electricity via a generator (Kaygusuz 2016). These plants can produce a significant amount of electricity and provide a reliable and stable power supply. However, the erection of large dams can have substantial environmental and societal impacts, such as habitat disruption and relocation of communities. In contrast, small-scale hydropower systems, also known as micro-hydro systems, utilize river currents or tidal movements to generate electricity. These systems are less invasive and can be deployed in remote areas to provide local power supply. They are particularly useful in regions without entrée to a centralized power grid.

Geothermal energy exploits the Earth's interior heat, found in rocks and fluids beneath the surface, to produce electricity and can offer direct heating. Geothermal-based power plants generally access this heat through wells drilled into geothermic reservoirs. Hot water or steam from these reservoirs is utilized to drive turbines, which generate electricity (Jolie et al. 2021). Geothermal power plants can be classified into three, namely, dry steam, flash steam, and binary cycle. Dry steam plants utilizes steam directly from the geothermal reservoir, flash steam plants converts high-pressure hot water into steam, and binary cycle plants transfers heat from geothermal water to a secondary fluid with a lower boiling point, which then vaporizes and drives the turbine. Geothermal energy can also be used directly for heating buildings, greenhouses, and industrial processes (Lund and Boyd 2016). Direct use applications are more efficient and cost-effective as they bypass the need for electricity generation. Geothermal heating systems are increasingly being adopted for residential and commercial use, providing a sustainable and reliable heating source.

Biomass energy source is hydrocarbon-based materials such as plant matter, agricultural waste, and even animal manure. These organic materials can be used to produce heat, electricity, or biofuels. Biomass can be burned directly to produce heat, a method that has been used for centuries in various forms, from burning wood in stoves to using agricultural residues in industrial boilers (Lim et al. 2015). Direct combustion is simple and cost-effective, but it can produce emissions if not properly managed. Biomass can also be transformed into biofuels such as ethanol and biodiesel (Demirbas 2011). Ethanol is commonly made from corn or sugarcane and used as a fuel additive or substitute for gasoline. Biodiesel is produced from vegetable oils or animal fats and can be utilized in diesel engines (Avagyan and Singh 2019). Advanced biofuels, made from nonfood crops or waste materials, are being developed to improve sustainability and reduce competition with food resources (Prasad et al. 2019). Additionally, biogas is produced through the anaerobic digestion of organic materials in the absence of oxygen (Singh et al. 2014). This process generates a mixture of methane and carbon dioxide, which can be deployed for electricity generation, heating, or as a vehicle fuel. Biogas systems are particularly useful for managing agricultural waste and reducing greenhouse gas emissions (Adegbeye et al. 2020).

15.3 Technological Advancements and Innovations in Renewable Energy

Renewable energy technologies have undergone significant advancements and innovations, ushering in an era of increased efficiency, reliability, and scalability. These developments are crucial in addressing global energy demands while mitigating environmental impacts. This chapter explores the latest technological breakthroughs across various renewable energy sectors: solar, wind, hydropower, geothermal, and biomass, highlighting their implications and potential for shaping a sustainable energy future.

15.3.1 Solar Energy

Perovskite Solar Cells: It has emerged as a favourable alternative to conventional silicon-based cells imputable to their high efficiency and lower production costs (Bati et al. 2023). These cells offer potential for flexible and lightweight applications, making them suitable for diverse environments and applications.

Bifacial Panels: Bifacial solar panels have revolutionized energy capture by harnessing sunlight from both the front and rear sides of the panel (Gallegos et al. 2024; Verma 2024). This dual-sided absorption significantly increases overall energy generation, especially in environments with reflective surfaces or varying light conditions.

Solar Tracking Systems: Advanced solar tracking systems adjust the orientation of solar panels throughout the day to optimize sunlight capture (Hasan et al. 2023). By following the sun's path, these systems maximize energy output, enhancing the efficiency and productivity of solar installations (Nadia et al. 2018).

Floating Solar Farms: Floating solar farms represent a creative solution to land scarcity issues. By deploying solar panels on water bodies such as reservoirs, lakes, or ponds, these installations not only conserve land but also benefit from the cooling effect of water, which can enhance panel efficiency (Kowsar et al. 2023).

15.3.2 Wind Energy

Larger Turbines: Innovations in turbine design have led to the development of larger and more efficient wind turbines.(Jackson et al. 2005) These turbines capture stronger and more consistent winds at higher altitudes, thereby generating more electricity per unit and reducing overall costs of wind energy production.

Offshore Wind Farms: Technological advancements have made offshore wind farms increasingly viable. Floating turbine platforms enable installations in deeper waters and areas with higher wind speeds, expanding the potential for offshore wind energy production and reducing visual and environmental impacts on land (Sun et al. 2012).

Advanced Materials: The adoption of innovative advanced materials like, carbon fibre composites for turbine blades, enhances efficiency and durability (Hussain and Immanuel 2022). Lightweight materials reduce structural loads and allow for larger blade designs, optimizing energy capture and operational lifespan.

Smart Grids and Energy Storage: Integration with smart grid technologies and energy storage systems is crucial for enhancing the reliability and stability of wind energy supply (Ahmed et al. 2020). These systems enable efficient management of electricity distribution, balancing fluctuations in wind power output and improving grid resilience.

15.3.3 Hydropower

Pumped Storage Hydropower: Pumped storage hydropower facilities play a pivotal role in energy storage and grid stabilization (Hunt et al. 2020). These systems store excess electricity by pumping water to a higher elevation during periods of low energy demand and release it to generate electricity during highest demand, enhancing grid flexibility and reliability (Wu et al. 2024).

Small and Micro-Hydropower Systems: Advancements in small-scale hydropower technologies make it feasible to harness energy from smaller rivers and streams, providing renewable energy solutions to remote and rural areas (Nautiyal et al. 2011). These systems support local energy independence and contribute to sustainable development goals.

Fish-Friendly Turbines: The development of fish-friendly turbine designs minimizes environmental impacts on the aquatic ecosystems (Koukouvinis and Anagnostopoulos 2023). These turbines features innovative designs that reduce fish mortality and improve passage efficiency, ensuring hydropower generation remains environmentally sustainable.

Tidal and Wave Energy: Technological advancements in tidal and wave energy capture harnesses the kinetic energy of ocean currents and waves, expanding the scope of hydropower as a renewable energy source (Chowdhury et al. 2021). These technologies hold promise for coastal regions with strong tidal currents and wave patterns.

15.3.4 Geothermal Energy

Enhanced Geothermal Systems (EGS): EGS technologies enhance the productivity of geothermal reservoirs by creating or enhancing permeability through hydraulic fracturing (Lu 2018). This technique improves heat extraction efficiency from deeper, less permeable rock formations, expanding the geographic viability of geothermal energy production.

Binary Cycle Power Plants: Binary cycle power plants utilize lower-temperature geothermal resources by circulating a secondary fluid with a lower boiling point than water (Tomarov and Shipkov 2017). This technology allows for efficient electricity generation from geothermal fluids with temperatures as low as 100 °C, enhancing resource utilization.

Co-production with Oil and Gas: Co-production of geothermal energy with oil and gas extraction operations optimizes resource use and reduces environmental impacts. By utilizing waste heat and fluids from oil and gas reservoirs, this approach enhances the overall efficiency and sustainability of geothermal energy production (Céspedes et al. 2022).

Direct Use Applications: It finds diverse applications beyond electricity generation, including district heating, greenhouse heating, and industrial processes. Direct use applications utilize geothermal heat directly without conversion to electricity, providing efficient and sustainable heating solutions (Lund and Boyd 2016).

15.3.5 Biomass

Advanced Biofuels: Advancements in biofuel production focuses on second- and third-generation biofuels derived from nonfood crops, agricultural residues, and waste materials (Kumar et al. 2020). These advanced biofuels offer higher energy yields, lower carbon emissions, and reduced competition with food resources compared to first-generation biofuels (Prasad et al. 2019).

Biogas Production: Innovations in anaerobic digestion technology enhance the efficiency of biogas production from organic waste materials. Biogas, primarily composed of methane, serves as a renewable energy source for electricity generation, heating, and transportation fuels, contributing to waste management and energy sustainability (Jeremiah et al. 2022).

Biomass Gasification: Biomass gasification processes convert solid biomass into syngas, a versatile fuel that can be used for electricity generation, heating, or as a feedstock for bio-based chemicals (Akbarian et al. 2022). This technology offers a flexible and efficient approach to utilizing biomass resources for energy production.

Integrated Biomass Systems: Integrated biomass systems combine biomass energy production with agricultural practices and other renewable technologies (Wang et al. 2021). These integrated approaches create synergistic systems that maximize resource efficiency, reduce environmental impacts, and enhance overall sustainability.

Renewable energy resources such as solar, wind, hydropower, geothermal, and biomass are being revolutionized by these technological advancements and innovations. Each source offers unique advantages and faces distinct challenges, but their continued development is vital for the evolution to a cleaner, more sustainable energy future. These innovations not only improve the efficiency and cost-effectiveness of renewable energy systems but also enhance their reliability and adaptability, making them more viable for widespread adoption and integration into global energy markets.

15.4 Global Perspectives on Renewable Energy Adoption and Deployment

Renewable energy has made significant strides over recent years, with its share in net global energy demand reaching 30% by 2023 (IEA 2023), up from 19% in 2020. This increase marks a notable achievement in the quest of sustainable energy solutions. However, the journey towards a predominantly renewable energy future remains challenging but urgent. Renewable energy adoption is accelerating globally, reflecting current trends driven by technological advancements, decreasing costs, and supportive policies across diverse regions. Solar energy is witnessing substantial growth worldwide, with robust expansions in residential, commercial, and utility-scale installations, particularly evident in emerging markets across Asia, Africa, and Latin America. Innovations such as community solar projects and solar-plus-storage systems are enhancing flexibility and resilience, making solar power more accessible and reliable in both developed and developing economies. The wind energy sector, especially offshore wind, is experiencing rapid global expansion, led by European countries and its increasing adopted in regions like China and the United States. Onshore wind continues to grow steadily across various continents, with hybrid projects integrating wind with solar and storage becoming mainstream to optimize efficiency and grid reliability. While large-scale hydropower faces environmental constraints and limited growth opportunities in certain regions, small and micro-hydropower systems are gaining popularity, particularly in remote and mountainous terrain globally. Countries with substantial hydropower resources are investing in modernization and exploring innovative technologies such as tidal and wave energy to diversify their renewable energy portfolios. Geothermal energy deployment is expanding globally, notably in countries rich in geothermal resources such as the United States, Indonesia, and Kenya, with advancements in enhanced geothermal systems and direct use applications unlocking new potential. Biomass energy is increasingly getting integrated into global energy strategies, with advanced biofuels becoming prevalent in transportation sectors and biogas and biomass power generation being incorporated into energy grids worldwide. The rising use of agricultural and industrial waste as biomass feedstock underscores global efforts towards sustainable waste management and renewable energy utilization. These global trends highlight a collective shift towards renewable energy sources as integral components of national energy policies and global climate action initiatives. As countries continue to invest in building renewable energy infrastructure and integrate these technologies with smart grid system and energy storage solutions, the global renewable energy landscape is equanimous to play a pivotal role in achieving energy security, mitigating climate change impacts, and fostering a sustainable future for generations to come.

15.5 Challenges in Renewable Energy Development

The quest for renewable energy, heralded as a beacon of hope amidst the looming climate crisis, is not without its trials and tribulations. As humanity grapples with the imperative to shift from fossil fuels to sustainable alternatives, the journey is fraught with technical, economic, social, and political challenges that intertwine in complex ways, shaping the landscape of renewable energy development.

15.5.1 Technical Challenges

At the heart of renewable energy's technical challenges lies the issue of intermittency. Unlike conventional power sources, such as coal or natural gas plants, renewable energy generation—whether from wind, solar, or hydropower—is inherently variable. The sun does not always shine, the wind does not always blow, and water flows fluctuate with seasons and weather patterns. This variability poses a fundamental challenge to the stability and reliability of the grid.

Addressing intermittency requires innovative storage solutions capable of keeping excess energy during peak hours of production periods, when energy demand exceeds supply (Parra et al. 2017). Battery storage technology has made significant strides in recent years, but scaling these solutions to match the capacity of large-scale power grids remains a formidable task (Farakhor et al. 2023). Additionally, integrating renewable energy sources into existing grid infrastructures—often designed with centralized, fossil fuel-based power generation in mind—presents technical hurdles related to grid management and balancing supply and demand in real time.

15.5.2 Economic Challenges

Economics forms another critical barrier to widespread renewable energy adoption (Seshadri 2012). While the costs of upcoming renewable technologies in particular solar photovoltaics (PV) and onshore wind have plummeted over the past decade, achieving cost competitiveness with fossil fuels remains a challenge in many regions (Luthra et al. 2015). The upfront capital costs of establishing renewable energy infrastructure, like solar farms or wind powered turbines, can be substantial, despite long-term operational cost savings.

Financing these projects is further complicated by uncertainties surrounding government policies and incentives, which can impact investor confidence and project viability (Polzin et al. 2019). Moreover, the economics of renewable energy development is intricately linked to global market dynamics, including fluctuations in commodity prices and geopolitical factors that influence supply chains and project financing options.

15.5.3 Social and Political Challenges

Perhaps the most nuanced challenges in renewable energy development are social and political in nature (Siamanta 2021). Public acceptance of renewable energy projects—whether wind farms dotting rural landscapes or solar arrays sprawling across desert plains—can be contentious. Local communities may express concerns about visual impact, noise pollution, or potential effects on property values, highlighting the importance of effective community engagement and stakeholder consultation processes in project planning and development (Jami and Walsh 2014).

Regulatory frameworks also play a pivotal role in shaping the renewable energy landscape. Policies that incentivize renewable energy deployment (Simpa et al. 2024), such as feed-in tariffs or renewable portfolio standards, are critical for driving investment and market growth. However, navigating complex and often fragmented regulatory environments across different jurisdictions can pose barriers to project development and grid integration. In some cases, political priorities and partisan divides may influence the stability and longevity of supportive policies, creating uncertainty for investors and developers alike.

Despite these multifaceted challenges, the trajectory of renewable energy development is undeniably positive. Technological advancements keep on pushing the boundaries of what is possible, from next-generation solar cells to innovative energy storage solutions and smart grid technologies that enhance system flexibility and resilience. Economic pressures, driven by both market forces and climate imperatives, are increasingly favouring renewable energy investments over fossil fuel alternatives in many parts of the world. The declining costs of renewable technologies, coupled with growing recognition of the economic benefits of clean energy transitions—including job creation, energy security, and reduced environmental impacts (Frondel et al. 2010)—are bolstering momentum towards a renewable energy future. Addressing social and political challenges requires a holistic approach that fosters inclusivity, transparency, and trust among stakeholders. Effective public engagement, informed by robust research and proactive communication, can help mitigate concerns and build consensus around the benefits of renewable energy development for local communities and broader society.

Moreover, enhancing policy coherence and stability at national and international levels is crucial for unlocking the full potential of renewable energy deployment. Collaborative efforts between governments, businesses, civil society, and academia are essential for overcoming regulatory hurdles, fostering innovation, and scaling up renewable energy solutions that are equitable, affordable, and sustainable (Scheffran and Froese 2016).

15.6 Policies and Initiatives Supporting Renewable Energy

Policies and initiatives supporting renewable energy plays a crucial role in driving global efforts towards sustainability and mitigating climate change (Charles Rajesh Kumar and Majid 2020). These efforts are enacted equally at international and national levels, encompassing a range of agreements, targets, and incentives aimed at fostering the adoption and development of renewable energy technologies (Pitelis et al. 2020).

At the forefront of global efforts are the international agreements like the Paris Agreement and the Kyoto Protocol (Breidenich et al. 1998; Moomaw et al. 1999). As per the 2015 United Nations Framework Convention on Climate Change (UNFCCC), the Paris Agreement aims to curb global temperature rise to well below 2 °C above pre-industrial temperature levels, with an aspiration to keep it to 1.5 °C (Rogelj et al. 2017). Central to this agreement are nationally determined contributions (NDCs), through which countries outline their climate action plans, including targets for renewable energy deployment and emissions reductions (Rogelj et al. 2017). The Kyoto Protocol, primarily focused on greenhouse gas emissions reductions, laid foundational frameworks for international cooperation on renewable energy initiatives and clean development mechanisms (Breidenich et al. 1998). The concepts of "net zero" and discussions on renewable energy were pivotal at COP26, the 26th United Nations Climate Change Conference held in Glasgow, Scotland, in 2021 (The National Archives 2021). This gathering marked a critical moment in global efforts to combat climate change, convening world leaders, policymakers, scientists, and activists to negotiate and commit to actions aimed at limiting global warming to well below 2 degrees Celsius, with an aspiration to achieve 1.5 °C. At COP26, countries and organizations pledged to achieve net zero emissions by mid-century or shortly thereafter, outlining ambitious targets and strategies that involve transitioning away from fossil fuels to renewable energy sources, enhancing energy efficiency, and implementing carbon capture and storage technologies. Renewable energy took centre stage in these discussions as a crucial solution for reducing greenhouse gas emissions and achieving sustainability goals. Initiatives announced included scaling up renewable energy deployment, improving energy efficiency measures, and phasing out coal-fired power plants. The conference highlighted the integral role of renewable energy not only in environmental preservation but also in stimulating economic growth, job creation, and fostering sustainable development worldwide. COP26 underscored the imperative for global cooperation and collective action to address climate change, with renewable energy recognized as a fundamental element in paving the way towards a sustainable and resilient future.

Numerous countries have implemented ambitious renewable energy policies and targets to align with international commitments and bolster domestic energy security. Case studies illustrate diverse approaches:

- Germany established the Renewable Energy Sources Act (EEG), which incentivizes renewable energy production through feed-in tariffs, spurring substantial growth in wind and solar power (Böhringer et al. 2017).
- China launched the National Renewable Energy Law, setting targets for renewable energy capacity expansion and implementing policies like preferential grid access and financial incentives to support solar, wind, and hydropower development (Schuman and Lin 2012).
- India aims to achieve 450 GW of renewable energy capacity by 2030 (PIP 2024) under its National Solar Mission and National Wind Mission, National Green Energy Mission, supported by policies such as generation-based incentives and viability gap funding to attract investment in renewable energy projects.
- In the United States, various states have renewable portfolio standards (RPS), mandating utilities to source a percentage of their total electricity from renewable resources. Federal tax incentives, such as the production tax credit (PTC) and investment tax credit (ITC) (Stokes and Breetz 2018), have been critical in supporting wind and solar energy development (Mcelroy and Chen 2017).
- Japan's feed-in tariff (FiT) scheme, introduced in 2012, guarantees fixed prices for electricity from renewable resources like solar, wind, and biomass (Dong and Shimada 2017). This policy promotes investment by providing long-term contracts and stable returns. Under the scheme, utility companies must buy electricity from renewable producers at set rates, offering financial security and encouraging development. The FiT scheme has significantly expanded Japan's solar power capacity, increased investment in wind projects, particularly offshore, and benefited biomass projects (Mizuno 2014). Overall, it has diversified Japan's energy mix, reduced dependence on the fossil fuels, and promoted a sustainable energy future, supporting energy security and environmental goals.

- Brazil's renewable energy policy, known as the National Program for the Production and Use of Biodiesel (PNPB), was launched in 2004 (Pousa et al. 2007). This policy provides incentives and loans for renewable energy projects, particularly focusing on bioenergy and hydropower. The goal is to diversify Brazil's energy matrix and enhance energy security by promoting the development and use of renewable energy sources. The program's financial support encourages the implementation of renewable energy technologies, contributing to a more sustainable and reliable energy system.
- The European Union's EU Emissions Trading System (EU ETS) places a price on carbon emissions, offering economic incentives for industries to invest in renewable energy and reduce emissions (The EU Emissions Trading System (EU ETS), n.d.). Among the top five countries in the EU for their participation and impact in the EU ETS are Germany, France, Italy, Spain, and the Netherlands. These countries have implemented significant measures to reduce carbon emissions and promote renewable energy development, contributing to the overall success of the EU ETS (Cansino et al. 2010).
- Australia's Renewable Energy Target (RET) mandates that a certain proportion of the nation's electricity must come from renewable sources (Cludius et al. 2014). RET aims to significantly reduce greenhouse gas emissions and foster a sustainable energy future for Australia. Implemented as part of the country's broader climate policy, the RET mandates that a specific percentage of Australia's electricity must come from renewable sources, such as wind, solar, hydro, and biomass. By setting ambitious goals, the RET incentivizes the development and deployment of renewable energy projects across the country.
- South Africa's Renewable Energy Independent Power Producer Procurement Program (REIPPPP) leverages a competitive bidding process to secure renewable energy projects. Launched in 2011, the program attracts investors by offering government guarantees and tax incentives, thereby supporting the development of renewable energy (Eberhard et al. 2014). This initiative has significantly boosted South Africa's renewable energy capacity, improved energy security, and reduced carbon emissions.

These examples highlight how national policies vary in scope and approach but share common goals of reducing dependence on fossil fuels, enhancing energy security, and mitigating environmental impacts.

Incentives and subsidies for renewable energy development are pivotal in overcoming economic barriers and accelerating the transition to renewable energy across sectors, from residential solar installations to large-scale wind farms and innovative biomass projects. Governments worldwide use measures such as feed-in tariffs (FiTs) to guarantee renewable energy producers a fixed price for electricity fed into the grid, thereby encouraging investment and providing revenue certainty. Tax incentives like investment tax credits (ITCs) and accelerated depreciation allowances reduce upfront costs for renewable energy projects, while grants, rebates, and green bonds leverage private sector investment with favourable terms. These incentives and subsidies collectively contribute to advancing renewable energy technologies, supporting global sustainability goals, and fostering a resilient energy future.

15.7 Building Bridges: International Collaboration for Renewable Energy

Collaboration and knowledge sharing between countries are essential to achieve sustainable energy goals and solve difficult problems. Developing countries have much to gain from international cooperation and a deeper understanding of technology, governance, and good practices. International cooperation and knowledge sharing within the framework of energy security policy have many advantages and opportunities.

Developing country collaboration is a powerful incentive to share best practices and important lessons learned in implementing sustainable energy policy. By sharing knowledge of successful initiatives, policy frameworks, and advanced systems, these countries can develop unique solutions to address their unique challenges. Developing countries with renewable technologies can also benefit (Bisbey et al. 2020) through capacity-building initiatives and technology transfers. By providing training programs and access to cutting-edge technologies, they can enhance the technical capabilities of developing nations, expediting the transition to sustainable energy solutions.

Through cross-border collaboration, sustainable energy projects can attract international financial support and private investment. By encouraging cooperation between countries, these projects become attractive to international investors and financial institutions (Ragosa and Warren 2019). Programs that emphasize teamwork and regional cooperation are more likely to receive funding from international organizations, financial institutions, and donors because they have a greater chance of success and impact. Cooperation and knowledge exchange can also promote policy and cooperation among developing countries. Countries can advance domestic and international goals by working together for sustainable energy development by learning from each other's successes and challenges. Cooperation between neighbouring countries can help promote the use of electricity to jointly solve energy security problems and improve energy in the region (Vasconcelos et al. 2022). Regional energy cooperation can improve cross-border energy use and provide backup power in emergencies. Cooperation can address environmental impacts of energy production, such as air and water pollution. Developing countries can jointly develop environmental standards and jointly monitor and control the environmental impact of energy production. South-South Cooperation is a way for countries to enhance cooperation through knowledge sharing and technology transfer. The strategy strengthens relations among developing countries and enables them to make the energy transition. Developing countries can collaborate with international organizations such as the United Nations and regional banks, which can help coordinate, provide assistance, and provide direct financing for renewable energy initiatives (Centobelli et al. 2020).

Collaborative research and innovation projects hold the potential to create significant breakthroughs in renewable energy technology. By pooling resources and expertise, developing nations can collectively address complex energy challenges and foster the advancement of sustainable energy solutions (Corrocher and Mancusi 2021). This collaborative approach is crucial, particularly for countries like India, which encounter specific obstacles in innovating offshore wind energy to overcome policy constraints and technological spaces. Knowledge sharing is another vital aspect of collaboration. By sharing best practices, case studies, and lessons learned, developing countries can gain valuable insights into overcoming their unique challenges (Irani et al. 2017). This exchange of information can significantly accelerate the learning curve and help avoid common pitfalls.

In contrast, developed countries have successfully navigated these challenges by establishing adequate policies, access to advanced technology, development of energy resources, and creation of good financial solutions. These elements collectively contribute to the successful deployment of offshore wind projects. For instance, countries in Europe have made significant strides in offshore wind energy by leveraging these strengths, thereby setting a benchmark for others to follow. To bridge the gap between developing and developed nations, it is essential to establish collaborative models that facilitate the growth of offshore wind energy in developing countries. One of the critical components of such collaboration is providing technical assistance. This includes transferring knowledge and expertise related to the design, installation, and maintenance of offshore wind turbines. Capacity-building initiatives can further enhance the local workforce's skills, ensuring they are equipped to handle the demands of this emerging industry. India's challenges in the offshore wind energy sector are multifaceted. The country struggles with inconsistent policies that hinder the smooth implementation of projects. Additionally, there is a notable lack of technical expertise required for the specialized field of offshore wind energy. The infrastructure necessary to support such projects is also inadequate, further complicating efforts. Moreover, financing difficulties pose significant hurdles, making it challenging to secure the substantial investment needed for these projects.

Policy guidance is also crucial in supporting developing nations. Developed countries can offer strategic advice on creating and implementing policies that foster a conducive environment for offshore wind energy projects. This guidance can help streamline regulatory frameworks, making it easier to navigate the bureaucratic landscape. Investment support and financial partnerships play a pivotal role in overcoming funding obstacles. Developed nations, along with international financial institutions, can provide the necessary capital and financial instruments to support large-scale offshore wind projects. This support can come in the form of grants, loans, or equity investments, reducing the financial burden on developing countries. Public-private partnerships (PPPs) and multilateral organizations are instrumental in facilitating joint ventures and resource-sharing. These entities can bring together stakeholders from various sectors, including government, industry, and academia, to work towards common goals. PPPs can effectively combine the strengths of both the public and private sectors, driving innovation and improving efficiency.

Finally, collaboration with developing countries can accelerate the deployment of offshore wind in developing countries by providing valuable insights, expertise, and resources, these collaborations can help developing nations overcome their challenges and achieve sustainable energy goals. This not only contributes to global energy sustainability but also supports economic growth and environmental protection in developing regions.

International collaboration and knowledge sharing offer a great opportunity for developing countries to make sustainable energy transitions. By sharing best practices, technology, and knowledge, these countries can use their collective resources to solve problems, develop regional energy, and move towards the common goal of a strong and low-carbon economy. Cooperation will help countries expand their influence, accelerate development, and coordinate the world's response to climate change and achieve their development goals. This cooperation not only supports the development of technology and new policies but also supports international cooperation to prevent climate change and ensure a safer and more sustainable future for all. Through international cooperation, developing countries can unlock new potential, stimulate economic growth, and take the lead in building strength and stability.

References

Adegbeye MJ, Ravi Kanth Reddy P, Obaisi AI et al (2020) Sustainable agriculture options for production, greenhouse gasses and pollution alleviation, and nutrient recycling in emerging and transitional nations – an overview. J Clean Prod 242:118319. https://doi.org/10.1016/J.JCLEPRO.2019.118319

Ahmed SD, Al-Ismail FSM, Shafiullah M et al (2020) Grid integration challenges of wind energy: a review. IEEE Access 8:10857–10878. https://doi.org/10.1109/ACCESS.2020.2964896

Akbarian A, Andooz A, Kowsari E et al (2022) Challenges and opportunities of lignocellulosic biomass gasification in the path of circular bioeconomy. Bioresour Technol 362:127774. https://doi.org/10.1016/J.BIORTECH.2022.127774

Avagyan AB, Singh B (2019) Biodiesel: feedstocks, technologies, economics and barriers. Assessment of environmental impact in producing and using chains. https://doi.org/10.1007/978-981-13-5746-6

Batel S (2020) Research on the social acceptance of renewable energy technologies: past, present and future. Energy Res Soc Sci 68:101544. https://doi.org/10.1016/J.ERSS.2020.101544

Bati ASR, Zhong YL, Burn PL et al (2023) Next-generation applications for integrated perovskite solar cells. Commun Mater 4:1–24. https://doi.org/10.1038/s43246-022-00325-4

Bisbey J, Li L, Gu Q, Chu CY (2020) Attracting private financing in cross-border infrastructure investments. J Infras Policy Dev 4:180–216. https://doi.org/10.24294/JIPD.V4I2.1199

Böhringer C, Cuntz A, Harhoff D, Asane-Otoo E (2017) The impact of the German feed-in tariff scheme on innovation: evidence based on patent filings in renewable energy technologies. Energy Econ 67:545–553. https://doi.org/10.1016/J.ENECO.2017.09.001

Breidenich C, Magraw D, Rowley A, Rubin JW (1998) The Kyoto protocol to the United Nations framework convention on climate change. Am J Int Law 92:315–331. https://doi.org/10.2307/2998044

Cansino JM, Pablo-Romero MDP, Román R, Yñiguez R (2010) Tax incentives to promote green electricity: an overview of EU-27 countries. Energy Policy 38:6000–6008. https://doi.org/10.1016/J.ENPOL.2010.05.055

Centobelli P, Cerchione R, Esposito E (2020) Pursuing supply chain sustainable development goals through the adoption of green practices and enabling technologies: a cross-country analysis of LSPs. Technol Forecast Soc Change 153:119920. https://doi.org/10.1016/J.TECHFORE.2020.119920

Céspedes S, Cano NA, Foo G et al (2022) Technical and environmental feasibility study of the co-production of crude oil and electrical energy from geothermal resources: first field trial in Colombia. PRO 10:568. https://doi.org/10.3390/PR10030568

Charles Rajesh Kumar J, Majid MA (2020) Renewable energy for sustainable development in India: current status, future prospects, challenges, employment, and investment opportunities. Energy Sustain Soc 10:1–36. https://doi.org/10.1186/S13705-019-0232-1/TABLES/37

Chowdhury MS, Rahman KS, Selvanathan V et al (2021) Current trends and prospects of tidal energy technology. Environ Dev Sustain 23:8179–8194. https://doi.org/10.1007/S10668-020-01013-4/FIGURES/5

Cludius J, Forrest S, MacGill I (2014) Distributional effects of the Australian renewable energy target (RET) through wholesale and retail electricity price impacts. Energy Policy 71:40–51. https://doi.org/10.1016/J.ENPOL.2014.04.008

Corrocher N, Mancusi ML (2021) International collaborations in green energy technologies: what is the role of distance in environmental policy stringency? Energy Policy 156:112470. https://doi.org/10.1016/J.ENPOL.2021.112470

Dada M, Popoola P (2023) Recent advances in solar photovoltaic materials and systems for energy storage applications: a review. Beni Suef Univ J Basic Appl Sci 12:1–15. https://doi.org/10.1186/S43088-023-00405-5/FIGURES/4

Demirbas A (2011) Competitive liquid biofuels from biomass. Appl Energy 88:17–28. https://doi.org/10.1016/J.APENERGY.2010.07.016

Dong Y, Shimada K (2017) Evolution from the renewable portfolio standards to feed-in tariff for the deployment of renewable energy in Japan. Renew Energy 107:590–596. https://doi.org/10.1016/J.RENENE.2017.02.016

Eberhard A, Leigland J, Kolker J (2014) South Africa's renewable energy IPP procurement program. World Bank Publications

Enevoldsen P, Valentine SV (2016) Do onshore and offshore wind farm development patterns differ? Energy Sustain Dev 35:41–51. https://doi.org/10.1016/J.ESD.2016.10.002

Farakhor A, Wu D, Wang Y, Fang H (2023) Scalable optimal power management for large-scale battery energy storage systems. IEEE Trans Transp Electrif. https://doi.org/10.1109/TTE.2023.3331243

Frondel M, Ritter N, Schmidt CM, Vance C (2010) Economic impacts from the promotion of renewable energy technologies: the German experience. Energy Policy 38:4048–4056. https://doi.org/10.1016/J.ENPOL.2010.03.029

Gallegos J, Arévalo P, Montaleza C, Jurado F (2024) Sustainable electrification—advances and challenges in electrical-distribution networks: a review. Sustain For 16:698. https://doi.org/10.3390/SU16020698

Hasan MM, Hossain S, Mofijur M et al (2023) Harnessing solar power: a review of photovoltaic innovations, solar thermal systems, and the dawn of energy storage solutions. Energies 16:6456. https://doi.org/10.3390/EN16186456

Hassan Q, Viktor PJ, Al-Musawi T et al (2024) The renewable energy role in the global energy transformations. Renew Energy Focus 48:100545. https://doi.org/10.1016/J.REF.2024.100545

Hunt JD, Zakeri B, Lopes R et al (2020) Existing and new arrangements of pumped-hydro storage plants. Renew Sust Energ Rev 129:109914. https://doi.org/10.1016/J.RSER.2020.109914

Hussain I, Immanuel RJ (2022) Composite materials and its advancements for a cleaner engine of future. In: Energy, environment, and sustainability, pp 169–191. https://doi.org/10.1007/978-981-16-8337-4_9

IEA (2023) Electricity – Renewables 2023 – Analysis – IEA. https://www.iea.org/reports/renewables-2023/electricity. Accessed 4 July 2024
Irani Z, Kamal MM, Sharif A, Love PED (2017) Enabling sustainable energy futures: factors influencing green supply chain collaboration. Prod Plan Control 28:684–705. https://doi.org/10.1080/09537287.2017.1309710
Jackson KJ, Zuteck MD, Van Dam CP et al (2005) Innovative design approaches for large wind turbine blades. Wind Energy 8:141–171. https://doi.org/10.1002/WE.128
Jaiswal KK, Chowdhury CR, Yadav D et al (2022) Renewable and sustainable clean energy development and impact on social, economic, and environmental health. Energy Nexus 7:100118. https://doi.org/10.1016/J.NEXUS.2022.100118
Jami AAN, Walsh PR (2014) The role of public participation in identifying stakeholder synergies in wind power project development: the case study of Ontario, Canada. Renew Energy 68:194–202. https://doi.org/10.1016/J.RENENE.2014.02.004
Jeremiah M, Kabeyi B, Olanrewaju OA (2022) Biogas production and applications in the sustainable energy transition. J Energy 2022:8750221. https://doi.org/10.1155/2022/8750221
Jolie E, Scott S, Faulds J et al (2021) Geological controls on geothermal resources for power generation. Nat Rev Earth Environ 2:324–339. https://doi.org/10.1038/s43017-021-00154-y
Kaygusuz K (2016) Hydropower as clean and renewable energy source for electricity generation. J Eng Res Appl Sci 5:359–369
Koukouvinis P, Anagnostopoulos J (2023) State of the art in designing fish-friendly turbines: concepts and performance indicators. Energies 16:2661. https://doi.org/10.3390/EN16062661
Kowsar A, Hassan M, Rana MT et al (2023) Optimization and techno-economic assessment of 50 MW floating solar power plant on Hakaluki marsh land in Bangladesh. Renew Energy 216:119077. https://doi.org/10.1016/J.RENENE.2023.119077
Kumar N, Sonthalia A, Pali HS (2020) Next-generation biofuels—opportunities and challenges. In: Green energy and technology, pp 171–191. https://doi.org/10.1007/978-981-13-9012-8_8
Lim MT, Phan A, Roddy D, Harvey A (2015) Technologies for measurement and mitigation of particulate emissions from domestic combustion of biomass: a review. Renew Sust Energ Rev 49:574–584. https://doi.org/10.1016/J.RSER.2015.04.090
Lu SM (2018) A global review of enhanced geothermal system (EGS). Renew Sust Energ Rev 81:2902–2921. https://doi.org/10.1016/J.RSER.2017.06.097
Lund JW, Boyd TL (2016) Direct utilization of geothermal energy 2015 worldwide review. Geothermics 60:66–93. https://doi.org/10.1016/J.GEOTHERMICS.2015.11.004
Luthra S, Kumar S, Garg D, Haleem A (2015) Barriers to renewable/sustainable energy technologies adoption: Indian perspective. Renew Sust Energ Rev 41:762–776. https://doi.org/10.1016/J.RSER.2014.08.077
Mcelroy MB, Chen X (2017) Wind and solar power in the United States: status and prospects. CSEE J Power Energy Syst 3:1–6. https://doi.org/10.17775/CSEEJPES.2017.0002
Mizuno E (2014) Overview of wind energy policy and development in Japan. Renew Sust Energ Rev 40:999–1018. https://doi.org/10.1016/J.RSER.2014.07.184
Moomaw W, Ramakrishna K, Gallagher K, Freid T (1999) The Kyoto protocol: a blueprint for sustainable development. J Environ Dev 8:82–90. https://doi.org/10.1177/107049659900800106/ASSET/107049659900800106.FP.PNG_V03
Nadia AR, Isa NAM, Desa MKM (2018) Advances in solar photovoltaic tracking systems: a review. Renew Sust Energ Rev 82:2548–2569. https://doi.org/10.1016/J.RSER.2017.09.077
Nautiyal H, Singal SK, Varun SA (2011) Small hydropower for sustainable energy development in India. Renew Sust Energ Rev 15:2021–2027. https://doi.org/10.1016/J.RSER.2011.01.006
Nikolaidis P (2023) Solar energy harnessing technologies towards de-carbonization: a systematic review of processes and systems. Energies 16(17):6153
O'sullivan M, Sandalow D (2017) The geopolitics of renewable energy
Parra D, Swierczynski M, Stroe DI et al (2017) An interdisciplinary review of energy storage for communities: challenges and perspectives. Renew Sust Energ Rev 79:730–749. https://doi.org/10.1016/J.RSER.2017.05.003

PIP (2024) India set to achieve 450 GW renewable energy installed capacity by 2030: Ministry of New and Renewable Energy (MNRE). https://pib.gov.in/Pressreleaseshare.aspx?PRID=1762960. Accessed 6 July 2024

Pitelis A, Vasilakos N, Chalvatzis K (2020) Fostering innovation in renewable energy technologies: choice of policy instruments and effectiveness. Renew Energy 151:1163–1172. https://doi.org/10.1016/J.RENENE.2019.11.100

Polzin F, Egli F, Steffen B, Schmidt TS (2019) How do policies mobilize private finance for renewable energy? A systematic review with an investor perspective. Appl Energy 236:1249–1268. https://doi.org/10.1016/J.APENERGY.2018.11.098

Pousa GPAG, Santos ALF, Suarez PAZ (2007) History and policy of biodiesel in Brazil. Energy Policy 35:5393–5398. https://doi.org/10.1016/J.ENPOL.2007.05.010

Prasad S, Venkatramanan V, Kumar S, Sheetal KR (2019) Biofuels: a clean technology for environment management. In: Sustainable green technologies for environmental management, pp 219–240. https://doi.org/10.1007/978-981-13-2772-8_11

Ragosa G, Warren P (2019) Unpacking the determinants of cross-border private investment in renewable energy in developing countries. J Clean Prod 235:854–865. https://doi.org/10.1016/J.JCLEPRO.2019.06.166

Rogelj J, Fricko O, Meinshausen M et al (2017) Understanding the origin of Paris agreement emission uncertainties. Nat Commun 8:1–12. https://doi.org/10.1038/ncomms15748

Scheffran J, Froese R (2016) Enabling environments for sustainable energy transitions: the diffusion of technology. In: Innovation and investment in low-carbon societies, pp 721–756. https://doi.org/10.1007/978-3-319-43884-9_34

Schuman S, Lin A (2012) China's renewable energy law and its impact on renewable power in China: progress, challenges and recommendations for improving implementation. Energy Policy 51:89–109. https://doi.org/10.1016/J.ENPOL.2012.06.066

Seshadri S (2012) Sustainability in clothing supply chains: implications for sustainability in clothing supply chains: implications for marketing, p 453–465

Siamanta ZC (2021) Conceptualizing alternatives to contemporary renewable energy development: community renewable energy ecologies (CREE). J Polit Ecol 28(1):47–69

Simpa P, Eze Ekechukwu D, Oduro P (2024) Renewable energy expansion: legal strategies for overcoming regulatory barriers and promoting innovation. World J Adv Eng Technol Sci 2024:168–186. https://doi.org/10.30574/wjaets.2024.12.1.0208

Singh VK, Kumar R, Teja PR (2014) Performance evolution of various models of bio-gas technologies in household environment. Int J Sci Technol Res 3:50–54

Stokes LC, Breetz HL (2018) Politics in the U.S. energy transition: case studies of solar, wind, biofuels and electric vehicles policy. Energy Policy 113:76–86. https://doi.org/10.1016/J.ENPOL.2017.10.057

Sun X, Huang D, Wu G (2012) The current state of offshore wind energy technology development. Energy 41:298–312. https://doi.org/10.1016/J.ENERGY.2012.02.054

The EU Emissions Trading System (EU ETS) (n.d.)

The National Archives (2021) [ARCHIVED CONTENT] COP26 Goals – UN Climate Change Conference (COP26) at the SEC – Glasgow. https://webarchive.nationalarchives.gov.uk/ukgwa/20230311034236/https://ukcop26.org/cop26-goals/. Accessed 6 July 2024

Tomarov GV, Shipkov AA (2017) Modern geothermal power: binary cycle geothermal power plants. Therm Eng 64:243–250. https://doi.org/10.1134/S0040601517040097/METRICS

Tyagi VV, Kaushik SC, Tyagi SK (2012) Advancement in solar photovoltaic/thermal (PV/T) hybrid collector technology. Renew Sust Energ Rev 16:1383–1398. https://doi.org/10.1016/j.rser.2011.12.013

Vasconcelos C, Silva J, Calheiros CSC et al (2022) Teaching sustainable development goals to university students: a cross-country case-based study. Sustain For 14:1593. https://doi.org/10.3390/SU14031593

Verma V (2024) Advancement in solar technology: evolution, generation, future prospective, and challenges – a review. https://doi.org/10.20944/PREPRINTS202407.0003.V1

Wang J, Dong F, Ma Z et al (2021) Multi-objective optimization with thermodynamic analysis of an integrated energy system based on biomass and solar energies. J Clean Prod 324:129257. https://doi.org/10.1016/J.JCLEPRO.2021.129257
Wu Z, Feng Y, Li X, Yuan B (2024) Flexibility definition and improvement of pumped hydro storage: a techno-economic analysis. J Energy Storage 93:112078. https://doi.org/10.1016/j.est.2024.112078

Chapter 16
Environmental Aspects and Electrical Energy Generation

Arun Kumar Choudhary

Abstract This chapter explores the correlation between the production of electrical energy and environmental factors, focusing on sustainable approaches. The text examines conventional energy sources such as coal, natural gas, and nuclear power, emphasizing their ecological consequences, particularly the release of greenhouse gases and pollution. The discussion then transitions to renewable energy sources, namely, solar, wind, hydro, and bioenergy, and explores their capacity to mitigate environmental damage. This chapter comprehensively elucidates significant strategies such as carbon capture, enhanced land use, and optimized distribution networks. The need for sustainable practices in the energy industry to ensure an environmentally friendly and resilient energy future is highlighted, along with the discussion of policy frameworks and international initiatives for sustainable development.

Keywords Electrical energy · Environmental · Sustainable · Ecological · Greenhouse gases · Bioenergy, Carbon capture.

16.1 Introduction

The generation and utilization of electrical energy are essential to contemporary civilization because they support economic expansion, improve living standards, and permit technical innovation. Electricity fuels various sectors, enterprises, and vital utilities, propelling efficiency and advancement. Dependable electricity is crucial for houses since it supplies illumination, warmth, cooling, and energy to operate equipment that enhances living conditions (Farghali et al. 2023). Additionally, it contributes to promoting public health by providing energy to medical institutions for health treatment activities and guaranteeing access to clean water via

A. K. Choudhary (✉)
Ministry of New and Renewable Energy, New Delhi, New Delhi, India

S. Kumar, V. K. Singh (eds.), *Renewable Energy Development: Technology, Material and Sustainability*, Clean Energy Production Technologies,
https://doi.org/10.1007/978-981-97-9626-7_16

purification systems. Electricity plays a decisive role in education by facilitating the use of digital learning platforms and ensuring the smooth functioning of academic institutions. Telecommunication networks, such as the Internet and broadcasting services, depend on a consistent and reliable electrical source to enable the flow of information and continuous connection. Moreover, electricity is vital in facilitating research and driving technical advancements, providing the necessary power for scientific labs and cutting-edge technologies such as artificial intelligence (AI) and the Internet of Things (IoT). Nevertheless, conventional means of energy production, such as coal and natural gas, have substantial environmental consequences, including releasing greenhouse gases (GHG) and deteriorating ecosystems. This phenomenon requires transitioning toward sustainable energy sources such as solar, wind, and hydroelectric power to promote sustainability. Expanding the range of energy sources also improves energy security by decreasing dependence on fossil fuels and minimizing the dangers of supply disruptions. Gaining a comprehensive understanding of the fundamental importance of electricity and effectively tackling the obstacles associated with its production are essential to attain a sustainable and prosperous future. Sustainable growth requires a careful balance between energy production and environmental conservation (Biedermann and Höhlein 1995). The increasing energy demand has resulted in substantial environmental degradation due to the heavy dependence on fossil fuels, including air and water pollution, habitat loss, and climate change. Transitioning to renewable energy sources (RES), such as solar, wind, and hydro, may reduce these effects while fulfilling energy requirements. Implementing sustainable methods in energy production conserves ecosystems and biodiversity improves public health and promotes long-term economic stability. Achieving this equilibrium is crucial in safeguarding the environment for future generations while simultaneously promoting advancement and ingenuity (Usman and Balsalobre-Lorente 2022). By analyzing the environmental effects of various energy sources, we can ascertain the advantages and disadvantages of each approach regarding sustainability. Although fossil fuels are dependable and have a well-developed infrastructure, they present substantial risks to air quality and the stability of the climate around the globe. Similarly, nuclear energy provides a low-emission alternative but requires meticulous management of radioactive materials and robust safety protocols. RES offers sustainable alternatives with low emissions, but they come with a cost in terms of material and land resources. They may harm the ecological balance of nearby species. Hydropower, a renewable alternative, may influence aquatic ecosystems and local populations. By thoroughly studying these issues, energy strategies that strike a strong balance between maximizing benefits and minimizing adverse environmental effects may be developed. This will lead us toward a more sustainable and resilient energy future (Saint et al. 2019).

This chapter covers the current, comprehensive state of several energy sources concerning their potential, production, consumption, per capita demand, etc. Additionally, it describes the environmental footprints and implications of various energy sources. Also, it explores the trade-offs associated with selecting one energy source over another. Furthermore, the worldwide environmental regulatory and policy framework, climate change, and greenhouse gas emissions in India are discussed. Finally, the conclusion and future directions were also covered.

16.2 Types of Electrical Energy Generation

Energy-producing technologies exhibit significant diversity, characterized by unique principles and use patterns. Fossil fuels, such as coal, oil, and natural gas, remain the predominant energy source worldwide, accounting for around 80% of total energy consumption (IEA 2023). These technologies use the combustion of fuels to create heat, which in turn powers turbines for electricity production. However, they are significant contributors to the release of GHG and air pollutants. Figure 16.1 depicts a global distribution of electricity generation in 2023, categorized by type of energy. Nuclear energy, which accounts for around 9.11% of worldwide power production, utilizes nuclear fission to generate heat to generate electricity. It has the advantage of producing minimal emissions but also raises problems over managing radioactive waste and safety precautions. Solar power, using sunlight via photovoltaic cells or solar thermal systems, is expanding significantly, accounting for around 5.53% of worldwide energy generation. It is highly regarded for its reduced ecological footprint and decreasing expenses. Wind power, which harnesses the kinetic energy of wind via turbines, accounts for about 7.82% of the world's electricity and is particularly noteworthy for its minimal operating emissions. Hydropower, which harnesses the kinetic energy of moving water to generate electricity via turbines, constitutes around 14.3% of the world's total electricity production.

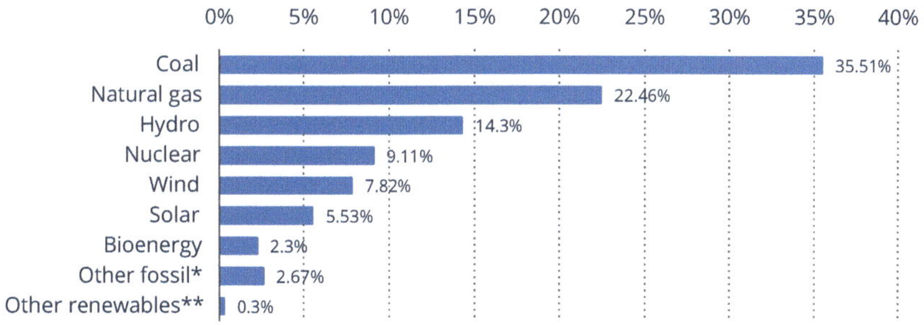

Fig. 16.1 Global distribution of electricity generation in 2023, categorized by type of energy. (Reproduced and adopted from: https://www.statista.com/statistics/269811/world-electricity-production-by-energy-source/, accessed on Jan 26 2024, Creative Commons license (usually Attribution-Only, or CC). *Encompasses oil, petroleum products, and produced gasses. **Encompasses geothermal, tidal, and wave energy sources)

While it provides a dependable and constant source of power, it also has the potential to affect aquatic ecosystems and local populations adversely. In developing countries, biomass energy, produced from organic materials such as plant and animal waste, is used to generate heat and power. However, if mismanaged, it may result in deforestation and other environmental problems. Geothermal energy harnesses the Earth's interior heat. It makes a modest but consistent contribution to the worldwide energy supply, particularly in areas with notable geothermal activity like Iceland, Indonesia, the Philippines, Turkey, and some sections of the United States (Gutiérrez-Negrín et al. 2024). The adoption and growth of each technique are contingent upon several variables, including the availability of resources, technical progress, economic feasibility, and environmental effects. An equitable approach to energy policy is required, considering both short- and long-term sustainability objectives (Osman et al. 2023).

16.3 Environmental Impacts of Different Energy Sources

Every energy-generating technique has unique environmental effects that differ in kind and scope, including human health, ecosystems, and patterns of the world's climate. Electricity production that relies on fossil fuels, namely, coal, oil, and natural gas, significantly contributes to environmental damage. Burning these fuels emits pollutants such as sulfur dioxide (SO_2), nitrogen oxides (NOx), and particulate matter, resulting in acid rain, smog, and respiratory diseases in people. In addition, releasing carbon dioxide (CO_2) from burning fossil fuels is the primary human-caused factor contributing to climate change. This worsens global warming and leads to rising sea levels and more frequent severe weather events (Anser et al. 2022). To tackle this, most countries have installed RES to mitigate these challenges.

The nations with the most installed renewable energy capacity as of 2023 are China, the United States, Brazil, India, Germany, and Japan. China is the global leader in installed power capacity, with over 1500 GW. This achievement is primarily due to substantial investments in solar, wind, and hydro energy sources. Figure 16.2 depicts that the United States has an installed capacity of around 404 GW, mostly from wind and solar energy sources. Brazil has around 194 GW of electricity generation capacity, mostly derived from hydroelectric sources. India now has an approximate capacity of 180 GW, specifically emphasizing solar and wind energy sources. Germany, boasting a capacity of 154 GW, and Japan, with a capacity of 149 GW, prioritize using solar and wind energy, as mentioned in Fig. 16.2. These countries jointly propel the worldwide expansion of renewable energy as they shift toward more environmentally friendly and sustainable energy (IEA 2024).

The global installed renewable energy capacity exceeded 3870 GW as of 2023, primarily due to significant investments and technological improvements in solar, wind, hydro, and biofuels. The rise is mainly driven by solar and wind energy, with solar photovoltaics reaching over 1419 GW and wind generation surpassing 1017

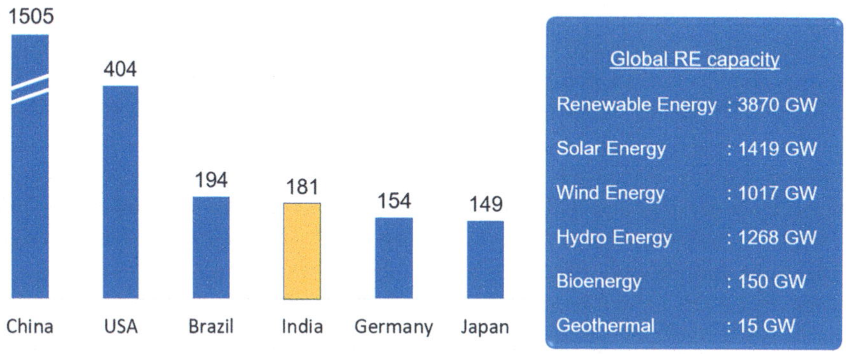

Fig. 16.2 Cumulative Global RE Capacity in gigawatts (GW) till December 2023

GW. Hydropower continues to be the most significant individual source, contributing more than 1268 GW. Bioenergy has almost 150 GW, and geothermal energy has nearly reached 15 GW. This development contributes to achieving climate objectives, decreases the release of greenhouse gases, and strengthens energy independence, signifying a crucial shift in the worldwide energy outlook.

Although promoted as a low-carbon substitute, nuclear energy presents notable environmental hazards linked mainly to radioactive waste disposal and reactor accidents. Although nuclear power facilities do not release GHGs during their operation, extracting and treating uranium ore and containing radioactive waste provide enduring environmental difficulties (Murray and Holbert 2020). The Chernobyl catastrophe (26 April 1986) and the Fukushima Daiichi nuclear disaster (11 March 2011) have revealed the severe and far-reaching effects that nuclear accidents may have, such as the discharge of radiation and the poisoning of land, water, and living species (Okada et al. 2021). Renewable energy sources provide cleaner options than fossil fuels and nuclear power but still have environmental consequences. Solar energy, produced by solar photovoltaic (SPV) panels and concentrated solar power (CSP) systems, necessitates substantial land use and may adversely affect local ecosystems and habitats. The manufacturing and disposal of PV panels also include using hazardous substances like cadmium and lead; however, recycling initiatives are enhancing. Wind power generated by wind turbines is usually regarded as ecologically harmless in terms of pollutants. However, collisions with the turbines' blades may hurt bird and bat populations. Additional problems include habitat disturbance and noise pollution, but efforts to mitigate these consequences are being made via turbine design and siting procedure advancements. Hydropower, harnessed from water movement in rivers or dams, is a sustainable energy source that ensures reliable electricity production but may have notable ecological consequences. Large dams modify river ecosystems, interrupt fish migrations, and submerge terrestrial areas, resulting in biodiversity loss and the relocation of populations. Biomass energy from organic substances such as wood, agricultural leftovers, and animal waste may provide a sustainable energy source. However, it is essential to acknowledge that environmental drawbacks are associated with its use. Deforestation and changes in land use related to biomass production may result in biodiversity depletion and the release of carbon emissions. In addition, the burning of biomass emits pollutants that are comparable to those released by fossil fuels. However, the intensity of these emissions may vary depending on the kind of biomass used and the combustion method used. Geothermal energy, derived from the heat stored under the Earth's surface, has low emissions during operation but may influence the surrounding geology and hydrology. Geothermal power facilities need the digging of wells to reach hot water or steam reservoirs, which may disrupt subterranean rock formations and possibly discharge hazardous pollutants like hydrogen sulfide. If geothermal fluids are not adequately handled after usage, water pollution is dangerous (IRENA 2023).

Every method of generating energy involves an intricate range of environmental factors that need to be thoroughly assessed and controlled to avoid adverse effects and maximize the advantages of energy production. Implementing sustainable energy regulations and using environmentally conscious technology are crucial for reducing these effects and successfully transitioning to a more sustainable and robust energy future.

The amount of water used differs significantly between renewable and nonrenewable energy methods. Nonrenewable sources such as coal, natural gas, and nuclear power plants need a significant amount of water to cool and generate steam, which often results in elevated rates of water extraction and consumption. Renewable technologies like wind and solar photovoltaic systems have meager water needs, unlike other forms of energy. Nevertheless, renewables like bioenergy and geothermal facilities need significant water for processing and cooling. The water use of hydropower is dependent on the unique circumstances and mainly involves managing the flow of water rather than consuming it. However, it may influence surrounding ecosystems. Therefore, the shift toward renewable energy sources often alleviates water scarcity but requires careful evaluation of the implications of each technology. Table 16.1 shows the water consumption factors for renewable and nonrenewable technologies (Macknick et al. 2011).

Table 16.1 Water consumption factors for renewable and nonrenewable technologies (gal/MWh)

Energy source	Cooling	Technology
Solar photovoltaics	Not required	Utility scale PV
Wind energy	Not required	Wind turbine
Concentrated solar power	Tower/dry/hybrid	Trough/power tower/fresnel
Bio power	Tower/pond/dry	Steam/biogas
Hydropower	Not required	Aggregated in-stream and reservoir
Geothermal energy	Tower/dry/hybrid	Steam/flash/binary/EGS
Nuclear energy	Tower/pond/dry	Generic
Natural gas	Tower/pond/dry	Combined cycle/steam
Coal	Tower/pond	Generic/subcritical/supercritical

16.4 Comparative Analysis of Environmental Effects

16.4.1 Comparison and Contrast the Environmental Footprints of Different Energy Sources

There are notable differences in the environmental footprints of various energy sources regarding how they affect human health, ecosystems, and climate change. Fossil fuels have significant environmental impacts because of their extensive extraction, processing, and burning procedures. Coal-fired power stations release significant amounts of carbon dioxide (CO_2), sulfur dioxide (SO_2), nitrogen oxides (NOx), and particulate matter into the atmosphere (Nieuwlaar 2013). These emissions contribute to global warming, acid rain, the creation of smog, and respiratory ailments. Oil extraction and refining activities emit pollutants and present the danger of oil spills, which harm marine ecosystems and coastal habitats. While burning natural gas releases less CO_2 than coal, it is essential to note that methane, a potent greenhouse gas, may escape during extracting and transporting natural gas, contributing to climate change.

Conversely, renewable energy sources such as solar, wind, and hydropower provide cleaner options with less environmental impact. SPV panels and CSP systems use sunlight to create energy without emitting greenhouse gases during operation. Nevertheless, the manufacturing and disposal of SPV panels include energy-intensive procedures and using hazardous substances such as cadmium and lead, necessitating appropriate recycling and disposal methods to mitigate environmental consequences. Wind power, created via turbines that capture wind energy, does not emit any direct pollutants and has little land required for each unit of electricity produced. However, wind farms may negatively affect bird and bat populations due to collisions with turbine blades and alter the surrounding landscapes and local ecosystems. Hydropower harnesses the kinetic energy of flowing water in rivers or dams to produce electricity, offering a dependable and sustainable energy resource. Massive dams can modify river ecosystems, obstruct fish movement, and flood land habitats, resulting in the loss of habitats and decreased biodiversity. Hydropower plants may cause population displacement and negatively affect downstream water quality and sediment movement. Biomass energy, obtained from organic substances such as wood, agricultural leftovers, and animal waste, is classified as a renewable source. However, it gives rise to apprehensions about deforestation, alterations in land use, and the release of pollutants, including particulates and nitrogen oxides, during the burning process. Geothermal energy harnesses the heat held under the Earth's surface and has a low emission of greenhouse gases during operation. However, it necessitates drilling and resource extraction, which might influence the local geology and hydrology. Effectively controlling geothermal fluids and mitigating subsurface pollution are essential to minimize environmental consequences. Although nuclear energy is low in carbon throughout its operation, it poses difficulties in the disposal of radioactive waste. It has potential catastrophes, as seen in the Chernobyl and Fukushima disasters. These incidents underscore the long-term environmental and health hazards connected with nuclear energy (Zhang and Wellalage 2022; Sohag et al. 2024).

Figure 16.3 illustrates the distribution of global CO_2 savings across several measures in the net zero scenario from 2022 to 2050. This data presents the anticipated worldwide GHG emissions and the expected difference from the most cost-effective strategies for limiting global warming to specific levels between 2030 and 2050. The figures are measured in billion metric tons of CO_2 equivalent and are categorized by scenario.

There is still a large gap between expected emissions and what is needed to limit global warming to 2 °C or 1.5 °C in 2030. Based on current unconditional nationally determined contributions, the 2 °C objective has a 14 GtCO2e gap, and the 1.5 °C goal has a 22 GtCO2e deficit. Conditional NDCs could close these disparities by 3 GtCO2e. The emissions gap is the difference between estimated global greenhouse gas (GHG) emissions if all countries fully implement their NDCs and emissions from the most economically efficient routes that meet the Paris Agreement's long-term temperature goals. These routes assume tight emission reductions starting in 2020, which have not happened. Thus, present emissions are higher, and the carbon budget is much lower. This implies a commitment to global warming unless emissions are reduced faster. The emissions gap estimates may be understated because they do not account for emissions since 2020. Table 16.2 shows that the 2030

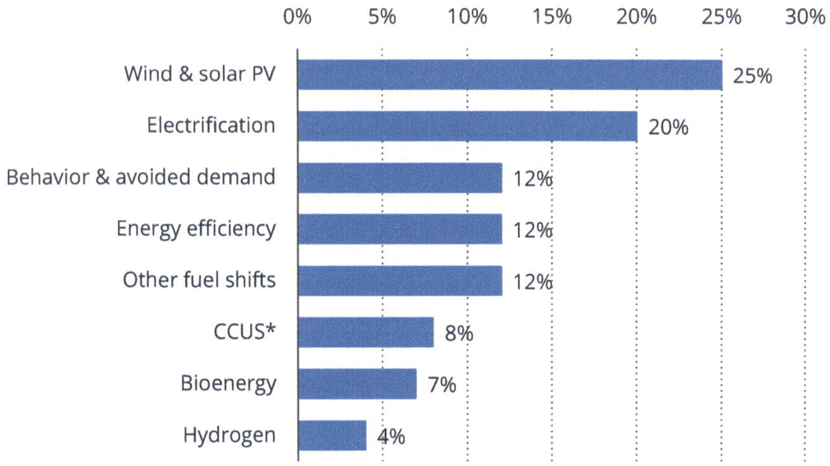

Fig. 16.3 Distribution of carbon dioxide (CO_2) emissions reductions in cumulative savings worldwide in the net zero scenario from 2022 to 2050, by mitigation measures. (Reproduced and adopted from: https://www.iea.org/data-and-statistics/charts/co2-emissions-reductions-in-cumulative-savings-by-mitigation-measures-in-the-net-zero-scenario-2022-2050, accessed on June 29, 2024, Creative Commons license (usually Attribution-Only, or CC). *Includes bioenergy with carbon capture and storage (BECCS) and direct air carbon capture and storage (DACCS))

Table 16.2 Global GHG emissions under different scenarios and the emissions gap 2030–2050 (Emissions Gap Report 2023, UNEP)

Timeframe and context	GHG emissions	Estimated gap for below 2 °C	Estimated gap for below 1.8 °C	Estimated gap for below 1.5 °C
2030				
Present policies	56	16	22	24
Unconditional NDCs	55	14	20	22
Conditional NDCs	52	11	17	19
2035				
Present policies continued	56	20	29	31
Unconditional NDCs continued	54	18	27	29
Conditional NDCs continued	51	15	24	26
2050				
Present policies continued	55	35	43	46
Unconditional NDCs and net zero pledges	44	24	32	36
Conditional NDCs and all net zero pledges	21	1	9	12

emissions gap is unchanged from last year. Full implementation of unconditional NDCs leads to a 14 GtCO2e gap for 2 °C and a 22 GtCO2e gap for 1.5 °C, while conditional NDCs lower these gaps to 11 and 19, respectively. Global greenhouse gas (GHG) emissions must be reduced by 28% for a 2 °C temperature increase and 42% for a 1.5 °C increase to follow the most cost-effective paths. This highlights the need to make considerable GHG emission reduction efforts. While fossil fuels continue to be the primary source of global energy, they present substantial environmental issues such as air and water pollution, habitat degradation, and climate change. Renewable energy sources provide cleaner alternatives with different environmental effects, ranging from low emissions and land use in solar and wind power to worries about habitat disturbance in hydropower and biomass. To achieve a balance between energy security, economic development, and environmental stewardship in the transition to a sustainable energy future, the environmental footprint of each energy source needs to be carefully assessed and managed through technological innovation, policy measures, and sustainable practices (IRENA 2020).

The global CO_2 emissions from fossil fuels and industry reached 37.15 billion metric tons ($GtCO_2$) in 2022 and have risen by 1.1% in 2023, reaching a record high of 37.55 $GtCO_2$ as depicted in Fig. 16.4. Since 1990, global CO_2 emissions have increased by over 60%. The primary sources of these emissions include the combustion of coal, oil, and gas, industrial activities, the release of gas through flaring, and the production of cement.

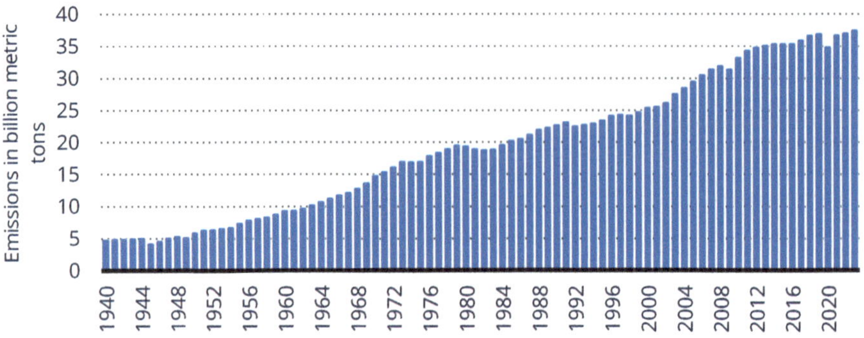

Fig. 16.4 Global carbon dioxide (CO_2) emissions every year from 1940 to 2023, measured in billion metric tons. (Reproduced and adopted from https://www.statista.com/statistics/276629/global-co2-emissions/, accessed on September 23, 2023, Creative Commons license (usually Attribution-Only, or CC))

Figure 16.5 shows the comparative analysis of the various energy sources. This analysis focuses solely on the primary power sources, excluding oil, since it is mainly used for transportation. Their land use is measured in square meters per megawatt-hour of power generated annually. This considers the varying capacity factors of various sources, explicitly relying on the actual output generated by intermittent technologies such as solar or wind. Initially, there are substantial disparities across the various sources. Nuclear energy is in the lowermost part of the figure. Wind power is the most land-efficient form of energy, requiring 50 times less area than coal and 18 to 27 times less land than on-ground solar PV per unit of electricity.

Furthermore, significant variations exist within a single energy technology. This is seen by the enormous variation in the land footprint, ranging from the least to the highest. This demonstrates that land use is heavily influenced by how technology is implemented and the specific circumstances of the area. The setting and kind of material have a significant role in the case of solar energy. Cadmium-based solar panels have higher energy and material efficiency than silicon panels, resulting in a lower land need per unit. The location where solar panels are installed, whether on roofs or the ground, significantly impacts their performance. Rooftop solar requires far less extra land since it utilizes an already occupied area, especially on the tops of existing structures. Nevertheless, during their lifespan, solar panels need a certain amount of land due to the necessity for mining the elements required for their production and the energy, primarily electricity, used during the silicon refining process. Ultimately, the density and spacing of the panels have a significant impact.

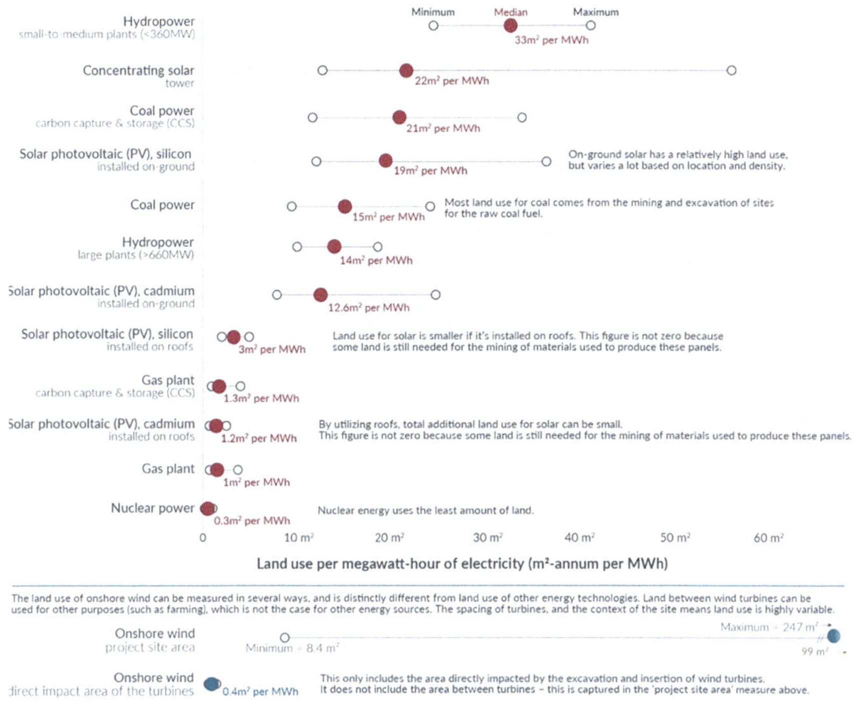

Fig. 16.5 Land use of energy sources per unit of electricity. (Adopted from https://ourworldindata.org/land-use-per-energy-source, accessed on June 11, 2023, Creative Commons license (usually Attribution-Only, or CC))

16.4.2 Trade-Offs Involved in Choosing One Energy Source Over Another

Selecting an energy source requires balancing a few intricate trade-offs that consider the environment, economy, and society. Fossil fuels provide a dependable energy source and well-established infrastructure but also have substantial environmental drawbacks. When they burn, they emit GHGs and air pollutants, contributing to climate change, air quality deterioration, and public health threats. Shifting toward RES, such as solar and wind, helps alleviate these environmental consequences by decreasing emissions and dependence on limited resources. Nevertheless, using renewable energy technologies sometimes demands significant land utilization and encounters intermittent operation obstacles, demanding the implementation of backup systems or energy storage solutions. Hydropower offers dependable energy generation, although it may disturb river ecosystems and result in the displacement of populations because of dam building. Biomass energy presents a sustainable substitute, although it elicits apprehensions about deforestation, depletion of biodiversity, and competition with food crops. Although nuclear energy is a low-carbon source, it has inherent dangers related to managing radioactive waste, reactor accidents, and public perception. Geothermal energy has a low greenhouse gas emission rate but requires meticulous handling of underground resources and geothermal fluids. The trade-offs of any energy source go beyond only environmental implications and include economic expenses, energy security, and cultural acceptability. Fossil fuels remain economically viable because of their cheap production costs and existing infrastructure. However, their negative impact on the environment requires strict restrictions and initiatives to reduce emissions. RES are becoming more economically competitive and provide long-term advantages such as energy self-sufficiency and the creation of employment in the environmentally friendly sector.

Nevertheless, using these technologies necessitates financial expenditures in grid modernization and storage systems to guarantee the dependability and consistency of the deployment. To accomplish a sustainable energy transition that fulfills current and future energy needs while minimizing environmental consequences and promoting social fairness, it is necessary to have integrated energy planning, informed decision-making, and stakeholder involvement. This involves carefully considering the trade-offs involved and making well-informed choices. Emissions stem from various sources, including fossil fuel use through combustion and flaring, industrial processes such as cement, steel, chemicals, urea production, and product use, excluding short-cycle carbon CO_2. Key sectors contributing to emissions are the power industry through power and heat generation plants, industrial combustion for manufacturing and fuel production, small-scale nonindustrial stationary combustion in buildings, mobile combustion in transport (road, rail, ship, and aviation), and fuel exploitation involving the production, transformation, and refining of fuels as shown in Fig. 16.6.

The globe has an immense capacity to harness solar energy. Solar energy is a nascent but rapidly expanding notion, outpacing all other forms of renewable energy globally. To give an integrated and consistent assessment of solar resources and the

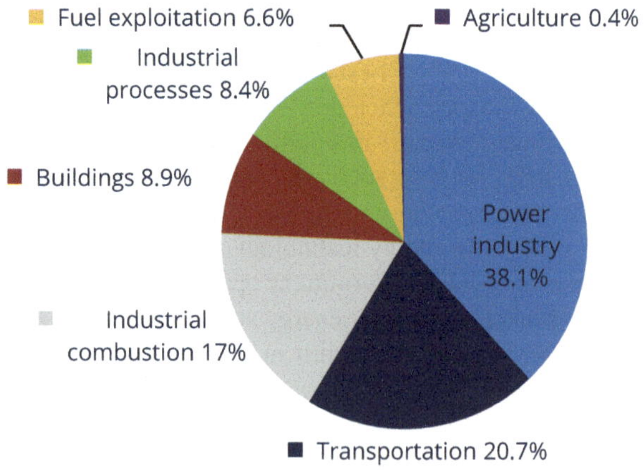

Fig. 16.6 Distribution of CO_2 emissions worldwide in 2022, by sector. (Adopted from https://www.statista.com/statistics/1129656/global-share-of-co2-emissions-from-fossil-fuel-and-cement/, accessed on April 10, 2024, Creative Commons license (usually Attribution-Only, or CC))

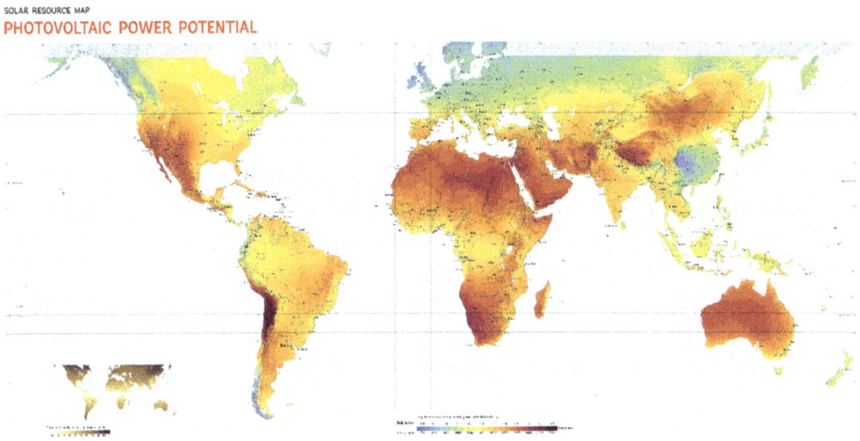

Fig. 16.7 Global photovoltaic power potential. (Adopted from https://globalsolaratlas.info/global-pv-potential-study, accessed on April 10, 2024, Creative Commons license (usually Attribution-Only, or CC))

potential for developing utility-scale photovoltaic (PV) power plants from the viewpoint of nations and regions, the World Bank produced research titled Global Photovoltaic Power Potential by country, as shown in Fig. 16.7.

16.5 Mitigation Strategies

Various technological and strategic solutions are implemented to reduce the environmental effects of energy production. These solutions are applied to different energy sources and strategic areas of concentration.

Carbon capture and storage (CCS) technologies are crucial in the field of fossil fuels. CCS is the process of collecting CO_2 emissions from power plants and industrial sources, so avoiding their escape into the atmosphere. The captured emissions are then stored underground. Moreover, by using state-of-the-art turbines and boilers, the combustion processes may be enhanced, resulting in a decrease in emissions per unit of energy generated. Renewable energy sources, such as solar and wind power, provide naturally cleaner alternatives but encounter obstacles such as intermittency. Technological advancements such as grid-scale energy storage devices, such as batteries and pumped hydro, contribute to stabilizing their energy production, hence improving dependability and facilitating integration into current power networks. Furthermore, advancements in the efficiency of solar panels and the design of wind turbines increase the amount of energy produced while simultaneously decreasing the impact on the environment (Mehmood et al. 2024). Despite its contentious nature, nuclear power offers a low-carbon energy alternative that produces little air pollution. Current research is centered on new reactor designs that provide improved safety, efficiency, and less waste production. Digitalization and smart grid technologies, on the other hand, offer a promising future for energy efficiency. They improve energy distribution, reduce transmission losses, and facilitate demand response tactics, resulting in increased energy efficiency and decreased environmental effects across all energy sectors. Strategic efforts also include regulatory frameworks that support the use of renewable energy, methods for pricing carbon emissions, and international partnerships to encourage the transfer of knowledge and the spread of innovation. To summarize, effectively dealing with the environmental effects of energy production requires a comprehensive strategy that incorporates technical progress, regulatory interventions, and global collaboration. By using these solutions across different energy sources and areas of concentration, society may advance toward a more sustainable and resilient energy future. Various technological ways and their strategic solutions along with focus areas shown in Table16.3.

By employing these technological innovations, strategic approaches, and policy measures, stakeholders can effectively mitigate the environmental impacts of energy generation. These solutions reduce emissions and environmental degradation, contribute to sustainable development goals, enhance energy security, and promote economic growth in a carbon-constrained world.

16.6 Regulatory and Policy Framework

Global and national policies governing environmental standards in energy generation are crucial in promoting sustainable energy practices, reducing emissions, and protecting ecosystems. In 2015, all United Nations Member States adopted the 2030 Agenda for Sustainable Development, which offers a unified vision for the planet's future and its inhabitants, ensuring peace and prosperity. The initiative's core consists of the 17 Sustainable Development Goals (SDGs), which serve as a pressing demand for action from all developed and developing nations in a worldwide collaboration. The SDGs acknowledge the need to simultaneously address poverty and other forms of deprivation via policies that enhance health and education, diminish inequality, stimulate economic development, and combat climate change while also striving to protect our seas and forests.

The SDGs are based on the substantial endeavors undertaken by governments and the United Nations, particularly the UN Department of Economic and Social Affairs, over numerous decades. At the Earth Summit held in Rio de Janeiro, Brazil, in June 1992, more than 178 countries agreed to adopt and execute Agenda 21. This comprehensive approach seeks to form a global partnership to advance sustainable development, improve human welfare, and protect the environment. The Millennium Declaration was overwhelmingly supported by the Member States of the United Nations during the Millennium Summit in September 2000, which took place at the United Nations Headquarters in New York. This summit formulated 8 Millennium Development Goals (MDGs). The Johannesburg Declaration on Sustainable Development and the Plan of Implementation, endorsed during the 2002 World

Table 16.3 Technological and strategic solutions to mitigate the environmental impacts of energy generation

Technological ways	Focus areas	Strategic solutions
Renewable energy technologies	Solar PV	Third-generation solar cell technologies
	Wind power	Wind turbine design and strategic placement of wind farms minimize impact on wildlife and habitats
	Hydropower	Implementing fish-friendly turbine designs and fish ladders in dams allows fish to migrate, reducing ecological disruptions. Utilizing run-of-river systems avoids the need for large reservoirs
	Bioenergy	Advanced feedstock selection and biomass supply chains, efficient conversion technologies, sustainability standards, and certification water and waste management
	Geothermal energy	Enhanced geothermal systems (EGS) and closed-loop geothermal designs minimize surface impacts and fluid disposal issues
	Marine energy	Advanced turbine design, environmental monitoring systems, biofouling prevention and ecosystem-based management
Energy storage and grid integration	Battery storage	Improving battery management systems (BMS) and advanced chemistry cell (ACC) for RE storage reduces intermittency issues and enhances grid stability
	Smart grids	Implementing smart grid technologies allows for efficient management of renewable energy sources (RES), demand response programs, and integration of distributed generation
Environmental monitoring and impact assessment		Conducting in-depth environmental impact assessments (EIAs), emissions trading, clean development mechanism (CDM) Utilization of advanced monitoring technologies, such as drones and satellite imaging, to track environmental changes and ensure compliance with regulations
Carbon capture, utilization, and storage (CCUS)		Developing and deploying CCUS technologies in fossil fuel power plants Utilization of captured CO_2 for enhanced oil recovery or industrial processes reduces emissions
Policy and regulatory frameworks		Establishing stringent emission standards and carbon pricing mechanisms to incentivize cleaner energy generation and penalize high-emission sources Implementation of renewable energy (RE) targets and contributions, feed-in tariffs (FiTs), renewable purchase obligation (RPO) trajectory, RE generation along with energy storage obligation, green open access and green markets, financial assistance to accelerate the adoption of RE technologies
Research and innovation (R&I) ecosystem		Investing in R&I for next generation RE technologies Active support to start-ups and entrepreneurs Promotion of innovation in clean energy technologies through public-private partnerships (PPP) and government funding initiatives

(continued)

Table 16.3 (continued)

Technological ways	Focus areas	Strategic solutions
Public engagement, awareness, and skill development		Educating the public about the benefits of RE and energy efficiency measures to foster support for sustainable energy policies and projects Engaging local communities in energy planning processes to address concerns, ensure transparency, and promote social acceptance of energy projects

Fig. 16.8 The UN Sustainable Development Summit in New York embraced 17 Sustainable Development Goals (SDGs). (Adopted from https://sdgs.un.org/goals, accessed on 10 April, 2024, Creative Commons license (usually Attribution-Only, or CC))

Summit on Sustainable Development in South Africa, reconfirmed the worldwide dedication to eliminating poverty and protecting the environment. The agreements expanded upon Agenda 21 and the Millennium Declaration by placing greater emphasis on collaborative partnerships among multiple nations (GSDR 2023). Figure 16.8 showed the UN Sustainable Development Summit in New York embraced 17 Sustainable Development Goals (SDGs).

As part of the outcome document "The Future We Want," which was accepted by Member States at the United Nations Conference on Sustainable Development (Rio + 20) in June 2012, a decision was made to develop a set of Sustainable Development Goals (SDGs) to expand on the Millennium Development Goals (MDGs). In 2013, the General Assembly established a 30-member Open Working Group to formulate a proposal about the SDGs. The culmination of this process occurred when the 2030 Agenda for Sustainable Development was adopted at the UN Sustainable Development Summit in September 2015. The agenda consists of 17 SDGs that form its core. In 2015, significant progress was made in global cooperation and the development of international policies. Several important agreements were adopted during this year, including the Sendai Framework for Disaster Risk Reduction in March, the Addis Ababa Action Agenda on Financing for Development in July, the Paris Agreement on Climate Change in December, and the "Transforming our World: the 2030 Agenda for Sustainable Development" with its 17 SDGs adopted at the UN Sustainable Development Summit in New York in September. The High-level Forum on Sustainable Development, convened on a yearly basis, serves as the principal United Nations platform for the oversight and assessment of the advancements made toward achieving the Sustainable Development Goals (SDGs). The Division for Sustainable Development Goals (DSDG) in the United Nations Department of Economic and Social Affairs (UNDESA) presently provides significant support and resources for the SDGs and their related thematic issues. The topics encompassed are water, energy, climate, oceans, urbanization, transport, research and technology, the Global Sustainable Development Report (GSDR), collaborations, and Small Island Developing States. The Department of Sustainable Development Goals (DSDG) plays a vital role in evaluating the overall advancement of the United Nations in executing the 2030 Plan.

Additionally, it is responsible for promoting and raising awareness about the SDGs via various outreach efforts. To effectively execute the global objectives of the 2030 Agenda, all stakeholders must demonstrate a strong commitment, ensuring that the SDGs are embraced and acted upon by a wide range of individuals and organizations. DSDG's objective is to assist in facilitating this interaction.

These organizations and frameworks are essential in setting and advancing environmental standards in the energy-generating industry, guaranteeing energy production processes' sustainability and environmental friendliness. Table 16.4 showed the global frameworks and governing environmental standards along with deliverables.

16.7 Climate Change and GHG Emissions in India

The increase in emissions in India in recent decades may be linked to causes such as a rapidly growing population, a rapidly rising economy, and an increasing demand for fossil energy. South Asian countries' contribution to global GHG emissions has increased to over 7%, positioning them as the third largest GHG polluters worldwide, behind the United States and China. However, India has the lowest CO_2 emissions per person among the G20 countries. It is just 1.9 tCO_2 per person, which is only one-eighth of the average annual emissions of an American. Its historical total CO_2 emissions contribution is much smaller compared to other big emitters, amounting to just 3% (Crippa et al. 2022).

India's GHG emissions have virtually quadrupled since the beginning of the twenty-first century. The nation emits over 4 billion metric tons of carbon dioxide equivalent ($GtCO_2e$) annually. India's fossil CO_2 emissions have almost quadrupled throughout this period, hitting an all-time high of 2.7 gigatonnes of CO_2 in 2022. The country's escalating emissions may be attributed mainly to the growing reliance

Table 16.4 Global frameworks and governing environmental standards along with deliverables

Global frameworks	Key deliverables and mechanism	Year of start	Governing body
Environmental Health Criteria (EHC) program	Offer guidelines and recommendations for managing environmental health risks	1970	World Health Organization (WHO)
Air quality guidelines	Provide global standards for air quality, focusing on minimizing the public health impacts of air pollution	1977	World Health Organization (WHO)
Assessment Reports (ARs)	Provide scientific assessments on climate change, its impacts, and potential future risks, as well as options for adaptation and mitigation	1990	Intergovernmental Panel on Climate Change (IPCC)
Basel convention	Controls the transboundary movements of hazardous wastes and their disposal	1992	Conference of the Parties (COP)
ISO 14001	Specifies the requirements for an environmental management system that an organization can use to enhance its environmental performance	1996	International Organization for Standardization (ISO)
Kyoto protocol	Commitment of industrialized countries to reduce GHG emissions to combat global climate change impacts	1997	United Nations Framework Convention on Climate Change (UNFCCC)
GRI standards	Provide the first global standards for sustainability reporting, including environmental impacts	1997	Global reporting initiative (GRI)
Rotterdam convention	Promotes shared responsibilities in relation to the importation of hazardous chemicals	1998	Conference of the Parties (COP)
Stockholm convention	The objective is to eradicate or limit the creation and utilization of persistent organic pollutants (POPs)	2001	United Nations (UN)
EU Emissions Trading System (ETS)	A key tool for reducing greenhouse gas emissions cost-effectively	2005	European Union (EU)
Performance standards on environmental and social sustainability	Provide a framework for managing environmental and social risks	2006	International Finance Corporation (IFC)
Green economy initiative	Aims to assist governments in "greening" their economies by reshaping and refocusing policies, investments, and spending	2008	United Nations Environment Programme (UNEP)
Renewable Energy Directive (RED)	Sets targets for renewable energy use in the EU	2009	European Union (EU)

(continued)

Table 16.4 (continued)

Global frameworks	Key deliverables and mechanism	Year of start	Governing body
Directive on Industrial Emissions (IED)	Aims to prevent and reduce pollution from industrial activities	2011	European Union (EU)
Sustainable Energy for All (SE4All)	A global initiative to ensure universal access to modern energy services, improve energy efficiency, and increase the use of renewable energy	2011	United Nations (UN)
ISO 50001	This specifies the requirements for establishing, implementing, maintaining, and improving an energy management system	2011	International Organization for Standardization (ISO)
Paris agreement	To limit global warming to well below 2 °C above pre-industrial levels and pursue efforts to limit the temperature increase to 1.5 °C. Its mechanism is based on NDCs where nations outline their plans to reduce GHG emissions	2015	United Nations Framework Convention on Climate Change (UNFCCC)
United Nations Sustainable Development Goals (SDGs)	Sustainable development goal 7 promotes affordable and clean energy and its targets under SDG 7.1 is access to energy, SDG 7.2 is renewable energy, SDG 7.3 is energy efficiency and goal 13 focuses on climate action	2015	United Nations General Assembly
Environmental and Social Framework (ESF)	Sets out the World Bank's commitments to sustainable development through policy and practice	2018	World Bank Group

on coal-fired energy production, which presently constitutes 75% of India's power supply and is known for its significant pollution. In 2022, the electricity industry in India accounted for 32% of all GHG emissions, excluding land use, land-use change, and forestry (LULUCF). It was the most significant contributor to emissions, with the agricultural and industrial sectors following behind. The Indian government has declared several goals it hopes to accomplish to combat climate change. These goals consist of decreasing the carbon intensity of the economy by a minimum of 45% by 2030 in comparison to the levels in 2005, fulfilling 50% of energy demand using energy sources that are not derived from fossil fuels by 2030, and attaining a state where carbon emissions are entirely neutral by 2070 (MoEFCC, 2022). India has achieved significant progress in its energy transition, establishing itself as a prominent player in the global renewable energy (RE) industry, especially in solar capacity for generation. However, India's ambitious climate objectives are compromised by its persistent reliance on coal despite its advancements and dedication to renewable energy. India is actively strategizing to boost its coal production to fulfill the escalating energy requirements from rapid industrialization, urbanization, and rising purchasing power. Consequently, the nation with the highest population in the world today confronts the task of harmonizing economic expansion with reducing carbon emissions.

India has a substantial legal foundation for environmental preservation that includes several essential laws and regulations. The Environment Protection Act of 1986 functions as a comprehensive law, granting the power to set benchmarks for air, water, and soil quality. The Water (Prevention and Control of Pollution) Act, 1974, and the Air (Prevention and Control of Pollution) Act, 1981, explicitly target the issues of water and air pollution. The Species Protection Act of 1972 protects species and their habitats. The National Green Tribunal (NGT) Act of 2010 guarantees the efficient and prompt resolution of legal matters concerning environmental protection and forest conservation. These statutes together provide a comprehensive framework for environmental regulation in India. Table 16.5 presents a comprehensive and resilient legislative structure aimed at safeguarding the environment in India.

Table 16.5 Environmental laws, regulations, and acts in India

Regulations	Title	Subject	Notification body
Constitutional provisions	Article 48A	State's duty to protect and improve the environment and to safeguard forests and wildlife	Republic of India
	Article 51A(g)	Protect and improve the natural environment including forests, lakes, rivers, and wildlife and to have compassion for living creatures	
Key acts	The Water (Prevention and Control of Pollution) Act, 1974	Prevention and control of water pollution	Parliament of India
	The Wildlife Protection Act, 1972	To protect wildlife and their habitats	
	The Forest (Conservation) Act, 1980	Conservation and protection of forests	
	The Air (Prevention and Control of Pollution) Act, 1981	Prevention, control, and abatement of air pollution	
	The Environment (Protection) Act, 1986, amended 1991	Framework for environmental protection	
	Energy Conservation Act, 2001	Promotes energy efficiency	
	The National Green Tribunal (NGT) Act, 2010	Disposal of legal cases related to environmental protection and conservation of forests	

(continued)

Table 16.5 (continued)

Regulations	Title	Subject	Notification body
Notable rules and notifications	Environment (Protection) Rules, 1986	Protection and improvement of the quality of the environment	Ministry of Environment, Forest, and Climate Change (MoEFCC), Govt. of India
	The Municipal Solid Wastes (Management and Handling) Rules, 2000	Guidelines and procedures for the management and handling of municipal solid waste (MSW)	
	Forest (Conservation) Rules, 2003	Govern the diversion of forest land for non-forest purposes	
	Environmental Impact Assessment (EIA) Notification, 2006	Legal framework that mandates the assessment of potential environmental impacts of certain industrial and developmental projects	
	Hazardous Waste Rules, 2016	Managing and handling hazardous wastes	
	The E-Waste (Management) Rules, 2016	Managing and handling electronic waste (e-waste)	
	The Plastic Waste Management Rules, 2016	Manage and control the generation, storage, collection, transportation, processing, and disposal of plastic waste	
	The Solid Waste Management Rules, 2016	Processing, recycling, treatment, and disposal of solid waste	
	The Coastal Regulation Zones (CRZ) Notification, 2019	This notification declares coastal stretches in India as CRZ and regulates certain activities within the CRZ	
	Plastic Waste Management Amendment Rules, 2024	Focus on strengthening the regulation and management of plastic waste	

16.8 Conclusion and Way Forward

The energy industry is steered toward environmentally friendly and sustainable practices that employ international and national legislation. These policies seek to diminish the environmental consequences of energy production, alleviate climate change, and encourage using clean energy technology by establishing guidelines, offering rewards, and promoting global collaboration. Practical environmental preservation in energy production requires close coordination between global efforts and country-specific activities. This chapter details the environmental impacts of different energy resources and their mitigation strategies, comparative evaluation of environmental effects, regulatory and policy framework, and climate change and renewable energy in the Indian context. With the rapid advancements in next-generation solar cells, energy storage technologies, smart grids, and hydrogen as an energy carrier, clean energy technology is fast changing the global energy environment. Hydrogen is being more recognized as a flexible and environmentally friendly energy carrier. The advancements in electrolysis have enhanced its production efficiency and cost-effectiveness, especially when using renewable sources such as wind and solar power. Hydrogen fuel cells are increasingly used in several sectors, such as transportation and grid storage since they provide emission-free alternatives to fossil fuels. Perovskite and tandem solar cells are considered the next generation of solar cells and demonstrate improved efficiencies and reduced manufacturing costs. These advanced cells could capture a wider range of radiation and may be used in flexible applications. The widespread use of solar energy worldwide relies on these technological breakthroughs. Advanced communication technologies are being integrated into power distribution networks to improve efficiency and reliability, leading to a revolution in smart grids. By implementing real-time monitoring and control of energy flows, the integration of renewable energy sources may be improved.

Demand response technologies and automated energy management systems can also optimize energy usage, minimize waste, and decrease expenses. Smart meters and IoT devices enable users to get comprehensive information about their energy consumption, enhancing efficiency. Energy storage innovations are crucial for developing a reliable and green electrical system. Progress in lithium-ion, sodium-ion, and solid-state batteries improves the amount of energy they can store, their longevity, and their safety. Flow batteries, on the other hand, provide flexible methods for storing vast amounts of energy, which helps to overcome the intermittent nature of renewable energy sources. In addition, thermal and compressed air energy storage provide alternate approaches to storing surplus energy. These technological advancements fuel the transition toward a more sustainable and reliable energy future. The world is on the verge of a substantial transition toward sustainable energy systems through advancements in hydrogen technology, solar cell efficiency, smart grid infrastructure, and energy storage solutions. This transition will guarantee a robust and environmentally friendly energy source for the future.

Competing Interest Declaration The authors have no conflicts of interest to declare that are relevant to the content of this book.

References

Anser MK, Usman M, Sharif M et al (2022) The dynamic impact of renewable energy sources on environmental economic growth: evidence from selected Asian economies. Environ Sci Pollut Res 29:3323–3335. https://doi.org/10.1007/s11356-021-17136-8

Biedermann P, Höhlein B (1995) On the environmental impact of energy conversion systems. In: Lapicque F, Storck A, Wragg AA (eds) Electrochemical engineering and energy. Springer, Boston. https://doi.org/10.1007/978-1-4615-2514-1_1

Crippa MO, Guizzardi DI, Banja MA, Solazzo EF, Muntean MA, Schaaf ED, Pagani F, Monforti-Ferrario F, Olivier JG, Quadrelli R, Risquez Martin A (2022) CO2 emissions of all world countries. JRC Science for Policy Report, European Commission, EUR. 2022;31182

Farghali M, Osman AI, Chen Z et al (2023) Social, environmental, and economic consequences of integrating renewable energies in the electricity sector: a review. Environ Chem Lett 21:1381–1418. https://doi.org/10.1007/s10311-023-01587-1

Global Sustainable Development Report (GSDR) (2023). https://sdgs.un.org/sites/default/files/2023-09/FINAL%20GSDR%202023-Digital%20-110923_1.pdf

Gutiérrez-Negrín LCA (2024) Evolution of worldwide geothermal power 2020–2023. Geotherm Energy 12:14. https://doi.org/10.1186/s40517-024-00290-w

IEA (2023) World Energy Outlook 2023. https://www.iea.org/reports/world-energy-outlook-2023

IEA (2024) Renewables 2023, IEA, Paris https://www.iea.org/reports/renewables-2023, Licence: CC BY 4.0

IRENA (2020) Global Renewables Outlook: Energy transformation 2050 (Edition: 2020), International Renewable Energy Agency, Abu Dhabi

IRENA (2023) World Energy Transitions Outlook 2023 (Vol. 1). Abu Dhabi: International Renewable Energy Agency. Retrieved from https://www.irena.org/Digital-Report/World-Energy-Transitions-Outlook-2023#page-3

Macknick J, Newmark R, Heath G, Hallett KC (2011) A review of operational water consumption and withdrawal factors for electricity generating technologies. In Proceedings of the U.S. Department of Energy's National Renewable Energy Laboratory. United States. https://doi.org/10.2172/1219149

Mehmood S, Zaman K, Khan S, Ali Z, Khan H (2024) The role of green industrial transformation in mitigating carbon emissions: exploring the channels of technological innovation and environmental regulation. Energy Built Environ 5(3):464–479. https://doi.org/10.1016/j.enbenv.2023.03.001

MoEFCC (2022) India's long-term low-carbon development strategy. Ministry of Environment, Forest and Climate Change, Government of India

Murray RL, Holbert KE (2020) Nuclear energy: an introduction to the concepts, systems, and applications of nuclear processes. Elsevier

Nieuwlaar E (2013) Life cycle assessment and energy systems. In: Reference module in earth systems and environmental sciences. Elsevier

Okada T, Cholii S, Karácsonyi D, Matsumoto M (2021) Communities in Fukushima and Chernobyl – enabling and inhibiting factors for recovery in nuclear disaster areas. In: Karácsonyi D, Taylor A, Bird D (eds) The demography of disasters. Springer, Cham. https://doi.org/10.1007/978-3-030-49920-4_11

Osman AI, Chen L, Yang M et al (2023) Cost, environmental impact, and resilience of renewable energy under a changing climate: a review. Environ Chem Lett 21:741–764. https://doi.org/10.1007/s10311-022-01532-8

Saint Akadiri S, Alola AA, Akadiri AC, Alola UV (2019) Renewable energy consumption in EU-28 countries: policy toward pollution mitigation and economic sustainability. Energy Policy 132:803–810

Sohag K, Husain S, Soytas U (2024) Environmental policy stringency and ecological footprint linkage: mitigation measures of renewable energy and innovation. Energy Econ 136:107721. https://doi.org/10.1016/j.eneco.2024.107721

United Nations Environment Programme (2023) Emissions gap report 2023: broken record – temperatures hit new highs, yet world fails to cut emissions (again). Nairobi. https://doi.org/10.59117/20.500.11822/43922

Usman M, Balsalobre-Lorente D (2022) Environmental concern in the era of industrialization: can financial development, renewable energy and natural resources alleviate some load? Energy Policy 162:112780

Zhang D, Wellalage NH (2022) Comparative analysis of environmental performance measures and their impact on firms' financing choices. J Clean Prod 375:134176. https://doi.org/10.1016/j.jclepro.2022.134176

Chapter 17
Environmental Aspects in Electrical Energy Generation: A Comprehensive Review

Dan Bahadur Pal and Amit Kumar Rathoure

Abstract This book chapter aims to provide a thorough examination of the environmental aspects associated with electrical energy generation. As the global demand for energy continues to rise, there is an increasing need to explore sustainable and eco-friendly methods of electricity production. This chapter covers a wide range of topics, including traditional and renewable energy sources, their environmental impacts, and emerging technologies that strive to minimize the ecological footprint of electricity generation. Additionally, it explores energy storage technologies, scrutinizing their environmental implications and their influence on policy formulation. Furthermore, it conducts an examination of global and regional policies and regulations pertaining to environmental considerations in the field of electrical energy generation.

Keywords Environmental sustainability · Green technologies · Environmental impact analysis · Clean energy solutions · Climate change mitigation

17.1 Introduction

In the quest for sustainable development and a greener future, the environmental impact of various energy sources has become a critical concern (Sekhar et al. 2024; Pramanik and Rathoure 2017). This chapter explores the intricate relationship between electrical energy generation and the environment, delving into the environmental aspects of different energy sources, their implications, and the strides made towards eco-friendly alternatives (Li et al. 2024). In the rapidly evolving landscape of energy production, the intersection between environmental sustainability and electrical energy generation has emerged as a critical focal point for the future

D. B. Pal · A. K. Rathoure (✉)
Department of Chemical Engineering, Harcourt Butler Technical University, Kanpur, Uttar Pradesh, India

S. Kumar, V. K. Singh (eds.), *Renewable Energy Development: Technology, Material and Sustainability*, Clean Energy Production Technologies,
https://doi.org/10.1007/978-981-97-9626-7_17

(Hassan et al. 2024b). As the world grapples with the challenges posed by climate change, the need for a harmonious coexistence between our growing energy demands and the health of our planet becomes increasingly evident (Olorunsogo et al. 2024). The chapters within this book navigate through the diverse facets of environmental impact, offering an in-depth analysis of the challenges and opportunities that lie at the nexus of energy and ecology (Bibri et al. 2024). Renewable energy sources, such as solar, wind, hydro, and geothermal, take center stage as promising alternatives, providing insights into their technological advancements, benefits, and potential drawbacks (Boadu and Otoo 2024).

17.2 Traditional Energy Sources

This section delves into the environmental impacts associated with conventional energy sources such as coal, air, and natural gas (Lin and Ullah 2024). The chapter evaluates the air and water pollution, greenhouse gas emissions, and land degradation linked to these sources. It discusses the regulatory frameworks in place and the technological advancements aimed at mitigating the environmental effects of traditional energy production (Sekhar et al. 2024).

17.2.1 Overview of Coal Combustion and Greenhouse Gas Emissions

Coal combustion, a predominant source of energy for decades, has played a central role in shaping the global energy landscape (Hassan et al. 2024b). However, its environmental consequences, particularly in terms of greenhouse gas (GHG) emissions, have raised significant concerns regarding climate change and air quality (Shah et al. 2024). This overview explores the intricacies of coal combustion and its associated greenhouse gas emissions, shedding light on the challenges and potential mitigation strategies.

17.2.1.1 Coal Combustion Process

Coal combustion involves the burning of coal to produce heat, which is then used to generate electricity or provide industrial heat (Kumar et al. 2024). The process typically occurs in power plants, where coal is pulverized into fine particles and burned in boilers to produce steam. The steam, in turn, drives turbines connected to generators, producing electricity. This process releases various by-products, including ash, sulfur dioxide (SO_2), nitrogen oxides (NO_x), and carbon dioxide (CO_2) (Schiavon

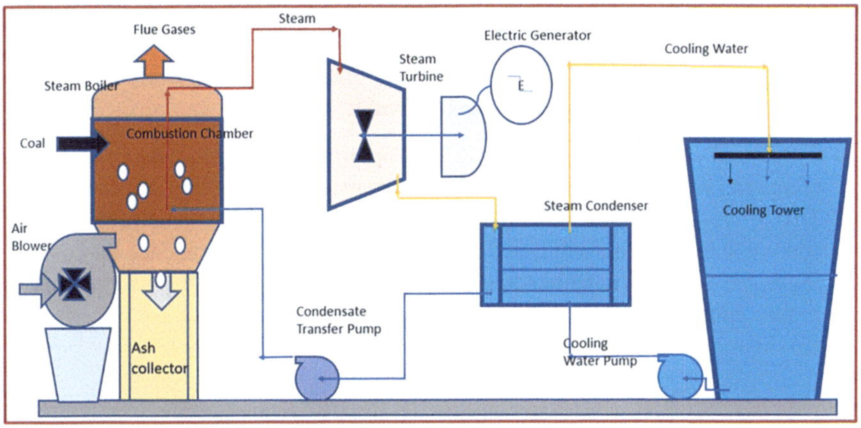

Fig. 17.1 Schematic diagram of coal-based electric power plant

et al. 2024). Figure 17.1 showed the schematic diagram of coal-based electric power plant.

17.2.1.2 Greenhouse Gas Emissions

(a) *Carbon Dioxide (CO_2)*: Coal combustion is a major contributor to CO_2 emissions, a primary greenhouse gas responsible for global warming. The carbon content in coal reacts with oxygen during combustion, releasing CO_2 into the atmosphere.
(b) *Methane (CH_4)*: While methane emissions from coal combustion are comparatively lower than those from natural gas, they still contribute to the greenhouse effect. Methane emissions often result from incomplete combustion and can be released during coal mining and transportation processes.
(c) *Nitrous Oxide (N_2O)*: Nitrous oxide is produced during coal combustion through the oxidation of nitrogen present in coal or from the release of nitrogen oxides (NO_x). N_2O is a potent greenhouse gas with a higher warming potential than CO_2.

17.2.1.3 Environmental Impacts

(a) *Air Quality*: Besides greenhouse gases, coal combustion releases pollutants like sulfur dioxide, nitrogen oxides, and particulate matter, contributing to air pollution and respiratory issues.
(b) *Climate Change*: The cumulative impact of CO_2 emissions from coal combustion significantly contributes to climate change by enhancing the greenhouse effect and trapping heat in the Earth's atmosphere.

17.2.1.4 Mitigation Strategies

To address the environmental impact of coal combustion, various mitigation strategies have been proposed, including the following:

(a) *Carbon Capture and Storage (CCS)*: Capturing CO_2 emissions from power plants and storing them underground to prevent their release into the atmosphere.
(b) *Transition to Cleaner Energy Sources*: Shifting towards renewable energy sources such as solar, wind, and hydropower to reduce dependence on coal.
(c) *Improved Efficiency*: Enhancing the efficiency of coal-fired power plants to reduce emissions per unit of electricity generated.

17.2.2 Air Pollution and Its Effects on Human Health and Ecosystems

Air pollution, a pervasive and complex environmental issue, poses significant threats to both human health and ecosystems (Adekanmbi et al. 2024). The release of pollutants into the atmosphere from various human activities has far-reaching consequences, impacting air quality and disrupting the delicate balance of ecosystems. This overview explores the sources of air pollution, common pollutants, and the multifaceted effects on human health and ecosystems.

17.2.2.1 Sources of Air Pollution

Air pollution originates from a diverse range of sources, including the following:

(a) *Combustion of Fossil Fuels*: Burning of coal, oil, and natural gas in power plants, vehicles and industrial processes.
(b) *Vehicle Emissions*: Exhaust gases from automobiles, contributing to the release of pollutants such as nitrogen oxides (NOx) and particulate matter.
(c) *Industrial Activities*: Emissions from factories and manufacturing processes, releasing pollutants like sulfur dioxide (SO_2) and volatile organic compounds (VOCs).
(d) *Agricultural Practices*: Release of ammonia (NH_3) from fertilizers and pesticides, as well as dust from plowing and harvesting.
(e) *Waste Management*: Disposal of solid waste in landfills and waste incineration.

17.2.2.2 Common Air Pollutants

Several pollutants are commonly associated with air pollution, including the following:

(a) *Particulate Matter (PM)*: Tiny particles suspended in the air, varying in size and composition (Nkuna et al. 2024).
(b) *Nitrogen Oxides (NOx)*: Gases produced during combustion processes, contributing to smog and acid rain (Nkuna et al. 2024).
(c) *Sulfur Dioxide (SO_2)*: Emitted from burning fossil fuels containing sulfur, leading to acid rain and respiratory issues (Nkuna et al. 2024).
(d) *Ozone (O_3)*: Ground-level ozone, a component of smog, formed through chemical reactions involving pollutants in the presence of sunlight (Nkuna et al. 2024).
(e) *Carbon Monoxide (CO)*: Colorless and odorless gas produced by incomplete combustion of fossil fuels (Nkuna et al. 2024).
(f) *Volatile Organic Compounds (VOCs)*: Organic chemicals released from various sources, contributing to ground-level ozone formation.

17.2.2.3 Effects on Human Health

(a) *Respiratory Issues*: Exposure to air pollutants can lead to respiratory problems, exacerbating conditions such as asthma and bronchitis (Peden 2024).
(b) *Cardiovascular Diseases*: Long-term exposure to certain pollutants is associated with an increased risk of heart diseases.
(c) *Cancer*: Prolonged exposure to specific air pollutants, such as benzene and formaldehyde, is linked to an elevated risk of cancer.
(d) *Neurological Effects*: Some pollutants may impact the central nervous system, affecting cognitive function and development.

17.2.2.4 Effects on Ecosystems

(a) *Acid Rain*: Emissions of sulfur dioxide and nitrogen oxides contribute to acid rain, harming aquatic ecosystems, soil quality, and vegetation.
(b) *Ozone Depletion*: Ground-level ozone can negatively affect plant growth and decrease crop yields.
(c) *Biodiversity Loss*: Air pollution, along with climate change, poses a threat to biodiversity by disrupting ecosystems and habitats.
(d) *Mining Impact*: Coal mining involves extensive surface and underground mining, leading to habitat destruction, soil erosion, and alteration of landscapes (Shobe et al. 2023). Mountaintop removal, a common coal mining practice, results in irreversible changes to the topography.

17.2.2.5 Mitigation and Prevention

(a) *Regulatory Measures*: Implementing and enforcing air quality standards and regulations.

(b) *Transition to Clean Energy*: Promoting renewable energy sources to reduce dependence on fossil fuels.
(c) *Green Transportation*: Encouraging the use of electric vehicles and public transportation.
(d) *Emission Control Technologies*: Implementing technologies to reduce emissions from industrial processes and vehicles.

17.2.3 Role of Natural Gas in Electricity Generation

Natural gas plays a crucial role in electricity generation and has become a cornerstone in the global energy mix (Hassan et al. 2024b). Its increasing prominence stems from its relatively lower environmental impact compared to other fossil fuels, as well as its flexibility and efficiency in power generation. Here, we explore the key aspects of the role of natural gas in electricity generation.

(a) *Combustion Efficiency*: Natural gas is a highly efficient fuel for electricity generation. When burned in modern, combined-cycle power plants, it can achieve high levels of combustion efficiency. Combined-cycle plants use both gas and steam turbines to generate electricity, maximizing the energy extracted from the fuel.
(b) *Flexibility and Reliability*: Natural gas power plants provide flexibility in responding to fluctuations in electricity demand. They can start up quickly and ramp their electricity output up or down efficiently, making them suitable for meeting peak demand periods or providing baseload power as needed. This flexibility enhances grid reliability and stability.
(c) *Lower Greenhouse Gas Emissions*: While natural gas is a fossil fuel, it produces fewer greenhouse gas emissions compared to coal and oil when burned for electricity generation. The combustion of natural gas releases lower levels of carbon dioxide (CO_2), as well as reduced emissions of sulfur dioxide (SO_2) and nitrogen oxides (NO_x). This makes natural gas a transitional fuel in efforts to reduce overall carbon emissions.
(d) *Complement to Renewables*: Natural gas complements renewable energy sources, such as solar and wind, by providing a stable and reliable source of electricity when weather-dependent renewables are not producing power (Oyewo et al. 2024). Gas-fired power plants can quickly adjust their output to balance fluctuations in renewable energy generation, contributing to grid stability.
(e) *Combined Heat and Power (CHP)*: Natural gas is often used in combined heat and power systems, also known as cogeneration. In these systems, natural gas is used to generate both electricity and useful heat simultaneously, increasing overall energy efficiency (Liang et al. 2024). CHP systems are particularly valuable in industrial settings and district heating applications.

(f) *Infrastructure and Distribution*: The existing infrastructure for natural gas extraction, transportation, and distribution makes it a readily available and accessible energy source. Utilizing this existing infrastructure can contribute to a reliable and cost-effective electricity generation system.
(g) *Transition Fuel*: Natural gas is often regarded as a "bridge" or transition fuel in the transition towards a low-carbon energy system. It can serve as a cleaner alternative to more carbon-intensive fossil fuels while facilitating the integration of renewable energy sources into the power grid (Hassan et al. 2024a).
(h) *Economic Considerations*: The cost-effectiveness of natural gas, driven in part by abundant reserves and efficient extraction technologies, makes it an economically viable option for electricity generation. This economic viability contributes to its widespread use in power plants globally.

17.2.3.1 Challenges

Despite its advantages, the role of natural gas in electricity generation is not without challenges. Concerns about methane emissions during extraction and transportation, as well as the need for continued efforts to enhance carbon capture and storage technologies, underscore the importance of considering the full environmental impact of natural gas use.

17.2.3.2 Methane Emissions and Their Impact on Climate Change

Methane emissions significantly contribute to climate change as a potent greenhouse gas, despite having a shorter atmospheric lifetime compared to carbon dioxide (CO_2) (Cotovicz Jr et al. 2024). Understanding the sources, impact, and mitigation strategies related to methane emissions is crucial for addressing the complex challenges posed by climate change.

(a) *Sources of Methane Emissions*: Fossil Fuel Production: Methane is released during the extraction, processing, and transportation of fossil fuels, including natural gas and oil. Leaks from wells, pipelines, and storage facilities contribute to emissions.
(b) *Agriculture*: Agricultural activities, particularly livestock digestion and manure management, are major sources of methane emissions. Rice cultivation in flooded paddies is also a significant contributor.
(c) *Waste Management*: Landfills produce methane as organic waste decomposes anaerobically. Inadequate waste management practices can lead to increased methane emissions.
(d) *Wetlands*: Natural wetlands are a natural source of methane, as the waterlogged conditions promote the anaerobic conditions necessary for methane-producing microorganisms.

(e) *Wildfires*: The combustion of biomass during wildfires can release significant amounts of methane into the atmosphere.

17.2.3.3 Impact on Climate Change

(a) *Greenhouse Gas Potency*: Methane is approximately 25 times more effective at trapping heat in the atmosphere than carbon dioxide over a 100-year period. In the short term (20 years), this potency increases to around 84–87 times that of CO_2.
(b) *Global Warming Potential (GWP)*: The GWP is a measure of how much heat a greenhouse gas traps in the atmosphere over a specific time frame compared to CO_2. Methane has a high GWP, making it a key driver of near-term global warming.
(c) *Feedback Loops*: Methane emissions can contribute to feedback loops that exacerbate climate change. For example, melting permafrost releases stored methane, amplifying warming trends.
(d) *Contribution to Ozone Formation*: Methane participates in atmospheric chemistry, leading to the production of ozone in the troposphere. While ozone is beneficial in the stratosphere, at ground level, it contributes to air pollution and has complex interactions with climate.

In summary, natural gas occupies a vital position in the global electricity generation landscape, offering a balance between reliability, efficiency, and lower environmental impact (Rani et al. 2024). Its versatility makes it a valuable component in the transition to a more sustainable and diverse energy portfolio.

17.3 Nuclear Power: A Controversial Solution

Environmental aspects of nuclear power in electrical energy generation involve both positive and negative factors (Dorji 2015). It's crucial to assess these aspects to make informed decisions about the role of nuclear power in the overall energy mix. Here are key environmental considerations:

17.3.1 Positive Environmental Aspects

17.3.1.1 Low Greenhouse Gas Emissions

Nuclear power plants produce electricity with minimal greenhouse gas emissions during operation. Unlike fossil fuels, nuclear power does not contribute directly to climate change through carbon dioxide (CO_2) emissions (Ramzan et al. 2024).

17.3.1.2 Land-Use Efficiency

Nuclear power plants generally have a smaller footprint compared to renewable energy sources like solar and wind farms, making them more land-use efficient.

17.3.1.3 Stable Energy Supply

Nuclear power provides a stable and reliable source of base load electricity, which can help maintain grid stability in conjunction with intermittent renewable sources.

17.3.1.4 High Energy Density

Nuclear fuel has a high energy density, meaning a small amount of fuel can produce a large amount of energy, reducing the need for extensive mining and transportation of fuel compared to other energy sources.

17.3.2 Negative Environmental Aspects

17.3.2.1 Radioactive Waste

One of the most significant environmental challenges is the management and disposal of radioactive waste generated during the nuclear fuel cycle (Banerjee et al. 2024). Safe disposal and long-term storage are critical to prevent environmental contamination.

17.3.2.2 Nuclear Accidents

Catastrophic events, such as nuclear accidents like Chernobyl and Fukushima, have raised concerns about the potential for long-term environmental damage and health risks. Strict safety measures and regulations are essential to minimize these risks.

17.3.2.3 Mining and Processing Impact

The extraction and processing of uranium for nuclear fuel can have environmental impacts, including habitat disruption, water pollution, and the release of other pollutants. Sustainable mining practices are essential to mitigate these effects.

17.3.2.4 Water Usage

Nuclear power plants require significant water for cooling purposes, which can have environmental consequences, especially in regions facing water scarcity. Implementing advanced cooling technologies and optimizing water usage is crucial.

17.3.2.5 Nuclear Proliferation

The spread of nuclear technology for energy generation raises concerns about nuclear weapons proliferation. Proper international regulations and safeguards are necessary to prevent misuse of nuclear materials.

17.3.2.6 High Initial Investment

The construction of nuclear power plants involves a high initial financial and environmental cost. The long construction timelines and associated resource use can impact ecosystems and contribute to emissions before the plant starts operating.

Nuclear power offers a low-carbon and efficient energy source, but it comes with significant challenges that need careful consideration. Advances in technology, stringent safety measures, and responsible waste management are crucial for mitigating the negative environmental aspects associated with nuclear power. Balancing the benefits and risks is essential for making informed decisions on its role in the future energy landscape.

17.4 Renewable Energy Sources

In the face of escalating climate change concerns and a growing need for sustainable energy solutions, the global focus has shifted dramatically towards renewable energy sources (Qing et al. 2024). This review explores the environmental aspects of key renewable energy technologies, including solar, wind, hydropower, and geothermal. Each source is examined in terms of its environmental benefits and challenges, utilizing life cycle assessments, land-use impacts, and potential ecological consequences as critical evaluation criteria.

17.4.1 Solar Energy

Solar energy, harnessed through photovoltaic (PV) cells and solar thermal technologies, stands out as a promising renewable resource (Raihan et al. 2023; Dorji 2015). Life cycle assessments indicate that solar panels produce minimal greenhouse gas emissions compared to conventional energy sources (Lima et al. 2024). However, challenges arise from the manufacturing process, which involves the use of rare materials and energy-intensive production methods (Muhammed et al. 2023). Additionally, land-use concerns, such as the impact on ecosystems and agricultural activities, require careful consideration.

17.4.2 Wind Energy

Wind power, derived from the kinetic energy of moving air, has gained popularity for its low carbon footprint during operation (Mastoi et al. 2023). Life cycle assessments suggest favorable environmental profiles for wind turbines. Challenges, however, include the visual and noise impact on local ecosystems, as well as the potential risks to bird and bat populations (Olabi et al. 2023). Responsible siting and technological advancements are essential to mitigate these concerns.

17.4.3 Hydropower

Hydropower has been a long-standing renewable energy source, utilizing the kinetic energy of flowing water to generate electricity (Dorji 2015). While it is a low-emission energy option during operation, the environmental impact of dam construction and alteration of river ecosystems cannot be overlooked. Striking a balance between energy generation and ecosystem preservation is crucial, emphasizing the importance of sustainable hydropower practices. Figure 17.2 showed the schematic diagram of tidal power plant.

17.4.4 Geothermal Energy

Geothermal energy, derived from the Earth's internal heat, is known for its reliability and low emissions. Life cycle assessments show that geothermal power plants have a relatively small environmental footprint (Paulillo et al. 2023). Challenges are primarily associated with the potential release of trace gases and chemicals from deep reservoirs during extraction. Rigorous monitoring and efficient extraction techniques are crucial to minimizing ecological consequences.

The transition to renewable energy sources is imperative for a sustainable future, but it requires a thorough understanding of the environmental impacts associated with each technology. Life cycle assessments, land-use considerations, and ecological consequences play pivotal roles in evaluating the overall sustainability of large-scale deployment. As technology advances and awareness grows, addressing these

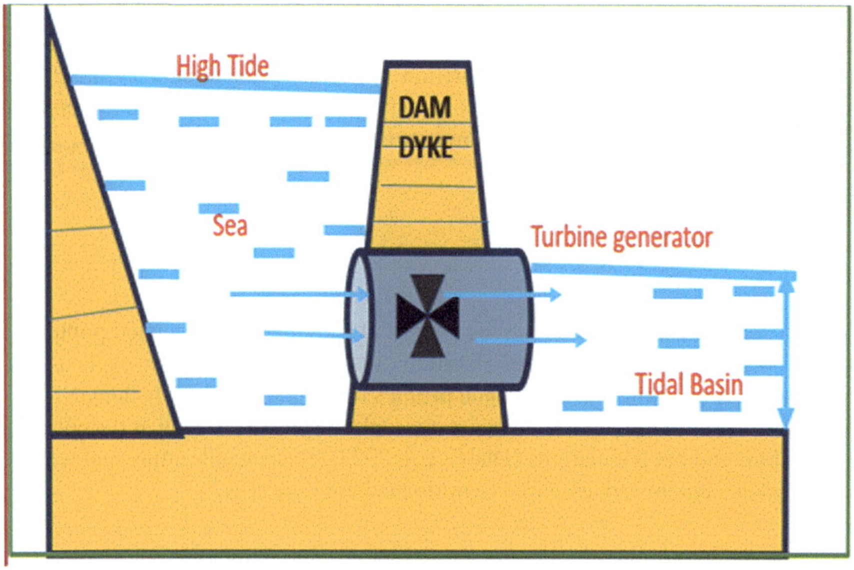

Fig. 17.2 Schematic diagram of tidal power plant

challenges will be essential for optimizing the positive environmental contributions of renewable energy sources while mitigating their potential drawbacks.

17.5 Energy Storage Technologies

As intermittent energy sources like solar and wind gain prominence, the need for effective energy storage becomes crucial (Tee et al. 2024). This section explores various energy storage technologies, such as batteries, pumped hydro storage, and thermal energy storage, discussing their environmental implications and potential contributions to sustainable energy systems.

17.5.1 Batteries

17.5.1.1 Types

Lithium-ion batteries, lead-acid batteries, sodium-ion batteries, etc.

17.5.1.2 Environmental Implications

The production and disposal of batteries can have environmental impacts. Mining for materials like lithium, cobalt, and nickel raises concerns. Recycling and responsible disposal methods are crucial to mitigate environmental damage (Mostaghimi and Behnamian 2023).

Contribution to Sustainability: Batteries play a vital role in renewable energy integration, providing grid stability, and enabling the widespread adoption of electric vehicles. Advancements in battery technology, such as solid-state batteries, aim to improve efficiency and reduce environmental impacts.

17.5.2 Pumped Hydro Storage

17.5.2.1 Operation

Involves pumping water to an elevated reservoir during times of excess energy and releasing it to generate electricity during high demand.

17.5.2.2 Environmental Implications

Construction may alter ecosystems, but once operational, pumped hydro storage has minimal direct environmental impact. Locations with suitable geography are limited.

17.5.2.3 Contribution to Sustainability

Pumped hydro storage is a mature and reliable technology that contributes to grid stability. It can store large amounts of energy for extended periods and plays a crucial role in balancing variable renewable energy sources.

17.5.3 Thermal Energy Storage

17.5.3.1 Types

Sensible heat storage, latent heat storage, and thermochemical storage.

17.5.3.2 Environmental Implications

The materials used for thermal storage and the efficiency of the storage process impact the environmental footprint. Some technologies may involve the use of high-temperature materials with associated production challenges.

Contribution to Sustainability: Thermal energy storage complements intermittent renewable sources by providing a dispatchable energy supply (Jacob et al. 2023). It finds applications in solar thermal power plants, industrial processes, and building heating, contributing to a more stable and sustainable energy system.

17.5.4 Flywheel Energy Storage

17.5.4.1 Operation

Kinetic energy is stored in a spinning rotor and converted back to electricity when needed.

17.5.4.2 Environmental Implications

The manufacturing and maintenance of flywheels have environmental impacts, but they are generally considered low compared to some other technologies.

17.5.4.3 Contribution to Sustainability

Flywheels offer fast response times and high cycling capabilities, making them suitable for grid stabilization and short-term energy storage applications. They can enhance the reliability and efficiency of the electricity grid.

17.5.5 Compressed Air Energy Storage (CAES)

17.5.5.1 Operation

Compresses air using excess energy and releases it to generate electricity.

17.5.5.2 Environmental Implications

The compression process may generate heat, and the choice of storage caverns is crucial to minimize leakage. Proper site selection and engineering can address environmental concerns.

17.5.5.3 Contribution to Sustainability

CAES is a promising technology for large-scale, long-duration energy storage. It can integrate with renewable sources, balancing supply, and demand on the grid.

In summary, each energy storage technology has its own environmental challenges (Ahmed et al. 2023), but they are essential for enabling a sustainable energy transition by improving the integration of renewable energy sources and enhancing the reliability and resilience of the energy grid. Continued research and development are critical to mitigating environmental impacts and improving the overall sustainability of these technologies.

17.6 Smart Grids and Demand-Side Management

The review addresses the role of smart grids and demand-side management in enhancing the efficiency of electrical energy systems (Saleem et al. 2023). It explores how these technologies can optimize energy consumption, reduce wastage, and contribute to a more sustainable and environmentally friendly electricity grid.

17.6.1 Smart Grids

Smart grids incorporate advanced communication, control, and monitoring technologies into the traditional electricity grid.

17.6.1.1 Role in Efficiency

Smart grids enable two-way communication between utilities and consumers, allowing for real-time monitoring, control, and optimization of electricity distribution. This improves overall grid efficiency by reducing transmission losses and enhancing system reliability.

17.6.1.2 Optimizing Energy Consumption

Through smart meters and sensors, smart grids provide detailed information on electricity usage. Consumers can make informed decisions about their energy consumption patterns, optimizing usage during periods of lower demand or higher renewable energy availability.

17.6.2 Demand-Side Management (DSM)

DSM involves the implementation of strategies to influence consumer electricity usage patterns, encouraging more efficient and flexible consumption.

17.6.2.1 Role in Efficiency

DSM helps balance electricity supply and demand by shifting or reducing energy consumption during peak periods. This reduces the need for additional power generation capacity and minimizes stress on the grid.

17.6.2.2 Optimizing Energy Consumption

Time-of-use pricing, incentive programs, and real-time feedback from smart meters empower consumers to adjust their electricity usage based on pricing signals or demand-response programs. This optimization leads to reduced energy wastage and lower overall costs.

17.6.3 Grid Flexibility and Integration of Renewable Energy

17.6.3.1 Smart Grids

Advanced sensors and automation in smart grids facilitate the integration of renewable energy sources, such as solar and wind. The grid can adapt to variable generation, ensuring a reliable power supply.

17.6.3.2 Demand-Side Management

DSM allows for better integration of intermittent renewables by aligning electricity consumption with periods of high renewable energy availability. Consumers can be incentivized to use electricity when renewable generation is abundant.

17.6.4 Reduction of Transmission and Distribution Losses

17.6.4.1 Smart Grids

Advanced monitoring and control capabilities in smart grids enable utilities to identify and address inefficiencies in the grid. This helps reduce transmission and distribution losses, contributing to overall energy conservation.

17.6.4.2 Demand-Side Management

By optimizing energy consumption patterns, DSM reduces the need for excess energy generation and, consequently, transmission losses.

17.6.5 Enhanced Reliability and Resilience

17.6.5.1 Smart Grids

Smart grids can quickly detect and respond to faults, improving the reliability of the electricity grid. This is crucial for minimizing downtime and ensuring a consistent power supply.

17.6.5.2 Demand-Side Management

Distributed energy resources and demand-side measures contribute to grid resilience. In times of high demand or emergencies, DSM can play a role in load shedding or shifting, maintaining grid stability.

In conclusion, smart grids and demand-side management play pivotal roles in optimizing energy consumption, reducing wastage, and enhancing the overall efficiency of electrical energy systems. These technologies contribute significantly to building a more sustainable and environmentally friendly electricity grid by promoting flexibility, integrating renewable energy, and empowering consumers to make informed choices about their energy usage.

17.7 Policy and Regulatory Framework

An analysis of global and regional policies and regulations related to environmental aspects in electrical energy generation is presented. This section discusses the effectiveness of current policies, potential areas for improvement, and the role of international cooperation in promoting sustainable energy practices.

17.7.1 *Global Policies*

17.7.1.1 Paris Agreement

The Paris Agreement, signed by numerous countries, sets targets to limit global warming and emphasizes the need for a transition to low-carbon energy systems. Nations commit to nationally determined contributions (NDCs) to reduce greenhouse gas emissions, influencing energy generation policies globally.

17.7.1.2 Sustainable Development Goals (SDGs)

SDG focuses on ensuring access to affordable, reliable, sustainable, and modern energy for all. This goal emphasizes the integration of renewable energy sources and increased energy efficiency in national policies.

17.7.2 *Regional Policies*

17.7.2.1 European Union (EU)

The EU has aggressive renewable energy targets and emissions reduction goals. Policies such as the Clean Energy Package aim to promote renewable energy, energy efficiency, and grid flexibility to create a more sustainable energy landscape.

17.7.2.2 California's Renewable Portfolio Standard (RPS)

California's RPS mandates a specified percentage of electricity sales to come from renewable energy sources. This policy has been influential in driving the adoption of renewable energy technologies in the region.

17.7.2.3 China's Renewable Energy Law

China has implemented policies promoting renewable energy deployment, grid connectivity, and technology innovation. These efforts align with the country's commitment to peak carbon emissions by 2030 and achieve carbon neutrality by 2060.

17.7.3 Effectiveness of Current Policies

17.7.3.1 Positive Impact

Global and regional policies have accelerated the deployment of renewable energy technologies, leading to a significant increase in renewable energy capacity. Many countries have witnessed a decline in the cost of renewable technologies, making them more competitive.

17.7.4 Challenges

Some policies face challenges related to enforcement, funding, and political will. Additionally, grid integration and storage technologies need further policy support to maximize the effectiveness of intermittent renewable sources.

17.7.5 Potential Areas for Improvement

17.7.5.1 Harmonization of Standards

There is a need for greater international cooperation to harmonize standards for renewable energy technologies, ensuring interoperability and facilitating cross-border energy trade.

17.7.5.2 Incentive Structures

Policymakers should continuously reassess incentive structures to encourage private investment in sustainable energy. Well-designed financial mechanisms and subsidies can drive innovation and adoption.

17.7.5.3 Transition Planning

Policymakers need to focus on creating comprehensive transition plans, considering the social and economic impacts of shifting from traditional to sustainable energy sources.

17.7.5.4 Technology-Neutral Policies

Policies should evolve to be technology-neutral, promoting the best mix of renewable energy sources based on regional characteristics. This approach fosters innovation and allows for adaptation to local conditions.

17.7.6 Role of International Cooperation

17.7.6.1 Technology Transfer

Collaborative efforts can facilitate technology transfer, allowing developing regions access to clean energy solutions and fostering a global transition.

17.7.6.2 Capacity Building

International cooperation can support capacity-building initiatives, aiding nations in developing and implementing effective policies, regulatory frameworks, and infrastructure for sustainable energy.

17.7.6.3 Financial Support

Developed nations can provide financial support to developing countries to overcome financial barriers associated with transitioning to sustainable energy, promoting global equity.

17.7.6.4 Knowledge Sharing

Sharing best practices and lessons learned internationally enhances the effectiveness of policies. Platforms for knowledge exchange contribute to the collective understanding of successful strategies. In conclusion, while global and regional policies have made significant strides in promoting environmental aspects in electrical energy generation, there are areas for improvement. A stronger emphasis on international cooperation, addressing challenges, and refining policies can further accelerate the transition to sustainable energy practices, contributing to the achievement of global climate and sustainability goals.

17.8 Future Outlook

The paper concludes with a discussion on the future prospects of environmentally sustainable electrical energy generation. It highlights the importance of continued research and development, collaboration among stakeholders, and the role of public awareness in fostering a cleaner and more sustainable energy future.

17.8.1 Advanced Technologies

Ongoing R&D is essential for the development of advanced technologies that can enhance the efficiency and reduce the environmental impact of electrical energy generation. Emphasis on next-generation renewable energy technologies, energy storage solutions, and smart grid advancements can significantly contribute to a more sustainable energy landscape.

17.8.2 Emerging Solutions

Continuous exploration of emerging solutions, such as advanced photovoltaics, offshore wind technologies, and breakthroughs in energy storage, can lead to more cost-effective and scalable options for sustainable energy generation. R&D efforts should focus on improving the performance, reliability, and affordability of these technologies.

17.8.3 Collaboration Among Stakeholders

17.8.3.1 Public-Private Partnerships

Collaborations between governments, private industries, and research institutions are crucial for accelerating the deployment of sustainable energy solutions. Public-private partnerships can facilitate the sharing of resources, expertise, and investments, fostering a conducive environment for the adoption of environmentally friendly energy technologies.

17.8.3.2 Global Cooperation

International collaboration is essential to address global energy challenges. Shared research initiatives and knowledge exchange can accelerate progress in developing and implementing sustainable energy solutions. Joint efforts in addressing climate change and promoting clean energy can lead to the establishment of global standards and best practices.

17.8.4 Public Awareness

17.8.4.1 Education and Advocacy

Increasing public awareness about the environmental impact of energy choices is crucial. Education campaigns can inform the public about the benefits of sustainable energy and empower them to make environmentally conscious decisions.

Advocacy for policy support and incentives can encourage governments to prioritize sustainable energy solutions.

17.8.4.2 Consumer Empowerment

Informed consumers play a vital role in driving demand for sustainable energy. Public awareness campaigns can highlight the benefits of using clean energy sources and energy-efficient appliances and adopting sustainable lifestyle practices. Promoting energy literacy empowers individuals and businesses to actively participate in the transition to a cleaner and more sustainable energy future.

17.8.5 Importance of Policy Support

17.8.5.1 Incentives and Regulations

Governments should provide financial incentives, tax credits, and supportive regulatory frameworks to encourage the adoption of sustainable energy practices. Stringent environmental regulations can drive industries towards cleaner energy alternatives and promote sustainable business practices.

17.8.5.2 Long-Term Planning

Governments and policymakers need to engage in long-term energy planning that prioritizes sustainability. This includes setting clear renewable energy targets, investing in research, and fostering an environment conducive to innovation and experimentation.

The future prospects of environmentally sustainable electrical energy generation are promising, provided there is a concerted effort from all stakeholders. Continued research and development, collaboration among governments, industries, and research institutions, along with increased public awareness, are integral to fostering a cleaner and more sustainable energy future. By embracing these principles, societies can not only mitigate environmental impacts but also pave the way for a resilient and sustainable energy landscape for future generations.

17.9 Conclusion

In summary, this comprehensive review paper provides a detailed examination of the environmental aspects of electrical energy generation, covering traditional, renewable, and emerging technologies. It aims to contribute to the on-going discourse on sustainable energy solutions and inspire further research in the pursuit of a greener and more environmentally friendly energy landscape.

References

Adekanmbi AO, Ani EC, Abatan A, Izuka U, Ninduwezuor-Ehiobu N, Obaigbena A (2024) Assessing the environmental and health impacts of plastic production and recycling. World J Biol Pharm Health Sci 17(2):232–241

Ahmed AA, Alsharif A, Yasser N (2023) Recent advances in energy storage technologies. Int J Electr Eng Sustain:9–17

Banerjee RH, Nelwamondo SM, Raje N, Parab H, Chaudhary N, Mathi P, Badenhorst J, Goede A, Sengupta P (2024) Radiation stability testing of hydrated and heat-treated magnesium potassium phosphates for back end nuclear fuel cycle applications. Ceram Int 50(1):2405–2414

Bibri SE, Krogstie J, Kaboli A, Alahi A (2024) Smarter eco-cities and their leading-edge artificial intelligence of things solutions for environmental sustainability: a comprehensive systematic review. Environ Sci Ecotechnol 19:100330

Boadu S, Otoo E (2024) A comprehensive review on wind energy in Africa: challenges, benefits and recommendations. Renew Sust Energ Rev 191:114035

Cotovicz LC Jr, Abril G, Sanders CJ, Tait DR, Maher DT, Sippo JZ, Holloway C, Yau YY, Santos IR (2024) Methane oxidation minimizes emissions and offsets to carbon burial in mangroves. Nat Clim Chang:1–7

Dorji G (2015) Environmental aspect of electric energy generation

Hassan Q, Algburi S, Sameen AZ, Salman HM, Jaszczur M (2024a) Green hydrogen: a pathway to a sustainable energy future. Int J Hydrog Energy 50:310–333

Hassan Q, Viktor P, Al-Musawi TJ, Ali BM, Algburi S, Alzoubi HM, Al-Jiboory AK, Sameen AZ, Salman HM, Jaszczur M (2024b) The renewable energy role in the global energy transformations. Renew Energy Focus 48:100545

Jacob R, Hoffmann M, Weinand JM, Linßen J, Stolten D, Müller M (2023) The future role of thermal energy storage in 100% renewable electricity systems. Renew Sustain Energy Transition 4:100059

Kumar V, Saxena VK, Kumar R, Kumar S (2024) Energy, exergy, sustainability and environmental emission analysis of coal-fired thermal power plant. Ain Shams Eng J 15(2):102416

Li Q, Zhao M, Hei P, Li F, Zhang K (2024) Driving sustainable development: exploring the nexus of financial inclusion, green mobility, and CO2 emissions in China's natural resource landscape. Resour Policy 89:104656

Liang P, Guo Y, Nutakki TUK, Agrawal MK, Muhammad T, Ahmad SF, Ahmad AYAB, Qin M (2024) Comprehensive assessment and sustainability improvement of a natural gas power plant utilizing an environmentally friendly combined cooling heating and power-desalination arrangement. J Clean Prod 436:140387

Lima GM, Belchior FN, Villena JEN, Domingos JL, Freitas MAV, Hunt JD (2024) Hybrid electrical energy generation from hydropower, solar photovoltaic and hydrogen. Int J Hydrog Energy 53:602–612

Lin B, Ullah S (2024) Modeling the impacts of changes in nuclear energy, natural gas, and coal in the environment through the novel DARDL approach. Energy 287:129572

Mastoi MS, Zhuang S, Haris M, Hassan M, Ali A (2023) Large-scale wind power grid integration challenges and their solution: a detailed review. Environ Sci Pollut Res 30(47):103424–103462

Mostaghimi K, Behnamian J (2023) Waste minimization towards waste management and cleaner production strategies: a literature review. Environ Dev Sustain 25(11):12119–12166

Muhammed NS, Gbadamosi AO, Epelle EI, Abdulrasheed AA, Haq B, Patil S, Al-Shehri D, Kamal MS (2023) Hydrogen production, transportation, utilization, and storage: recent advances towards sustainable energy. J Energy Storage 73:109207

Nkuna SG, Olwal TO, Chowdhury SD, Ndambuki JM (2024) A review of wastewater sludge-to-energy generation focused on thermochemical technologies: an improved technological, economical and socio-environmental aspect. Clean Waste Syst:100130

Olabi A, Obaideen K, Abdelkareem MA, AlMallahi MN, Shehata N, Alami AH, Mdallal A, Hassan AAM, Sayed ET (2023) Wind energy contribution to the sustainable development goals: case study on London array. Sustainability 15(5):4641

Olorunsogo TO, Ogugua JO, Muonde M, Maduka CP, Omotayo O (2024) Environmental factors in public health: a review of global challenges and solutions. World J Adv Res Rev 21(1):1453–1466

Oyewo AS, Aghahosseini A, Movsessian MM, Breyer C (2024) A novel geothermal-PV led energy system analysis on the case of the central American countries Guatemala, Honduras, and Costa Rica. Renew Energy 221:119859

Paulillo A, Cui X, Brown P, Striolo A, Lettieri P (2023) Geothermal energy impact estimator: a software application for estimating the life-cycle environmental impacts of geothermal energy. Open Res Eur 3

Peden DB (2024) Respiratory health effects of air pollutants. Immunol Allergy Clin 44(1):15–33

Pramanik H, Rathoure AK (2017) Electrooxidation study of NaBH4 in a membraneless microfluidic fuel cell with air breathing cathode for portable power application. Int J Hydrog Energy 42(8):5340–5350

Qing L, Abbas J, Najam H, Ma X (2024) Investment in renewable energy and green financing and their role in achieving carbon-neutrality and economic sustainability: insights from Asian region. Renew Energy 221:119830

Raihan A, Voumik LC, Rahman MH, Esquivias MA (2023) Unraveling the interplay between globalization, financial development, economic growth, greenhouse gases, human capital, and renewable energy uptake in Indonesia: multiple econometric approaches. Environ Sci Pollut Res 30(56):119117–119133

Ramzan M, Razi U, Usman M, Sarwar S, Talan A, Mundi HS (2024) Role of nuclear energy, geothermal energy, agriculture, and urbanization in environmental stewardship. Gondwana Res 125:150–167
Rani P, Parkash V, Sharma NK (2024) Technological aspects, utilization and impact on power system for distributed generation: a comprehensive survey. Renew Sust Energ Rev 192:114257
Saleem MU, Shakir M, Usman MR, Bajwa MHT, Shabbir N, Shams Ghahfarokhi P, Daniel K (2023) Integrating smart energy management system with internet of things and cloud computing for efficient demand side management in smart grids. Energies 16(12):4835
Schiavon M, Ravina M, Zanetti M, Panepinto D (2024) State-of-the-art and recent advances in the abatement of gaseous pollutants from waste-to-energy. Energies 17(3):552
Sekhar SJ, Samuel MS, Glivin G, Le T, Mathimani T (2024) Production and utilization of green ammonia for decarbonizing the energy sector with a discrete focus on sustainable development goals and environmental impact and technical hurdles. Fuel 360:130626
Shah IH, Manzoor MA, Jinhui W, Li X, Hameed MK, Rehaman A, Li P, Zhang Y, Niu Q, Chang L (2024) Comprehensive review: effects of climate change and greenhouse gases emission relevance to environmental stress on horticultural crops and management. J Environ Manag 351:119978
Shobe C, Bower S, Maxwell A, Glade R, Samassi NM, Campforts B (2023) The uncertain future of mountaintop-removal-mined landscapes. Geomorphology 445:108984
Tee WH, Gan CK, Sardi J (2024) Benefits of energy storage systems and its potential applications in Malaysia: a review. Renew Sust Energ Rev 192:114216

The manufacturer's authorised representative in the EU is Springer Nature Customer Service Centre GmbH, Europaplatz 3, 69115 Heidelberg, Germany. If you have any concerns regarding our products, please contact ProductSafety@springernature.com

Printed and bound by CPI Group (UK) Ltd, Croydon, CR0 4YY
01/10/2026
02225930-0011